AF900127

THE CURSUS ENIGMA

STUDIES IN THE BRITISH MESOLITHIC AND NEOLITHIC

Volume 3

Series Editors
David Jacques and Graeme Davis

PETER LANG
Oxford • Bern • Berlin • Bruxelles • New York • Wien

THE CURSUS ENIGMA

Prehistoric Cattle and Cursus Alignments

David Saunders

Edited by David Jacques

PETER LANG

Oxford • Bern • Berlin • Bruxelles • New York • Wien

Bibliographic information published by Die Deutsche Nationalbibliothek.
Die Deutsche Nationalbibliothek lists this publication in the Deutsche National-bibliografie; detailed bibliographic data is available on the Internet at http://dnb.d-nb.de.

A catalogue record for this book is available from the British Library.

Library of Congress Cataloging-in-Publication Data

Names: Saunders, David John, 1957- author.
Title: The cursus enigma : prehistoric cattle and cursus alignments / David John Saunders.
Description: Oxford ; New York : Peter Lang, [2021] | Series: Studies in the British Mesolithic and Neolithic, 2297-1068 ; vol. 3 | Includes bibliographical references and index.
Identifiers: LCCN 2020025326 (print) | LCCN 2020025327 (ebook) | ISBN 9781789977455 (hardback) | ISBN 9781789977547 (ebook) | ISBN 9781789977554 (epub) | ISBN 9781789977561 (mobi)
Subjects: LCSH: Neolithic period--Great Britain. | Monuments--Great Britain. | Mounds--Great Britain. | Earthworks (Archaeology)--Great Britain. | Cattle trails--Great Britain--History. | Cattle herding--Great Britain--History. | Great Britain--Antiquities.
Classification: LCC GN776.22.G7 S37 2020 (print) | LCC GN776.22.G7 (ebook) | DDC 936.1--dc23
LC record available at https://lccn.loc.gov/2020025326
LC ebook record available at https://lccn.loc.gov/2020025327

Cover design by Peter Lang Ltd.

ISSN 2297-1068
ISBN 978-1-78997-745-5 (print) • ISBN 978-1-78997-754-7 (ePDF)
ISBN 978-1-78997-755-4 (ePub) • ISBN 978-1-78997-756-1 (mobi)

© Peter Lang AG 2021

Published by Peter Lang Ltd, International Academic Publishers,
52 St Giles, Oxford, OX1 3LU, United Kingdom
oxford@peterlang.com, www.peterlang.com

David Saunders have asserted their right under the Copyright, Designs and Patents Act, 1988, to be identified as Author of this Work.

All rights reserved.
All parts of this publication are protected by copyright.
Any utilisation outside the strict limits of the copyright law, without the permission of the publisher, is forbidden and liable to prosecution.
This applies in particular to reproductions, translations, microfilming, and storage and processing in electronic retrieval systems.

This publication has been peer reviewed.

To Hannah, my partner and part-time sub-editor

Contents

Figures

Tables

Tables in Appendixes

List of Maps

Foreword

In the 100th edition of *Current Archaeology* magazine, a reader's letter was published that pleaded for help in providing a definition for the term 'cursus monument'. 'I would be most grateful if you could inform me precisely what is meant by a "cursus"' wrote Norah Leggett. She stressed that she had been unable to find out what a cursus monument was, even in an archaeological dictionary. In the forty years or so since this plaintive letter was written, archaeologists have made periodic efforts to answer this question, with mixed success. From time to time a piece of research comes along that makes a fascinating contribution to this discussion, and David Saunders' book contains just such a piece of research.

This kind of research is needed, because these large to enormous earthwork and timber post-defined enclosures of fourth millennium BC date may have dominated the lives and landscape of the first farmers in Britain, but cursus monuments have made substantially less impact on the pages of archaeological textbooks. Cursus are the kind of site that generally get a contractual obligation paragraph or two in synthesis of the Neolithic period, and prehistory, of Britain, a trend that goes back to Stuart Piggott's 1954 book *The Neolithic Cultures of the British Isles*. This paragraph is usually an apologetic interpretative shrug of the shoulders that concludes with the supposition that these were monuments of a ceremonial nature, probably involving processions, but what else is there to say?

The book that you hold in your hands is a rare example of a detailed, data-driven study that has as its central focus cursus monuments. Works of a similar depth have been few and far between, with notable exceptions being book-length contributions to solving the 'cursus enigma' or 'cursus problem' written by Roy Loveday (*Inscribed across the Landscape*, 2006) and myself (*Reading between the Lines*, 2015). Alongside an edited volume (*Pathways and Ceremonies*, edited by Barclay and Harding in 1999), a few major fieldwork projects, both research and developer-funded, have focused on a cursus-dominated place, from Drayton in Oxfordshire to the Cleaven Dyke in Perth and Kinross. Yet such studies do not afford the space to think or the words to express a detailed exploration of the cursus monument as either a prehistoric tradition or a curious archaeological piece of terminology.

David Saunders in this book presents an innovative, fascinating, provocative and intriguing answer to Norah Leggett's question. That is not to say that this is the definitive account, or that David has solved the cursus problem. Neolithic monuments tend to fall into roughly coherent typological groupings that we then burden with obscure and problematic names: cursus, henge, palisaded enclosure and so on. Yet these classifications often impose an artificial uniformity in terms of what the function of such monuments might have been within Neolithic society. Yet it is likely that enclosures served various different roles and purposes even to those who built, and used them, and that meaning may have changed from generation to generation.

The meaning of cursus monuments to us, archaeologists, has also changed from generation to generation, and David's ideas presented in this book offers a cursus theory for our time, when there is more concern than ever with thinking about the Mesolithic inheritance (as Denise Telford once put it) that influenced Neolithic farmers. By considering cursus monuments in terms of landscape, hunting practices and animal husbandry, we can start to consider how these enormous monuments cut across a range of aspects of the lives of those who built, used and encountered them, as opposed to fuzzy ritualistic roles. We may never truly know what cursus monuments were, but this book moves us one further step along the journey to achieving the goal of satisfying the curiosity of one Norah Leggett.

Kenneth Brophy, March 2020

Preface

Many have struggled to understand the reason behind the construction of cursus monuments and their possible purpose, which has led to them being often described as 'enigmatic'. Previous cursus monument studies have tended to focus on the construction or post-construction phases of the monument, rather than upon the reason a Neolithic community decided to locate and align these monuments where they did. Perhaps, at this current time, the sole use of an archaeological methodology has gone as far as it can with regard to explaining the 'cursus enigma'. This book takes a new approach to the cursus monument debate by incorporating a fresh methodology, previously unexplored in cursus monument investigation. It looks to combine previous archaeological methodologies with an understanding of the different styles of movement of various large herbivores, to identify whether any correlation exists between the initial placement and alignment of Neolithic cursus monuments and these types of movement. This has enabled the study to identify a strong correlation between the placement and alignment of Neolithic cursus monuments and the movement of domestic cattle within that landscape, thus making an original contribution to the cursus monument debate.

Growing up in West Sussex, living less than 50 m from the rivers, floodplains and meadows of the Pulborough Brooks, my early memories are of wandering those pastures, moving through fields of cattle with fishing rod in hand, the herds often coming within a few feet, sometimes less, as they tried to discover what I was up to. Little did I know that those childhood experiences of long summers spent interacting with cattle would play such an important role in my future archaeological career.

David Saunders

Summary

The research outlined in this book has identified a strong correlation between the movement of domestic cattle and the placement and alignment of Neolithic cursus monuments, thereby making an original contribution to the cursus monument debate. The possible correlation between these monuments and the movement of Neolithic pastoralist communities within the landscape had not been well explored in previous research. Previous cursus monument studies have tended to focus on the construction or post-construction phases of the monument, rather than on the reason a Neolithic community decided to locate and align the monuments where they did. The study on which this book is based used quantifiable data gathered through the fitting of GPS collars to American range cattle (George et al. 2007) to determine the terrain over which cattle move. This was combined with GIS elevation and slope data from a GIS software programme supplied by Environmental Systems Research Institute.

This enabled a quantifiable examination of fifty cursus monument sites on or adjacent to the English chalkland belt to determine the movement of cattle across the landscape at each individual monument site. Investigation into areas of natural restriction to the landscape and areas affected by winter flooding of pasture enabled the identification of areas that could have aided cattle movement and husbandry at prime points during the early spring. The inclusion of boots-on-the-ground field observations across each of the fifty monument sites has helped overcome issues associated with previous studies, where results from a few ideal examples appear to have then been extrapolated to the rest.

The linking of cattle movement data to mainstream archaeological research identified that cursus monument sites appear to align with the potential routes of domestic cattle across the valley profile. Further investigation into areas affected by the winter flooding of pasture identified a strong correlation with the precise location of cursus monument sites.

The study suggests that cursus monuments commenced life as droveways, perhaps identifying an initial practical function to the landscape prior to any possible later ritual importance of the sites. A case study was undertaken to re-evaluate these ideas using research undertaken in the Milfield Basin in Northumberland, an area that appears to have had a droveway which did not develop into a cursus monument.

Acknowledgements

I would like to thank the following people and establishments: David Jacques of the University of Buckingham for his enthusiasm and support throughout this process and for use of material relating to the Blick Mead site; The University of Buckingham, the University of Durham, the University of York and the University of Reading for the opportunity to analyse potential large herbivore movement throughout the Stonehenge landscape by making available their research undertaken at Blick Mead; Peter Rowley-Conwy and Bryony Rogers of the University of Durham for the opportunity to analyse potential aurochs movement throughout the Stonehenge landscape by making available their research on isotopic data undertaken from two aurochs' teeth discovered at Blick Mead.

I am especially grateful to the following people: Graeme Davis and Nick James for their continued support both during and after their supervision of my studies in Archaeology; Tim Darvill of the University of Bournemouth for his support and encouragement throughout this process and for his kind permission to critically dissect the cursus monument definition within his 2008 publication of *The Concise Oxford Dictionary of Archaeology*; Peter Rowley-Conwy of The University of Durham for giving me access to his chapter 'To the Upper Lake: Star Carr revisited – by birchbark canoe' as one of the fifty offprints within his latest book Economic Zooarchaeology; Chris Green of Oxford University for his explanation on the use of mass modelling data undertaken as part of the English Landscape and Identities (EngLaId) project and how to incorporate his research within the Mesolithic/Neolithic transition period; Steve Marshall for his extensive guidance with regards to the spring networks surrounding the Avebury monuments: Richard Crook, of Vineys Farm, Amesbury, a local farmer with fifty years' experience farming in the Stonehenge landscape, for explaining why cattle appear to relish eating grass within mist-filled valleys; and Mike Clarke, custodian of Blick Mead, Amesbury, for many hours discussing the countryside and country skill aspects of the Blick Mead area.

I am grateful to the following people who took the time to discuss this research in person, enabling me to incorporate many of their latest ideas: Richard Bradley, Roy Loveday, Chris Tilley, Jim Leary, Martin Green, Martin Bell, Mark Bowden, Gill Hey, Dave Field, Simon Banton, Carol Druce and Philip Aitkens. I am also grateful for Cllr Fred Westmoreland's commitment to myself and the first cohort of archaeological students from the University of Buckingham through the sponsorship of our publication *Stonehenge: A Landscape Through Time* by the Amesbury Archaeological Fund. I would also like to thank the following people and organizations for their permission to use their copyright material within this publication: Framework Archaeology including Wessex Archaeology and Oxford Archaeology, Cambridge Archaeological Unit, British Archaeological Reports Ltd, Proceedings of the Prehistoric Society, Historic England, Ordnance Survey, Digimap, British Geological Survey UKRI, OpenStreetMap Foundation and OpenTopoMap, Peter Lang Ltd and John Gover.

Due to the extent of the study area, I have been especially dependent on the help and advice of the staff of various libraries. I would especially like to thank the staff of Cambridge University Library, UCL Library and the University of Buckingham Library. But special thanks must go to Kenny Brophy, of the University of Glasgow, for agreeing to be my external PhD tutor and for the hours spent discussing the challenges of this study.

On a personal note, a final thanks must go to my partner Hannah, for the endless hours spent reading this research and listening to me, as she would put it, 'banging on' about cursus monuments, and without whom I would not have met David Jacques and would never have set out on this journey.

Abbreviations

ALS	Airborne laser scanning
DEM	Digital elevation model
DTM	Digital terrain model
EH	English Heritage
ERT	Earth resistivity tomography
Esri	Environmental Systems Research Institute
Fms	Fragrance materials in sludge-amended soils
GIS	Geographical information system
GPR	Ground-penetrating radar
GPS	Global positioning system
HE	Historic England
HER	Historic Environment Record
Lidar	Light detection and ranging
NISP	Number of identifiable specimens
OD	Ordnance datum
OSL	Optically stimulated luminescence
OS Grid Ref	Ordnance Survey grid reference
RCHME	Royal Commission on the Historical Monuments of England
SEP	Stonehenge Environs Project
SHLP	Stonehenge Hidden Landscapes Project
SRP	Stonehenge Riverside Project
UNESCO	United Nations Educational Scientific and Cultural Organization
WHS	World Heritage Site

Introduction

Several prominent cursus monument investigators have struggled to explain clearly the reasoning behind the physical construction and placement of cursus monuments. This has led Loveday (2006, p. 11) to describe them as an 'enigma', McOmish (2003, p. 1) to identify them as 'a 6,000-year-old puzzle' and Bradley (1986, p. 1) to describe them as 'weird', while Brophy (2016, pp. 3–13) dedicates a complete chapter within his book to them, which he titles 'The weirdest type of field monument in the country'. So perhaps to move the cursus monument debate forward, a different style of methodology will be required from that previously used.

Did Sir John Lubbock (1865) fully understand the dilemma he was instigating when he first coined the term 'Neolithic'? I think he probably did, for, while this appears initially to have been principally an attempt to separate the New Stone Age in which tools were ground and polished from earlier periods in which they were only chipped and flaked, in his publication *Pre-historic Times*, in which he first defines the period, he quickly moves from this simple technological definition to become one of the first people to suggest that the keeping of cattle might be an important characteristic for the start of the Neolithic.

Thomas (1999, p. 7) strongly contests that it was a full farming economic base that defined the start of the Neolithic, suggesting that 'a quite different set of economic practices prevailed during the Neolithic of Southern Britain' which appear to be more complex, messy and fragmented. An apparent lack of arable agriculture or sedentary settlement suggests that either Early Neolithic migrants undertook a lifestyle more pastoral than fully agricultural, moving with their herds along traditional paths and territories previously used by Mesolithic communities, or that essentially Mesolithic hunter-gatherer communities adopted some of the Early Neolithic material cultures while continuing to use the traditional routes and territories they had used for generations. Perhaps, as Ray and Thomas (2018, p. 84) propose, the birth of Neolithic Britain was a 'co-creation achieved through contact and interaction between Continental Neolithic people and the indigenous British population' drawing on both indigenous people and incomers from diverse points of origin.

It is my intention to use these theories as a starting point to explore the idea that there could be a correlation between the placement and alignment of Neolithic cursus monuments and the activities undertaken by earlier Mesolithic and Neolithic communities that lived within the area during the Mesolithic/Neolithic transition period. Little exploration has previously been undertaken with regard to the correlation between large herbivore movement and the placement and alignment of Neolithic cursus monuments. An important factor throughout the study was that only quantifiable data would be used to identify the way in which various large, wild and domestic herbivores would have moved throughout the immediate landscape of the cursus monuments found on or near to the English chalkland belt.

The idea started to form during my connection with excavations carried out beside an ancient spring at Blick Mead under the direction of David Jacques (Jacques et al. 2018) as part of the University of Buckingham's Archaeology programme, *Stonehenge: A Landscape Through Time*. This connection also gave access to the site's fauna (Peter Rowley-Conwy – University of Durham) and environmental (Nick Branch – University of Reading) data sets and has given rise to long discussions with Bryony Rogers regarding the isotopic analysis of cattle teeth from the region. This enabled a scientific element to be introduced to my own fieldwork which was included when analysing the probable movement of large herbivores throughout both the Stonehenge and the wider British landscape.

The sedimentary sequence within the springhead basin at Blick Mead revealed the remains of aurochs, wild boar and red deer. Analysis of bone remains (Durham University and the Natural

History Museum) reveals that over 2,430 pieces of animal bone fragments have been found at Blick Mead (Jacques et al. 2018, p. 127). Identifiable fragments made up 11 per cent of the total assemblage, while a further twelve fragments were identified using ZooMS (Charlton 2018 – In press), which indicated that between 57 and 59 per cent of the assemblage was from aurochs, the largest amount found nationally from a Mesolithic site (Jacques and Phillips 2014, p. 24).

It was Jacques' (Jacques et al. 2017, p. 20) suggestion that the natural vantage points of the wider Stonehenge environment could have been invaluable to the Mesolithic hunter-gatherers at Blick Mead where a natural funnel, possibly created by an earlier palaeochannel, slopes down from the King Barrow Ridge to a fording point in Stonehenge Bottom and that the Mesolithic posts were perhaps set up to mark the movement of aurochs through the relatively open landscape (Jacques and Phillips 2014, p. 24) that awakened my initial interest in studying animal movement across the Stonehenge knoll, leading eventually to the writing of my PhD thesis, 'An Investigation into Potential Correlations between the placement of Neolithic Cursus Monuments and Large Herbivore Movement'.

The fact that this landscape is the precise location for the Stonehenge Greater Cursus potentially enhances Thomas et al.'s (2009, p. 44) earlier suggestion that 'the Greater Cursus might reflect an earlier Mesolithic route between the River Avon and the River Till'. Although Thomas was clearly indicating an association with people, the movement could be construed in different ways. One interpretation that gains credence looking at parallels in both the cursus direction and its topography is that herds of large herbivores may have migrated along this route. Initial research highlights that this would appear to be mirrored across other cursus monuments, a factor supported by Brophy's (2016, pp. 138–39) suggestion that 'cursus monuments have for some time been linked with other traditions where the connections of alignment and location at the Dorset and Stonehenge Greater cursus monuments seem to have been both physical and conceptual'.

The identification of potential aurochs movement, and therefore presumably domestic cattle movement, by Bryony Rogers et al. (2018, pp. 127–52) through the analysis of strontium, carbon and oxygen isotopes carried out on two aurochs' mandibular M3 teeth found at Blick Mead and the resultant suggestion of a potential link between cattle movement and the chalkland belt of the British Isles (Jacques et al. 2018, pp. 136–46) established the range of my data set, as Rogers et al.'s (2018, p. 141) oxygen isotopic data ($\delta^{18}O$) values for the aurochs' teeth from Blick Mead (BM421 and BM422) appear to highlight that both individuals came from either the same region or from different regions with similar ($\delta^{18}O$) values – although there is some ambiguity as to which actual regions these were. The highest oxygen ($\delta^{18}O$) values indicate that the aurochs perhaps originated in parts of eastern England, while the carbon ($\delta^{13}C$) isotopic results clearly identify that the herds of aurochs remained on grassland throughout the spring and summer months and only split into smaller groups to head away from these grassland plains to winter within the forests on acorns and browse (Rogers et al. 2018, pp. 127–52).

The objective of the study was to expand upon my earlier evidence for large herbivore movement within the pre-Stonehenge ritual landscape which had been identified during *Stonehenge: A Landscape Through Time* (Saunders 2018) and to investigate potential correlations between a greater number of Neolithic cursus monuments and the movement of large herbivores. The suggested methodology for the movement of aurochs, and by association domestic cattle, across the Stonehenge Plain appears to be supported by the discovery of aurochs footprints along the route suggested by Saunders (2015, p. 59) during excavations of a Mesolithic layer adjacent to the springline at Blick Mead in October 2017. Although this clearly places aurochs within the vicinity of the landscape near to where the later cursus monument would be constructed and appears to identify at least one potential route that cattle took across the landscape, it obviously does not form a correlation between the two. To reach that point, the study will need to investigate a number of factors, such as whether the correlation was between cursus monument construction and large herbivores generally, or whether it is possible to be more specific and determine which large herbivores the correlation was with, such as deer, aurochs or domestic cattle.

The study area stretched along the English chalkland belt from the Yorkshire Wolds to the south coast. While the selection of this geographical area resulted in the omission of several large cursus monument sites, such as the Thornborough Cursus and the Scorton Cursus in North Yorkshire, the East Adderbury Cursus near Banbury and the Potlock Cursus in Derbyshire, it has been one of the largest studies of English cursus monuments yet undertaken, providing sufficient data to ascertain whether a combination of the natural topography together with the winter flooding of springs and rivers would have led to a concentration of cattle movement at relevant points of the landscape during the early spring, and whether a correlation could be identified between the movement of cattle, either domestic or wild, and the placement and alignment of Neolithic cursus monuments at these precise locations.

CHAPTER 1

Background

Archaeological History of Cursus Monuments

Before starting to investigate any potential reason behind the placement and alignment of cursus monuments, I feel it is prudent to spend a moment outlining their archaeological history. This will perhaps identify some of the problems associated with recognizing the majority of cursus monument locations through cropmark production and may ascertain some of the reasoning that lies behind the current accepted distribution pattern for these monuments.

On 6 August 1723 William Stukeley (1740), a Lincolnshire minister with strong antiquarian interests, discovered the first monument to be classified as a cursus, around 1,000 m north of the main Stonehenge monument. Although he appeared to trace the full length of its course and initially appreciated its square-ended enclosure form, by the time his findings were published (1740) he had convinced himself that this elongated enclosure was in fact a Roman chariot racetrack and thereby falsified his drawings to coincide with this later theory. He also decided to totally ignore his previous observations that the cursus monument terminated some 40 m short of a long barrow set across its alignment. However, it was these factors which led Stukeley (1740) to name this new style of monument a 'cursus', stating that the western end curved into an arch like the end of a Roman circus.

It was nearly a century later before Sir Richard Colt Hoare (1812, pp. 157–58) recognized the Stonehenge Lesser Cursus, the length of which he mistakenly extended to fall in line with Stukeley's original Roman chariot racetrack theory. After the death of William Cunnington, he also publicized Cunnington's recognition of 3 km of the Dorset Cursus on Cranbourne Chase (Colt Hoare 1819, p. 33), where he again believed it to be a Roman chariot racetrack, even though it crossed several watercourses and a bog (Brophy 2016, p. 14). However, these three monuments were to remain the sole cursus monument representatives for the next 124 years.

It would require the invention of the aeroplane and aerial photography before more buried cursus ditches were to come to light. In 1922, Air Commodore Clark Hall noticed strange marks on RAF aerial photographs of the downs near Winchester that turned out to be Celtic field systems. However, a further ten years passed before Major Allen, an interested observer, was to realize the full potential of this new knowledge. Allen brought his suggestion to the attention of Edward Thurlow Leeds at the Ashmolean Museum. At the time, Leeds (1934) was excavating a ditch that appeared to be either Early Bronze Age or Late Neolithic. However, after speaking with Major Allen, he identified that cropmarks in one of the photographs appeared to extend the line of his ditch. In the next edition of the *Antiquaries Journal*, Crawford (1935) responded to this idea by suggesting that Major Allen's rectangular enclosures were in fact 'cursuses', placing in print for the first time the Anglicized plural for the Latin 'cursus'.

Due to Crawford's suggestions, the number of cursus monument discoveries expanded rapidly. Increased understanding of the techniques of aerial photography supported by excavations at both the Dorset Cursus and the Dorchester-on-Thames Cursus (Atkinson 1951) enabled the number of cursus monuments to rise to fifteen by 1960, when the English Royal Commission published *A Matter of Time*. This increased to nineteen, when Webster and Hobley (Webster et al. 1964) used the same techniques to identify several smaller sites within the Warwickshire Avon Valley.

With each following decade, the number of identified cursus monuments increased, reaching a total of twenty-nine when Gordon Maxwell (1979) identified through cropmark production that pit-defined sites in Scotland also belonged to this class of monument, and further increased to forty-five when Roy Loveday (1985) completed a nationwide trawl of cursus monuments for his PhD thesis.

To date, over 100 possible cursus monument sites in Britain and Ireland have been identified. However, as Loveday points out in the opening page of his PhD thesis (1985, p. 1), 'it is difficult to produce a definitive list of cursus monuments from cropmarks alone due to their production being critically linked to the permeability of the soil, the type of crop growing in the field, seasonality, the amount of rainfall and the underlying geological solid'.

Heavy moisture, retentive soils and areas of forest or pasture can each act as blankets obscuring large areas of the British Isles, thereby limiting the ability of aerial photography to identify cropmark production. This is supported by the discovery of the Stadhampton Cursus which only occurred in 1986, even though the cursus monument is only situated 5 km from the great Dorchester-on-Thames complex in Oxfordshire, an area that had been carefully mapped by Major Allen some fifty years earlier, and by the recent discoveries of two cursus monuments at Stoke Hammond in Buckinghamshire in 2011 and a further two monuments at Clifton Reynes, also in Buckinghamshire, during the exceptionally dry summer of 2018.

Brophy (2016, pp. 8–12) went some way to explain this problem by identifying that although some areas, such as the chalk uplands of southern England and eastern Yorkshire, are open to aerial survey and provide areas of cropmark potential, other areas which appear to consist of similar material, such as the Chiltern chalk ridge, remain largely unresponsive to cropmark production due to their clay with flint capping, while other areas might be masked by modern farming methods, areas of urban development and even differing practices between aerial archaeologists.

Roger Palmer of the Cambridge Air Photo Service (1979) identified that even areas which respond well to cropmark production are extremely dependent upon the weather and require the archaeological flier to pass overhead at precisely the right time. However, the increased use of Google Earth and of aerial drones, combined with a methodology more in tune with the detection of potential cursus monument sites, could perhaps overcome some of these current difficulties, effectively allowing more air miles to be flown over high-potential sites.

The current cursus monument distribution pattern may therefore be a direct result of receptive subsoils and reflect no more than the limitations of the technique. This appears to have been supported by the mass data modelling exercise undertaken by Green (2017, personal communication) as part of the *English Landscape and Identities (EngLaid) Project* which identified areas that appear to be impaired to cropmark production and by Brophy and Cowley (2005) who put forward the idea that the lack of cursus monuments in upland locations, on the west coast of Scotland and on some Scottish islands may have been a product of either lower levels of reconnaissance or the non-receptiveness of the landscapes for cropmark production. Therefore, Loveday's (2006, p. 133) belief that there seems to be remarkable consistency in the siting of cursus monuments, in each case the location appearing either to be in a chalkland valley or on a flat expanse of river terrace gravel, could have arisen solely from the limitations of the technique.

Harding (Barclay and Harding 1999, pp. 30–38) believes that the siting of both the Rudston Cursus and Dorset Cursus appear to show a deliberate association with topography, illustrated by the way their boundaries resonate to regulate movement, suggesting these sites should not be removed from everyday social interactions but require to incorporate activities from within the surrounding landscape. Harding (1999, p. 34) further suggests that 'cursuses may have therefore acted as symbolic boundaries which constrained movement and interaction across the wider landscape', a concept supported by Hedges and Buckley et al.'s (2001, pp. 153–54) suggestion that the Springfield Cursus aligned with a strategic location where groups of people passed through to disperse up onto the boulder clay plateau.

Could Hedges and Buckley et al. (2001, pp. 153–54), Harding (1999, p. 34), Loveday (2006, p. 132) and Brophy (2016, p. 155) have each unknowingly stumbled upon the primary reason behind cursus monument siting? Could the reason behind the monuments' location lie in the fact that they aligned with naturally occurring open landscape that acted as a concentration point for large herbivore movement?

A Neolithic Economy

The initial theme for the classification of the British Neolithic – the introduction of agriculture – remained dominant throughout British archaeological research until recent changes in our understanding of the Neolithic. Ray and Thomas (2018, p. 20) suggest 'there is a continuing tension between approaches that address artefacts and monuments typologically and those that attempt to place the object into a specific context', where past tendencies have identified the Neolithic period primarily in terms of an economic phenomenon believed to lie within agricultural practice. However, Ray and Thomas (2018, p. 20) believe that 'these procedures lead to very different ways of representing the past, the former concentrating on the definition of particular kinds of entity at different scales of resolution while the second focuses instead on social practices and lived experiences'.

Although the domestication of various species appears to occur in the earliest Neolithic period, it should be noted that domesticated sheep and pigs could be kept in areas away from any actual living area, provided they were kept safe from predators. This has led Ray and Thomas (2018, p. 94) to suggest that 'it was the keeping of cattle that most tied those communities to their animals'. The owning of cattle started to provide a source of social and economic wealth that no Mesolithic community had previously enjoyed. However, Pryor (2015b, p. 78) believes that, apart from this, there would have been little difference between the last hunter-gatherers and initial Neolithic communities because 'Mesolithic hunters actually managed their game to such an extent that they were almost livestock farmers', through their controlling patterns of grazing management, such as using fire to clear woodland and shrub from around water holes to improve access. Manipulation of the environment would have allowed Mesolithic hunter-gatherers virtually to harvest both red deer and roe deer.

Although their herding and feeding patterns contrast significantly with those of wild cattle, Pryor (2015b, p. 78) suggests that 'when the first domesticated cattle, sheep and pigs arrived with Neolithic newcomers, they would have been readily accepted by the native British population', who could have adapted and extended the grazing areas already provided for wild prey. He (ibid., p. 78) therefore identifies any adoption of livestock farming as 'more a shift in lifestyle than a thoroughgoing revolutionary change'. Prior (2015b, p. 9) thinks that the assumption of earlier prehistorians that Britain's Mesolithic people moved around in a seemingly 'aimless fashion', following the migrations of game and the availability of other natural resources, is wrong, believing that (ibid., p. 9) 'Mongols, and indeed all other nomads did actually lead highly structured lives and there was no pointless wandering at all. Put another way, people followed known routes rather than their noses. And those routes would cover much the same areas of land from one year to the next', where each journey was planned to prevent interference with the migrations of other groups of nomads. Ray and Thomas (2018, p. 92) further suggest that 'if the indigenous component of this formation was modest in size, the number of incomers from north-east France, Belgium and other regions may also have been limited, so the composition of these groups may have changed over time'. They (ibid., p. 24) believe that 'directly or indirectly, these ideas would have had a profound impact on the way the British Neolithic was envisaged'. Especially how farming might have been adopted. Therefore, if farming was primarily a system that people could adopt or discard, rather than being an extension of cultural preferences, there was no reason to suppose that it might not have been adopted by indigenous hunter-gatherers who had been in contact with Neolithic communities from whom they acquired livestock.

Whittle's Bayesian statistical modelling (Whittle et al. 2011) appears to have eventually produced a consensus for the expansion of the Neolithic into the British Isles. Crane (2016, p. 80) believes the expansion followed the pattern of warmer summers and cooler winters that began around 4100 BC, peaking 'around 3800 BC when conditions reached their driest'. Therefore, for around 300 years the growing season had been getting gradually longer and lands suitable for grazing domesticated animals and cultivating crops had extended steadily northwards, followed by the Neolithic community's exploitation of these factors.

Zvelebil and Rowley-Conwy (1984, p. 74) suggest that the first Neolithic people in Britain left a totally invisible archaeological imprint upon the landscape, which would support Megaw and Simpson's (1979, p. 79) earlier suggestion that 'it would be reasonable to assume the earliest Neolithic monuments did not reflect the structures erected by the first, second or even the third generation of migrant agriculturists to settle in this country, a theory that would also hold true if Neolithic ideas had been taken up by the indigenous Mesolithic population. Thomas (1999, p. 13) does not deny the significance of the origins and spread of agriculture – he just does not believe that 'the cultural and social innovations were subsidiary to the inception of farming'.

Bailey (2007, p. 214) believes that this version of change, which appears to operate over different time spans across different parts of the country, raises important challenges with regards to achieving chronological precision, a factor further complicated by Griffiths' (2014, pp. 221–43) use of Bayesian statistical modelling that compares radiocarbon dates from rod microliths, which appears to indicate a significant overlap between the Late Mesolithic and the Early Neolithic material cultures. This would suggest that some form of ancestral Mesolithic lifestyle potentially persisted for at least the first 300 years of the Neolithic period. However, many of Griffith's (2014, p. 222) sites, such as March Hill Top, South Haw, Dean Clough, Dan Clough, Dan Bridge and Rocher Moss South in the Pennines and Lydstep on the western coast of Wales, are perhaps indicating the remnants of Mesolithic communities moving to the upland and coastal areas away from encroaching Neolithic populations.

This would appear to be supported by further investigations undertaken by Rowley-Conwy (2017, personal communication, December) on the isotopic results of fauna discovered during earlier excavations of the Coneybury Anomaly (Richards 1990, pp. 40–61). These appear to highlight that the groups of domestic cattle came from three distinct separate locations of cleared grazing ground around the Stonehenge Area, even though previous pollen records appeared to show that only minimal clearance had occurred at this time, while the wild animal fauna within the pit appears to have originated from forest cover near to the location of one of the domestic herds. Alex Bayliss's redating of the Coneybury Anomaly, giving a date range between 3760–3700 cal BC (Barclay 2014), has led Rowley-Conwy (2017, personal communication, December) to suggest it is highly probable this identifies a phased introduction of the Neolithic economy in the Stonehenge area where small Neolithic farming units were situated alongside indigenous Mesolithic populations.

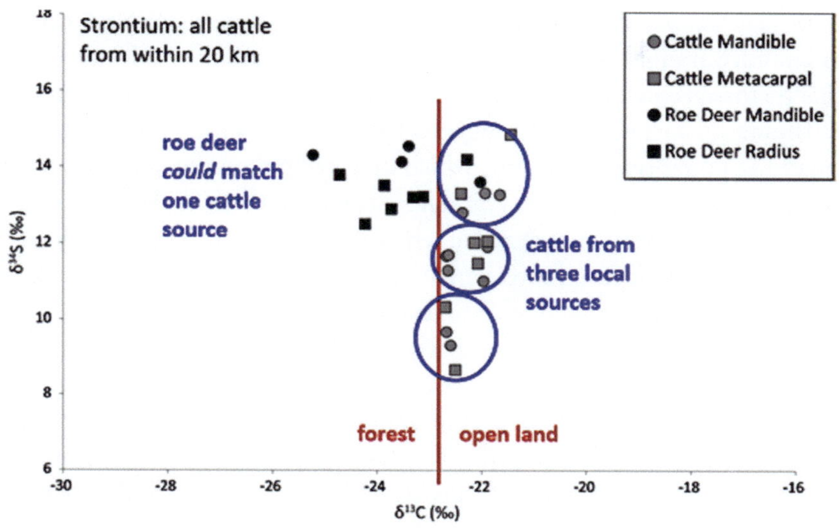

Figure 1: Sulphur and carbon isotopic results from fauna within the Coneybury Anomaly (After Peter Rowley-Conwy – Courtesy Proceedings of the Prehistoric Society, Open Access article)

The number of sites dating to the final period of the Mesolithic is growing, which is starting to result in the perceived chronological gap between Mesolithic and Neolithic sites being eroded. This could be due to the fact that a greater significance is now being given to tree throw holes across sites such as at March Hill in Northamptonshire and South Haw in the Pennines, which appear to contain Mesolithic flints that overlap with the earliest Neolithic presence in Britain. Recent excavations by Oxford Archaeology at the site at Stainton West, near Carlisle in Cumbria, also reveal a large occupation dating to the period immediately before the start of the Neolithic. However, Darvill's (1987, pp. 56–57) attempts to interpret Neolithic house structures from southern Britain during this period as farmstead settlements have been challenged by Herne (1988, p. 25) who suggests that this tends to overlook the un-domestic character of the deposits encountered within most of these structures. This has led Thomas (1999, p. 9) to suggest that although these may have been houses within the minimal sense of a domestic context, 'they may not have provided the year-round dwelling of domestic communities'. Thomas (ibid., p. 9) believes that although it was probable that 'some of these initial dwellings were lived in, at least on a temporary basis. But what is striking is how few can be unequivocally identified as permanent dwellings, of the sort associated with agricultural settlement in subsequent periods'. Thomas (1999, pp. 16) further suggests that this absence of substantial domestic architecture from the Early Neolithic period was a pattern maintained until well into the Bronze Age. He identifies that preserved Early Neolithic structures tended to be flimsy in the extreme and that the density of occupation material indicates a relatively short period of occupation.

Brophy (2016, pp. 175–76) suggests that initially, rather than cultivating large open fields 'the first farmers of mainland Scotland were probably planting crops amidst light woodland'. Yet within a few generations of their arrival the landscape was being radically changed. Brophy (2016, p. 176) uses the environmental pollen evidence from around the Early Neolithic timber hall at Warren Field to support his theory, which suggests that 'the building was located within a large clearance in the natural woodland' where cereal crops were grown nearby. Perhaps this started after harvested cereal crops were dropped within the vicinity of the settlement, germinating later to produce crop growth nearer the dwelling. The apparent lack of arable agriculture or sedentary settlement could potentially suggest that either Early Neolithic migrants were undertaking a lifestyle that was more pastoral than fully agricultural, moving with their herds along traditional paths and territories previously used by Mesolithic communities, or that communities that were essentially Mesolithic hunter-gatherers adopted some of these Neolithic material cultures while continuing to use the traditional routes and territories they had used for generations. Perhaps the actual transformation to a full Neolithic subsistence activity was a much more gradual event, an idea supported by Harding and Healy (2007, pp. 45–46) who suggest that historical continuity may have been as significant as cultural disjuncture during this period and by Thomas (2007, p. 73) who suggests that 'the Mesolithic populations had a dynamic role in the formation of the British Neolithic which involved interaction between the differing communities and that the Neolithic character could have been readily assimilated by local populations'.

Current evidence tends to suggest that the arrival of an agricultural economy was probably not the mainstay of an Early Neolithic culture, which could make it problematic to link the start of the Neolithic to this event. Earlier research, such as that by Case (1969, p. 181), appears to have clouded later views, perhaps due to his suggestion that Early Neolithic monuments required a phase of stable adjustment and it was only when a farming economy had been in place for some generations, ensuring the generation of sufficient surplus, that monuments could become a feature of the landscape. Looking at the number of worker hours needed to construct the various cursus monuments, Loveday (2006, pp. 144–46) also implied that a full agricultural economy would have been a requirement to increase population levels sufficiently to support the additional worker hours required. However, to date it has not been proven that complex monuments need to be based upon a system of sedentary agriculture. Dincauze and Hasenstab (1989, p. 73) identified that during the late Archaic period occurring at Cahokia in the lower Mississippi Valley between 1800 BC and 500 BC, a number of large enclosures were constructed by communities of sedentary hunter-gatherers. Hunter-gatherers and those practising garden horticulture or other simple forms of cultivation often had many hours of spare time when they were not engaged in productive labour, which suggests monument building need not have been dependent solely upon large agricultural surplus (Sahlin 1974). This is supported by Thomas (1999, p. 23) who believed that the case for monument building requiring the development of a resultant economic base due to the transition from hunting and gathering to farming is 'flawed'.

Thomas (1999, p. 24) further suggests that the current evidence for Early Neolithic domestic plants within southern England appears to represent only small-scale garden horticulture carried out on a sporadic basis. This would support the earlier research by both Evans et al. (1993, p. 188) who suggests that these plots could have been used for many years without soil decline or fall in yields, which would indicate a potential short-term tillage, rather than the foundation of permanent fields and by Burl (1987, p. 32) who suggests that Neolithic field systems are rare and that the Neolithic boundaries at Fengate, Cambridgeshire, may indicate paddocks solely for the control of cattle.

The earlier view appeared to cast monuments as some kind of optional extra or refinement to be indulged in only when the economic conditions allowed. However, Bradley (1984a, p.14) suggested that large monumental construction tended to be undertaken by dominant groups at a time when they were establishing their authority, or when they were under conditions of stress and instability. Thomas (1999, p. 38) went even further, believing these monuments had some degree of symbolic content, referring to things beyond themselves, where their influence had an active role in the actual process of social change.

Tilley (2010, p. 42) argues that the adoption of Neolithic traits was a highly localized and indigenous development and that Mesolithic places of significance and the pathways between them retained their significance into the Neolithic (Tilley 2017, personal communication, March). Darvill's (2006, p. 63) investigation into the Mesolithic postholes within the Stonehenge landscape could potentially support this theory that information was passed across generations. He casts doubt on the idea that all the posts within the Mesolithic postholes (Vatcher and Vatcher 1966) stood at the same time. This is because, although pine tends to rot within approximately fifty years, the radiocarbon dates from the various postholes show a date span of between 300 and 1,600 years. This suggests that no evidence from the previous post or posthole would have survived at the time of its replacement, yet future generations knew within a few feet where to position the new post. This strongly suggests that, through some form of communication (perhaps stories or song), ancient customs were being passed from generation to generation. This general dictum about the span and reach of memory has been supported by Bradley (2002, p. 8) who suggests that there appears to be a 200-year duration for the normal transmission of unaltered oral traditions. Ray and Thomas (2018, p. 52) suggest that the raising of substantial posts was used by Mesolithic communities to mark significant locations. This appears to be supported through the discovery in 2012 of an oak post bearing complex geometric markings, found buried on its side in peat at a site near Maerdy in the upper Rhondda Valley, in south Wales, which dated to around 4175 BC.

The elaborate designs have been compared to those on the large passage tomb of Gavrinis in the Gulf of Morbihan, perhaps indicating contact between the Mesolithic people from Wales and Neolithic communities in Brittany. Other apparent markers include the Down Farm Shaft at Fir Tree Field on Cranborne Chase, where a scatter of Mesolithic flints was discovered just below a layer of domesticated cattle and sheep bones, indicating a time span as short as 200 years between Mesolithic and Neolithic communities, while Mesolithic burnt flint, postholes and pits discovered at Perry Oak, within the boundary of the later Stanwell Cursus monument complex, would potentially indicate that the cursus monuments had been deliberately constructed to connect these earlier features. Ray and Thomas (2018, p. 58) suggest that what seems undeniable is that these locations were identified as places of enduring ancestral significance, perhaps attracting later activity of various kinds owing to the earlier occurrences and ceremonies that had occurred there, forming parts of the collectively remembered geography that enabled people to make themselves at home in a landscape.

Although this book is primarily about the alignment and location of cursus monuments, I feel it is important to spend some time discussing the Neolithic monuments that preceded them, to identify whether cursus monuments were potentially an extension of these earlier monuments. The Early Neolithic period within southern England first saw the construction of both earthen long barrows and causewayed enclosures. Initially, small wooden structures covered with mounds of earth, containing collections of disarticulated human bone, were constructed on the southern chalkland, to be followed by the first causewayed enclosures which appear to have been constructed amongst the slightly earlier long barrows.

The first Megalithic stage of Neolithic monument development appears to have been the construction of tombs and bank barrows. While earlier tombs may have had up to four chambers, each approached by a separate passage through the sides of the monument, in later monuments this chamber space was reached through a single entrance at the end of the mound, a factor that Thomas (1999, p. 48) believes identifies movement through a symbolically charged area. Paul Ashbee (1970) recognized the sites at Maiden Castle, Long Bredy, Broadmayne and Pentridge in Dorset as belonging to this bank barrow group. However, Loveday (2006, p. 89) suggests that from 'the cursus monument question, a number of narrow so called cursus cropmarks seem better interpreted as bank barrows (e.g. North Stoke and Llandegai)', drawing this conclusion from the remnants of bank barrow-like mounds which lay within the centre of the Scorton Cursus in Yorkshire and the Stanwell Cursus at Heathrow. Loveday (ibid., p. 89) argues that at the Maiden Castle bank barrow, three separate segments 'lay just off the crest of the ridge so the monument was sky lined when viewed from the Frome Valley'. This would support Stone's (1947, p. 11) earlier assumption that the Maiden Castle bank barrow could be viewed as 'a cursus monument on a small scale', which led Brophy (2016, p. 28) to suggest that these two types of monuments 'are often be discussed as two sides of the same coin'.

In Gathering Time, Whittle et al. (2011) were able to date twenty-seven of the seventy-four causewayed enclosure sites investigated. Nearly all these earliest enclosures, dating to around 3750 BC, occur in south-east England, the earliest dates appearing to have been obtained for sites overlooking the coastal plain of Sussex or located on and around the Thames estuary. This perhaps suggests that it was precisely the areas that maintained close contact with the continent where the development of enclosure first arose. Ray and Thomas (2018, p. 142) suggest that 'given the increasing contrasts in material culture that by this time existed between the continent and Britain, it is likely that the idea of enclosure may have travelled with individuals rather than representing a mass migration as such, and in these terms causewayed enclosures creation may have involved a deliberately adopted process'.

Crane (2016, p. 89) suggests that causewayed enclosures 'were an imported form, appearing first on the coasts closest to the continent'. Yet by the middle of the thirty-seventh century they had reached the chalk heartland and within seventy-five years open spaces had been enclosed in a contiguous region that extended from Lyme in the south to the Severn in the west and the Wash in the east with several more appearing in isolated spots as far north as the Firth of Tay. Explanations for the function of causewayed enclosures have ranged from enclosed settlements (Oswald 2001, 120–32), to cattle kraals (Piggott 1954, pp. 29–30), to exchange centres (Whittle et al. 1999, p. 354), to necropolis and cult centres (Pryor 1998, 363–71) – all of which suggest that established routeways already existed across the region prior to the commencement of monument construction (Whittle et al. 2011, p. 11). Earlier work by Bradley (1978, p. 103), Gardiner (1984, p. 21) and Evans et al. (1988, pp. 73–84) would appear to support Whittle's suggestion that causewayed enclosures existed at the edges of inhabited areas or groups of monuments. However, Legge (1981a, pp. 169–81) believed that faunal remains from causewayed enclosures would have represented only a small part of a much greater complex pastoral economy, probably also involving other sites elsewhere in the landscape.

Loveday (2006, p. 130) suggests a variation on this theme is that cursus monuments should be seen as an exaggerated representation of smaller rectangular houses. This would support Thomas's (2006) suggestion that the timber cursus monuments in Scotland appear to share a close affinity with the timber hall sites of eastern Scotland. However, Brophy (2016, p. 188) warns that although timber cursus monuments were constructed from posts, even if these do indicate separate bursts of post erection, they may have occurred only a few weeks or months apart. He suggests (ibid., p. 188) that 'the posts would have probably looked different from one another anyway, of different height and girth' and were in different states of preservation. This could support Pryor's (1998, p. 364) earlier interpretation that ditched cursus monuments, such as those at the Maxey Cursus near Peterborough, had their ditches constructed at different times producing a situation that appears more like a project in progress rather than a structure built to a predetermined plan. This could be identified as a deliberate act to continually disturb the landscape, perhaps to scare wildlife away from, or reduce the numbers of wildlife using, an area. Shearer and McLellan (2008, p. 66) have argued that the repetition and rhythm created through the sequences of timber replacement within Scottish timber cursus monuments could have been seen as being as fundamentally important as the original form of the monument. This has led Brophy (2016, p. 188) to suggest that these were creeping, moving monuments whose linear features consisted of fresh posts, rotted stumps and abandoned overgrowth. Perhaps, as Thomas (1999, p. 48) suggests, these boundaries served to constrain the conditions under which the space within the monument could be experienced. It was perhaps this desire to influence movement that can be seen in the development of the more linear aspect to these monuments in the period after 3700 BC (Whittle et al. 2011), such as in the development of passage tombs in Ireland.

Whether or not the birth of the British Neolithic was, as Ray and Thomas (2018) propose, a co-creation achieved through contact and interaction between Continental Neolithic people and the indigenous British population, what is known is that from around 4100 BC the improving climate allowed continental breeders and growers to enter the British Isles, to set up new economies and to leave a new footprint on the landscape. Crane (2016, p. 96) suggests this improvement continued, 'with the driest decades clustered around 3800 BC' allowing the new footprint to spread as far as northern Scotland. However, after this period of plenty, conditions became less favourable as solar intensity dropped, producing colder and stormier winters, where from around 3650–3600 BC cereal cultivation began to decline and the number of settlements fell. It therefore appears that around the time that northern communities were starting to develop their initial affinity with cursus monuments the resultant reduction in the nutritional value of previous high-yield grasslands would have been causing competition between domestic and wild animals. To survive, people had to forage for wild foods, moving with their domestic animals while becoming less dependent on permanently occupied sites and reverting to the more mobile lifestyle of their ancestors.

Types of Cursus Monument

Loveday (2006, p. 25) stressed the importance of placing the primary focus for cursus monument investigation upon sites which have been proven by excavation, as he believes this increases the security of data when trying to establish what constitutes a cursus monument. To comply with Loveday's suggestion, this study ensured that approximately 60 per cent of the cursus monuments investigated had either been physically excavated or had geophysical surveys undertaken. However, while I also agree with Loveday's other statement, on the necessity of cross-referring results from potential individual monument sites and comparing them with the numerous other cursus monuments, this does not appear previously to have occurred in a methodological manner. One of the first problems encountered during the investigation of cursus monuments was that of scale. Several of the largest cursus monuments extend for miles, such as the massive 10 km of the Dorset Cursus on Cranbourne Chase, while others such as the Stanwell 4 Cursus appear to be only 82 m in length. However, even with this vast size variation, Loveday (2006, pp. 25–26) believes that all cursus monuments can be defined no matter what their size, although he acknowledges that this does not resolve all the difficulties. This has led Loveday (1985, pp. 35–62) to suggest that appropriate divisions should be made within various groups of cursus monuments based upon their mathematical length. He (ibid., pp. 35–62) therefore categorized cursus monuments into four broad groupings:

> Long mortuary enclosures which extend up to 150 m in length and 25–30 m in width. Minor cursuses which extend 180–800 m in length. Major cursuses which extend 1,000–2,000 m in length and achieve 40–100 m in width. Mega cursuses which range from 2,700–5,640 m in length and demand distinction from the former group.

However, Loveday (2006, p. 26) found that, even within this framework, significant variation was still to be found. This caused him (ibid., pp. 28–31) to propose several subdivisions for these monuments. He noted that individual terminal forms appeared to differ according to their degree of curvature, and therefore devised five principle types – convex, flattened convex, precisely squared, irregularly squared and squared – although he suggests that there could be further degrees of variation within each of these forms. While each terminal end appears to take the same form across all individual recorded sites, he does accept that minor variations, appear to make one end more prominent. It should be noted that Loveday's (2006, p. 28) definition of rounded and squared terminals, normally referred to as Type A and Type B, and the further distinctive subgroup of rectangular sites, which appear to be laid out with geometric precision, termed Bi sites, give no implication of construction date.

Others, however, do not seem to have readily adopted Loveday's terminology, even though McOmish (2003, p. 9) suggests that 'the terminals appear to be the most important parts of the enclosure'. He (McOmish 2003, p. 9) indicates that on some sites, such as the Dorset Cursus, the terminal banks appear to be on a more massive scale than the banks on either side. Could it be that the terminal was the primary factor in changing the direction of some kind of movement along the parallel sides of the monument? Tilley's (1994, p. 175) earlier work appears to have supported this prominence of terminal theory when he suggested that the terminal end of the Dorset Cursus was further emphasized by the placement of a large barrow that outlined where the cursus terminal began. Tilley (1994, p. 178) also believes that the symbolic importance of this cursus terminal was marked out in the landscape through its relationship with four surrounding barrows and its intervisibility with a number of others that lay to the east. Brophy (2016, p. 64) later suggested that the terminal postholes at the timber cursus monuments at Dunragit, Castle Menzies Home Farm, Upper Largie and Douglasmuir were generally deeper, suggesting that 'the terminals may have been the visually dominant elements of these monuments, perhaps with posts graded in height towards one or both ends'.

Although most cursus ditches appear straight, closer inspection show some to be decidedly irregular. Anomalies are found in the Dorset Cursus, Rudston Cursus A in Yorkshire, the Drayton Cursus in Oxfordshire, the Springfield Cursus in Essex and at the Fornham All Saints Cursus in Suffolk, all of which have distinctive sinuous or angular sections. Richard Atkinson (1955, p. 9) also noted that one side of the Dorset Cursus ditch appeared inferior to the other – one side being prominent and fairly straight, while the opposite bank followed an irregular course. He explained the inferior side as potentially being built as a secondary ditch where less care had been taken during construction than with its neighbour. However, if the parallel ditches had been constructed to prevent or control movement across them, it would make sense to build the ditch closer to the item being restricted more prominent than the other ditch. In this manner, the design would operate similar to a First World War trench system, where frontline trenches were less prominent than second-line trenches – the initial obstruction acting as a deterrent, able to be breached but at a significant cost which made the second obstruction impenetrable. Loveday (2006, p. 120) further suggests that regarding earlier monuments, 'it often appears that the straighter cursus ditch is on the other side of the cursus'. He (ibid., p. 120) also proposed that the Stratford St Mary Cursus monument's continuous but adjusted construction from one terminal to the other suggests that the 'cursus monument width was determined in advance and that earlier monuments were not the primary aligning factor'. McOmish (2003, p. 12) suggests this ditch imbalance was intentional, with one side deliberately built on a more monumental scale. During Julian Richards' (1990, pp. 72–92) earlier excavations of several trenches at both the Greater and Lesser Stonehenge Cursus monuments, he (Richards 1990, p. 80) noticed that not only, as at other monuments, the terminal ditches appear larger than the side ditches, but also that the cursus ditch profile, width and depth varied as he moved along the monument.

This marked variability in cursus monument design also appears to exist when investigating the internal banks. While the presence of internal banks is illustrated at many excavated cursus monuments, such as the Greater Stonehenge Cursus (Christie 1963, pp. 370–82), the Dorset Cursus (Atkinson 1955, p. 7) and Rudston A in Yorkshire (Manby 1958), exceptions exist, such as at the Stanwell 1 Cursus (Lewis et al. 2010), the Scorton Cursus (Topping 1982) and the Cleaven Dyke (Richmond 1940), where a central bank or axial mound seems to replace the internal banks associated with most other cursus monument sites.

However, perhaps the greatest variability when investigating cursus monuments is the total replacement of the actual ditches themselves, as occurred with the use of pits or posts within lowland Scotland and Lincolnshire. Loveday (2006, p. 28) notes that excavation of these pits reveals that in most cases they once 'held posts', suggesting (ibid., p. 28) that perhaps they acted as 'mortuary platforms for the exposure of the dead'. He proposed that the in-turn of the side posts and the fact that the terminal and central division postholes were larger than the sides suggest that the site at Douglasmuir was 'originally a one-unit structure that was later almost exactly doubled in size' (ibid., p. 28).

Brophy (2016, p. 61) identifies this group of monuments as being 'a genuinely regional phenomenon with no convincing examples yet found outside Scotland', although Patrick Clay (2001, p. 9) has suggested that 'the fact no cursus monuments have been discovered within the Lincolnshire Wolds could be because linear pit or post alignment monuments were adopted as an alternative to ditched cursus monuments'. Clay (2001, p. 9) suggests 'examples of these monuments occur at Bag Enderby, Stenigot and Harlaxton', while other examples may also occur at both Ryhall and Oakham in Rutland and in the Nottinghamshire Trent Valley. Brophy (2016, p. 61) suggests that these monuments, which he terms 'timber cursus monuments', were some of the earliest monumental structures to have been built within the British Isles. Brophy (2016, p. 61) believes that these monuments, found mostly in eastern and south-western lowland Scotland, potentially embodied 'the germ of the cursus monument tradition that gradually expanded to the south'. These pit-defined monuments range between 60 m and approximately 200 m in length, however Brophy (2016, p. 63) believes that the width range appears to be the more important factor as the width of nearly all timber cursus monuments ranges between only twenty to 34 m. Brophy (2016, p. 63) suggests that 'in overall form, most of these monuments could be characterized as rectangular, with a few verging on trapezoidal' where the terminals range from almost perfectly square to rounded. However, at a number of sites the terminals have not yet been recorded. Using the accepted formula that a 1 m pit can support a 3.5 m post, there appear to have been variations in the height of the various monuments. At Bannockburn, the height of the posts appears to have been around 1.4 m while at Dunragit the side posts appear to have been up to 3 m in height while the terminal posts rose to 5 m. This suggests that, in some cases, ramps would have been required to enable construction to have taken place. Brophy (2016, p. 65) suggests that the construction of the larger sites would have required upwards of '300 such posts requiring a remarkable quantity of timber, probably accompanied by large-scale woodland clearance'.

Brophy (2016, pp. 186–87) believes that there was a dynamic nature to these timber cursus monuments which he classified as 'moving monuments'. He suggests that the wobbly boundaries and potentially segmented construction techniques hinted that 'various incarnations of these monuments occurred, where these relatively small enclosures were either replaced or extended through time or where similar enclosures were abutted onto one end'. Brophy (2016, p. 188) further believes that 'although we can identify specific settings of posts that may have formed a group for some reason or other, the overall plan of these monuments suggests there may be other ways of considering how they developed through time'.

It is important to proceed with care when investigating cursus monument construction dates. This is primarily due to the fact that many cursus monuments have yet to be excavated and therefore have not provided dating evidence. Of the fifty cursus monuments researched as part of this study, only 24 per cent have currently been dated. On one of these, the Eynesbury Cursus, the optically stimulated luminescence (OSL) methodology was used, while the Wolverton Cursus complex has only been dated through the discovery of Peterborough Ware pottery within the ditch fill. Although the dates for the remainder of the cursus monuments have been achieved through radiocarbon dating, the date range of between 300 to 500 years could appear problematic.

However, I think it is prudent to start any section that deals with the chronology of cursus monument construction with the earliest monuments that appear within the British Isles – the timber cursus monuments from Scotland. Brophy (2016, pp. 81–86) identifies that radiocarbon dates have been published for nine Scottish timber cursus monuments. However, he warns it is useful to bear in mind that in most cases the dates were retrieved from the analysis of charcoal, therefore the data could be biased towards the burnt elements of such monuments. He (ibid., p. 82) also suggests that 'the date may refer to heartwood from the centre of the tree that died long before the tree did' and thereby the dates could be up to several centuries too old. Loveday (2006, p. 28) also argues that the dates for the Scottish timber cursus monuments may be a little 'too early', which appears to be supported by the Bayesian statistical modelling analysis undertaken by Whittle et al. (2011, p. 830) on a suite of radiocarbon dates from Scottish timber cursus monuments from the eastern lowland region. Whittle et al. (2011, p. 830) believe that the outcome from this analysis suggests that the timber cursus monument tradition belongs to the middle rather than the early centuries of the fourth millennium BC. However, Brophy (2016, p. 84) suggests that this statistical approach suggested that Scottish timber cursus monuments were all built after 3700 BC. Using Whittle et al.'s (2011) potential expansion of the Neolithic, which identifies Scotland's Neolithic as commencing perhaps as late as 3800 BC, Brophy (2016, p. 84) believes that 'these monuments are still amongst the earliest large structures built by the first farmers in Scotland'.

Table 1: Scottish timber cursus monument construction dates (After Brophy 2016)

Cursus Site	**Date**	**Dating method**
Bannockburn	3400–3000 BC	Radiocarbon
Castle Menzies	4040–3660 BC	Radiocarbon
Douglasmuir	3930–3390 BC	Radiocarbon
Dunragit	3760–3630 BC	
Holm	3990–3660 BC	Radiocarbon
Holywood	3890–3370 BC	
Upper Largie	3800–3650 BC	Radiocarbon

However, Whittle et al. (2011, pp. 202–06) suggest, when looking at south Wessex, that we are still a long way from processing any reliable chronology for the first centuries of the Neolithic. This raises the question of how well the cursus monuments within my study area have been dated. The problem of dating cursus monuments through the use of radiocarbon is not new, being initially addressed by Bradley et al. (1983), although at that time very few relevant radiocarbon dates existed. Only twenty-four radiocarbon dates currently exist from English cursus monuments, plus a further thirty-two which are relevant through the dating of closely related monuments such as bank barrows and long enclosures (Barclay and Bayliss 1999, p. 11). The fact that cursus monuments are often relatively clean of finds has tended to result in a situation where material that is functionally or stratigraphically related to the construction of these monuments is virtually non-existent for radiocarbon analysis (ibid., p. 17). For example, antler tools found in the base or primary silts of cursus ditches provide a plausible suggestion that they were used to dig the ditch and then deposited. However, this cannot be proven with certainty (ibid., p. 17).

Barclay and Bayliss (1999, p. 23) also used Bayesian statistical modelling to provide dating evidence for a number of cursus monuments. The Dorchester-on-Thames Cursus has been dated via a red deer antler from the primary fill of the cursus monument ditch, where interpretation that the antler was used in the construction of the ditch provided a potential radiocarbon date for the construction of the cursus monument of 3380–2920 cal BC (ibid., p. 22).

The Dorset Cursus has been dated from another antler, this time recovered from the partially stabilized surface on top of the primary silt, rather than from the base of the ditch. This meant that the radiocarbon date for the Dorset Cursus of 3600–3030 cal BC was potentially later than the construction date, as the antler must have been deposited some time after construction (Barclay and Bayliss 1999, p. 23). However, two stratified radiocarbon measurements from beneath the bank and one from the primary silt gave an estimated date of construction of 3610–3380 cal BC.

The Springfield Cursus could only be dated from three samples of oak which were stratigraphically later than construction. Consequently, these samples only provide a *terminus ante quem* for the monument's construction, due to the fact that the samples could have a large age-at-death offset (Barclay and Bayliss 1999, p. 23). Although the Springfield Cursus has been dated to 3400–3000 cal BC, the uncertainty of this data caused Barclay and Bayliss (1999, p. 23) to omit the data from their tables.

The two phases of construction for the Stonehenge Lesser Cursus can be dated from antlers found at the base of the ditches for each phase of construction, which can be assumed to be related to the digging of the ditches (Richards 1990, p. 76). Barclay and Bayliss (1999, p. 23) suggested that the first phase of construction for the Stonehenge Lesser Cursus occurred at 3640–3130 cal BC while the second phase occurred at 3510–3030 cal BC. This initially suggested that it appeared likely that the Stonehenge Lesser Cursus had been constructed and remodelled prior to any initial construction of the Stonehenge Greater Cursus (ibid., p. 23), which had been radiocarbon dated (Richards 1990, p. 96) from material excavated during the 1940s (Stone 1947). This produced a date between 3015 and 2935 cal BC. However, as the material in question was thought perhaps to be intrusive, they (Barclay and Bayliss 1999, p. 23) warned that the date for the Stonehenge Greater Cursus could be far less reliable. This proved to be the case, for in 2007, as part of the Riverside Project, a new date was obtained for the Stonehenge Greater Cursus through the radiocarbon dating of an antler pick to 3630–3370 BC (Parker Pearson et al. 2009, p. 319).

At the Stanwell Cursus complex, Framework Archaeology (Lewis et al. 2010, p. 34) used a combination of optically stimulated luminescence (OSL) methodologies and the Bayesian statistical modelling of radiocarbon dates in an attempt to date the Stanwell C1 Cursus monument. The cleanliness of the site led to problems with the dating of the cursus monument, which resulted in them being able to infer only that the Stanwell C1 Cursus had been built within the span of other English cursus monuments, between 3640–3380 and 3260–2920 cal BC. They concluded that their attempt to create a calendrical Neolithic chronology of the Heathrow landscape at Terminal 5 through the use of absolute dates failed due to the poor preservation of suitable material, contamination by later material and the clean nature of the cursus monument.

At the Drayton North Cursus, Bayesian statistical modelling was carried out on radiocarbon dates that came from material within a tree throw hole that had been sealed beneath the cursus bank, providing reliable *termini post quem* measurements for the construction of the monument (Bayliss et al. 2008, p. 183). This provided an estimate for the construction date of the cursus monument of 3620–3390 cal BC. This date would be consistent with the theory that the Drayton long barrow aligned on the northern terminal of the cursus monument, as the long barrow was potentially constructed sometime in the mid fourth millennium BC (ibid., p. 184).

The Eyenesbury Cursus was dated from the late fifth to the mid fourth millennium BC (4830–3460 BC) where the calculation was based on a constant series of optically stimulated luminescence (OSL) measurements (Rhodes 2007, p. 61), although care must be taken when using OSL measurements due to the large correction factors of plus or minus 600 years that are involved with this methodology.

The remainder of dated English cursus monuments have only single radiocarbon dates available or the dates have not been subject to Bayesian statistical modelling. This has resulted in Barclay and Bayliss (1999, p. 25) stating that the dating evidence presented for cursus monuments is disappointing due to the fact that datable evidence is minimal.

Table 2: English cursus monument construction dates

Cursus Site	**Date**	**Dating method**
Aston-on-Trent	3700–3400 BC	Radiocarbon
Biggleswade	3340–3020 BC	Radiocarbon
Eynesbury	4830–3460 BC	OSL
Dorchester-on-Thames	3380–2920 BC	Radiocarbon
Dorset	3360–3030 BC	Radiocarbon
Drayton North	3610–3380 BC	Radiocarbon
North Stoke	3630–3340 BC	Radiocarbon
Stanwell	3600–3300 BC	Radiocarbon
Springfield	3400–3000 BC	Radiocarbon
Stonehenge Lesser	3640–3130 BC	Radiocarbon
Stonehenge Greater	3630–3370 BC	Radiocarbon
Wolverton	3500–3000 BC Late fourth Millennium	Peterborough Ware
Rudston	3750–3250 BC Second half of fourth Millennium	Radiocarbon

The Enigma of Cursus Monuments

As we saw earlier, potential difficulties exist when attempting to generate a definitive list of cursus monument locations through use of cropmark production alone. Heavy moisture, retentive soils and areas of forest or pasture can all act as blankets obscuring large areas of landscape, effectively limiting the application of this technique. However currently, except for a few examples in the Rhine and Mosel Valleys in Germany (Brophy 2016), cursus monuments appear to be exclusive to the British Isles and Ireland. Although they are found throughout the British Isles, ranging from the large mega cursus monuments of southern England to the pit-alignment monuments of Scotland, the question must be, what exactly constitutes the nature of these cursus monuments?

This is an area that many others have tried, and failed, to fully answer. McOmish (2003, p. 8) believes that few monuments, from any period in history or prehistory, are as enigmatic as cursus monuments. He sums it up by suggesting that cursus monuments 'have turned the concept of mystery into an art form'. Loveday (2006, p. 11) also uses the word 'enigma' when describing cursus monuments. He believed so strongly in their tendency to defy logic that he used the word in the title of his book, *Inscribed across the Landscape: The Cursus Enigma*. He suggests that the largest cursus monument, the Dorset Cursus, requiring approximately 500,000 worker hours to construct, 'appears to enclose nothing' (Loveday 2006, p. 11), further suggesting that as we follow cursus monuments down their huge linear scale, 'the same picture presents itself' time and time again'. This has led both McOmish (2003) and Loveday (2006, p. 11) to agree that the parallel lines of cursus ditches appear to lead nowhere and enclose nothing.

But perhaps that is precisely the point. Perhaps it is not the land enclosed within the cursus monuments that is important, but the land to one or both sides of them. Perhaps they do not lead anywhere, but are designed to either block or control movement occurring parallel to them or across them.

McOmish and Loveday suggest that the problem is further compounded by a lack of excavation of cursus monuments, apparently as a result of their large size and their reputation for barren ditches and interiors which produce pitifully little by way of internal features, artefacts or dating evidence. Even the extensive excavations at the Stanwell Cursus complex (Lewis et al. 2010) undertaken prior to the construction of Heathrow Terminal 5 appear to support McOmish (2003) and Loveday (2006, p. 11).

Bradley (1991, pp. 209–19) and Tilley (1994, pp. 143–201) endeavoured to overcome these problems using phenomenology. In an attempt to understand the meaning of these enclosures, they examined how ritual participation may have taken place within the interior of the Dorset Cursus, identifying potential relationships with its distinctive architecture, associated burial monuments and immediate topography. However, despite Bradley and Tilley both separately walking the entire length of the Dorset Cursus, recording spatial relationships that potentially offered evidence for a structured procession through the enclosure, their interpretations remain problematic, as it is evident that these interpretations are not immediately applicable to other sites – although this obviously does not invalidate the methodology. I personally believe that the use of phenomenology does much to carry forward the cursus monument debate and did indeed use it, alongside other methodologies, within my own fieldwork. However, it should be noted that at this moment there appears to be no concrete evidence that ritual procession ever took place within cursus monuments.

From the outset, the concept of linking cursus monuments with some sort of linear movement appears to have been centred on the idea of people moving along them in some form of ritual or ceremony. Initially this was seen as Romans in chariots (Stukeley 1740), but more recently as some form of processional pathway. Sir Norman Lockyer's (1906, p. 311) suggestion that as in Bronze Age stone rows in Dartmoor, astronomical analysis of the Stonehenge Cursus identified it to be a processional road appears to be the earliest expression of this idea. This was followed by Stone (1947, p. 18) who argued that 'the Stonehenge Greater Cursus was a processional way' and by Atkinson (1956, p. 9) who suggested that 'ceremonies of a processional kind took place along the Dorset Cursus'. In Atkinson's research it appears to be a lack of material evidence that led to this theory. Bradley (1983, pp. 15–20) believed that cursus monuments could potentially constrain movement towards a specified location, where activities took place which were associated with deposition, burning and perhaps the dead. This was supported by Tilley (1994, p. 197) who used phenomenology to suggest that 'movement occurred in a specific direction'.

It appears that hitherto the movement envisaged has always been imagined as human movement, normally in the form of some ceremonial procession. Yet I would question whether the movement consisted of humans moving through the landscape. Could it instead be the movement of large herbivores, or for that matter humans moving across the landscape with their domestic animals?

Nonetheless, even without any existing material evidence, the processional theory continues to be one of the main things associated with cursus monuments.

Barrett (1994, p. 24) argued that the structure of many other Neolithic monuments could be understood around the principle of processional movement towards a focal point, while Lucas (1995, p. 140) continued along this theme, suggesting that movement within the cursus monument might elaborate upon processions which culminated in the deposition or the removal of human remains from the mortuary structure. However, Loveday (2006, p. 126) suggests that 'Major Cursuses seem too wide for processions and minor ones too short'. This is supported by Brophy (2016, p. 30), who suggests that 'cursus monuments were not physically suited for the control and definition required for focused ceremonial processions'. Parker Pearson and Ramilisonina (1998) go some way further, putting forward a notion that cursus construction may have actually stopped people accessing the area so that they became a virtual empty path where the route was for ancestors only.

While prevalent theories continue to link the role of cursus monuments to this concept of an arena for ritual procession, this appears to be primarily due to their massive scale and prominence within the landscape. Pryor (1982), McOmish (2003) and Loveday (2006), however, believe that these theories don't add up, due to the fact that they cannot be generally applied across all cursus monuments. Although some cursus sites could well have been used for processions, in the same way as a modern church is used for various activities such as marriage and death, this model does not fit across the board. Both Loveday (2006) and McOmish (2003) accept that some cursus monuments may well have been used as processional routes, although McOmish (2003, p. 8) has suggested that 'what we would perceive to be orderly processions would be difficult, or even impossible at many sites due to the various ground surfaces and types of vegetation incorporated within them'. He (McOmish 2003, p. 12) also identified that flowing rivers actually cut across several cursus monuments, which led him to ask, 'how would a procession operate across a river?'

This would also bring into question, just who was undertaking these processions? The variability of cursus monument construction once again suggests that there appears to be no constant theme. Lewis et al. (2010) argue that the Stanwell Cursus, Scorton Cursus and Cleaven Dyke each appear to be socially inclusive monuments, which would have been in full view due to their raised central bank, potentially representing communal co-operation rather than exclusion; while McOmish (2003, p. 12)) identifies that ceremonies undertaken within the Dorset Cursus would have been obscured from outsiders due to its high flanking banks; and Loveday (2006) and Brophy (2016) have both noted the apparent minimal scale of some cursus earthworks. At this moment, the notion of processional movement along cursus monuments continues to be problematic, as movement along the entire length would be something of a trial, particularly during the winter months, although that could potentially have been the objective of the procession.

This suggests that cursus monuments were originally laid out upon the landscape for reasons other than those previously discussed. Francis Pryor's (1982) excavation of the Maxey Cursus near Peterborough and Richard's (1990) excavation of the Stonehenge Cursus show that they both appear to have been backfilled virtually straight after construction. This led Pryor (1998, p. 364) to suggest that the Maxey Cursus appeared more like a project in progress rather than a structure built to a predetermined plan. He (ibid., p. 364) believes that people may have visited the site on a seasonal basis, adding new bits to the structure year after year, indicating that the actual construction activity appears to be more important than the actual cursus architecture. If as Pryor suggests, the actual activity of disturbing the landscape was the important factor, perhaps this was to bring closure to the site's previous uses. Perhaps, by changing the landscape, people were also altering the way it would be used by wildlife in the future.

Looking at the actual design and construction of cursus monuments, McOmish (2003) notes that, despite the occasional wiggle, they were almost obsessively straight, which gives encouragement to the idea that movement occurred from a start point to a finishing point with an apparent visual focus upon one end of the monument. Bradley (1991, pp. 209–19) and Tilley (1994) suggest that movement along a cursus monument occurred in one direction only. If, as I am proposing, cursus monuments align with routes taken by large herbivores to traverse the natural topography, it could either suggest that animals moved in a circular route across the country or that hunting occurred in only one of the seasons where any form of migration took place.

Brophy (2016, p. 158) also suggests that many cursus monuments were planned, with topography in mind, due to the fact that the majority seem to occupy low-lying positions that run alongside or across valleys and streams, often resulting in the apparent linking of watercourses. His original thought during his earlier PhD investigations (Brophy 1999) was that, 'as rivers could potentially be seen as both life-giving and dangerous places, the cursus monument could be seen as a symbolic river built in response to this paradox of nature'. However, in Brophy's later book (2016, p. 159) his position appears to have changed slightly to one where the spatial connection is with the overall landscape upon which rivers, streams and springs flow. While Brophy's (1999) initial suggestion appeared to be in direct opposition to that previously put forward by Loveday during his PhD (1985), who noted that 'there was no obvious constancy between the alignment of cursus monuments and their relationship to rivers', this spatial connection to the overall landscape brings the two theories closer together. However, McOmish (2003, p. 13) added strength to Brophy's (1999) initial argument by noting that, at the Buscot Wick Cursus in the Thames Valley, 'the relationship between the two cursus enclosures, both in terms of relative size and orientation appear to mirror that between the River Thames and the River Cole'. When put together, this evidence brings to the foreground the notion that rivers, or perhaps winter flooding and springs, together with the overall topography of the landscape, may be significant factors in the actual location and alignment of cursus monuments.

At other cursus monuments, the actual direction of the monument alignment appears to be the significant feature. The Springfield Cursus appears to align on a smaller enclosure some 300 m away, while the Dorset Cursus alignment potentially implies movement from the north-east to the south-west, a direction which could reflect an interest in the midwinter sunset (Buckley et al. 2001, pp. 101–62). McOmish (2003, p. 11) also believes this preoccupation with the solar calendar could be seen at the natural chalk ridge situated at the western end of the South Dorset Ridgeway, where a concentration of Neolithic monuments including at least two cursus monuments were aligned on a knoll marking the end of the Ridgeway. McOmish (2003, p. 10) further suggests that many cursus enclosures could have represented 'the formalization of sections of long-established paths or routes carved out by thousands of years of either human or animal movement, inevitably restricting previous free access of the route for day-to-day tasks'. Could this explain the episodic construction, irregularities in alignment and disparities between the two sides of many monuments? Or might this just reflect necessary compromises between ancient pathways and the demands of monumental design? Loveday (2006, p. 13) sums it up by stating that 'this then is the nature of the challenge – to explain vast empty enclosures, often laid out with striking precision' for no apparent reason.

The lack of a coherent answer to the question 'What is the nature of a cursus monument?' suggests that either the wrong methodologies are being used or we have been asking the wrong questions. Atkinson (1955, p. 9) hints at a rite of passage for young men when he suggests that the Dorset Cursus may have had a religious or ceremonial rather than domestic function. He (ibid., p. 9) outlines the Roman festival of Lupercalia, in which young men armed with whips ran along a course striking at bystanders as they passed. Tilley (1994, p. 198) later turns this into a rite of passage theory when he suggests that novices are taken out of the mundane world into the bounded space of the cursus monument, while McOmish (2003, p. 13) suggests that such sites could have served as proving grounds for young men, where an association with artefacts such as arrowheads might suggest that hunting or archery was part of the test.

These various theories regarding the location and alignment of cursus monuments could each potentially have some degree of justification. However, whether looking out from the cursus monument as in phenomenology, acting as a testing ground for the rites of passage of young men or aligning with earlier monuments, distant aspects of the landscape or celestial events, each of these theories become problematic in that they do not appear to be readily transferable across all cursus monument sites and that quantifiable figures have not so far been produced.

I believe we have yet to discover the model that gives a consistent fit across cursus monuments, and we may have to proceed several stages further to make sense of the enigma. But could a methodology that combines the theories of cursus alignment with the natural topography of the landscape (Loveday 2006, pp. 114–30), together with the alignment associated with the natural concentration of large herbivore movement at river crossings (Brophy 2016, pp. 159–62) or springs (Marshall 2016, pp. 21–33), move the debate forward? Could it also be the case that these alignments led to the formalization of long-established paths (McOmish 2003, p. 10) through the concentration of large herbivore movement (Vera 2000, pp. 52–55)? Could Buckley et al.'s (2001, pp.153–54) suggestion that the Springfield Cursus aligns with the strategic location where groups of people pass through to disperse up on to the boulder clay plateau, perhaps in the spring and summer and returning in the autumn and winter, have picked on the wrong species? Could the alignment instead be symbolising the seasonal migration of large herbivore movement, either wild or domestic?

Potential Function of a Cursus Monument

The elongated shape and size of cursus monuments featured prominently when Atkinson (1955, p. 9), investigated the Dorset Cursus. He was the first to conclude that it was primarily an arena for religious or ceremonial processions. Although this explanation appears to have remained the favoured position since that time, it seems to be merely lack of evidence that led Atkinson to this theory. Bradley (1991, pp. 209–19) and Tilley (1994, p. 198) have attempted to reconsider the meaning of these enclosures by using phenomenology to interpret the Dorset Cursus as a path along which rites of passage could have been made tangible. Tilley suggests that it was bodily movement along the monument which made it meaningful to the participants (ibid., p. 198). However, it is evident that these interpretations are not immediately applicable to other sites.

This has led Johnston (1999, pp. 39–48) to challenge this view, believing that, if processions were to be seen as being generally ceremonial in nature, a number of conditions would be required to meet this concept, such as having participants and an audience, that it moved along a predetermined route from start to finish and that it had a purpose. However, Johnston (1999, p. 44) decrees that, with regards to it being a pathway, 'the cursus does not make sense when the earthwork is in place'. It appears that the construction of the monument actually infringes upon processional movement, suggesting that the construction of the bank and ditch occur when human processions no longer take place along the route, effectively creating a boundary from everyday use and passing it into the more sacred realms of the ancestors (ibid., p. 46). However, taking Johnson's suggestion one stage further, I would question who was undertaking the procession along the cursus monument? Was it necessarily a human procession, or could it instead have been a symbolic underlining of previous large herbivore movement, perhaps across a hunting zone in the landscape – or perhaps, as put forward by Jacques (Jacques et al. 2014, p. 29) even to memorialize or bring closure to this event. Or could it simply be concerned with the seasonal movement of domestic cattle by members of an initial Neolithic community?

Further studies (Bradley 1993; Chapman 2005, pp. 159–70; Loveday 2006, pp. 125–26; Tilley 1994, pp. 196–200) have introduced a number of other possible interpretations, such as the possibility cursus monuments could have demarcated an alignment with a celestial body. However, David Field's and Trevor Pearson's (2011, pp.35–40) investigations of the passage of the sun during the spring and autumn equinoxes at the Greater Stonehenge Cursus led them to suggest that, if the general function of a cursus monument was to act as some form of astronomical observatory, the anticipation might be that a greater degree of conformity in shape, size and orientation of cursus monuments would have occurred than is actually the case. They conclude that any alignment of the Stonehenge Greater Cursus was probably along already existing prominent man-made structures. Loveday (2006, p. 126) also believes that any notion of cursus monuments as celestial alignment devices is problematic, suggesting that 'they are too wide to have been usable with any approach to accuracy'.

Pearson and Field (2011, p. 36) suggest that cursus monuments linked earlier monuments and routes together. They (ibid., p. 36) suggest a plausible link with the Greater Stonehenge Cursus and the Amesbury 42 long barrow and the fact that the cursus monument terminated just short of the long barrow could suggest a deep measure of respect for the earlier monument. However, the lack of any significant man-made structure to the western end of the cursus has brought this theory into question, with Loveday (2006, p. 126) pointing out that 'although a very few cursuses link existing monuments, most do not'.

Francis Pryor (1998, p. 364), suggested that the Maxey Cursus near Peterborough was more like a project in progress rather than a structure built to a predetermined plan. This led him to believe that 'people visited the site on a seasonal basis, adding new bits to the structure year after year', indicating that that the actual construction activity was more important than the architecture. This was potentially supported by Pearson and Field's (2011, p. 37) excavations of the Greater Stonehenge Cursus ditches, which appear to provide evidence that the perimeter of the bank was originally formed through a series of conjoined mounds, suggesting periodic construction. However, they differ from Pryor in suggesting that this occurred for entirely practical reasons rather than a ritual process of construction.

Brophy (1999) developed the next cursus monument interpretation during his PhD thesis by arguing that Neolithic communities' lack of control over seasonal flooding resulted in cursus monuments being constructed within low-lying valley bottoms to create some form of symbolic river that was entirely under human control. However, this argument does not appear readily applicable within chalkland settings, although the ditches of low-lying cursus monuments probably did become water filled during wet periods. Brophy (2016, p. 158) later modified his earlier theory to suggest that many cursus monuments were planned, with topography in mind, due to the fact that the majority seemed to occupy low-lying positions running alongside or across valleys and streams, which often resulted in the apparent linking of watercourses. This later theory could potentially be supported by the fact that, when looking at the Stonehenge Greater Cursus, the possibility exists that Stonehenge Bottom either carried surface water or became marshland at certain periods (Saunders 2015 unpublished – after Jacques and Phillips 2014, pp. 7–27). This would provide some degree of connection with both Brophy's earlier and later theories.

Harding (1999, pp. 33–34), put forward the notion that the cursus monuments at Rudston created barriers to free movement along the Great Wold Valley, arguing that part of the process could have been the aim of dividing the landscape into fixed territories. Pearson and Field (2011, p. 37–39) support Harding's (1999 pp. 30–38) suggestion, arguing that the Stonehenge Greater Cursus could potentially have blocked an important north-south route through Stonehenge Bottom dividing the landscape into fixed territories. They (ibid., p. 39) also suggest a similar relationship between monument and valley system was demonstrated when they undertook a GIS catchment analysis of the Dorset Cursus. They suggest that, in both cases, the relationship with the local valley system is due to ritual significance attached to natural springs and watercourses (ibid., p. 39). However, I would question against whom, or against what, was the barrier to free movement intended? I shall be putting forward a theory that cursus monuments are potentially creating either a route for, or a barrier against, some form of large herbivore movement across either past hunting grounds or domestic cattle grazing areas, thereby acting as closure to an old lifestyle or the start of a new economy.

It appears that, while many of the traditional ideas about the function of cursus monuments fit across a few monuments, they fail to establish the reasons for placement and alignment across all cursus monuments. However, the 'best fit' appears to be a combination of the theories for cursus alignment with the natural topography of the landscape put forward by Loveday (2006) with cursus alignment associated with the natural concentration of herd movement at river crossings put forward by Brophy (2016). Therefore, the next sections of this book will focus on the questions: how open was the landscape environment at the point of the Mesolithic/Neolithic transition period and what evidence is there for animal movement throughout this period, especially across cursus monument sites?

CHAPTER 2

The Openness of the Landscape during the Mesolithic/Neolithic Transition Period

Traditional archaeological methodologies currently seem unable to establish the reasons why Neolithic communities chose the precise location and alignment for the placement of their cursus monuments, or for that matter the reason behind the placement of any Neolithic monument. Could further investigation into the theory that the cursus monument aligns with the natural topography of the landscape at points associated with the concentration of large herbivore movement move the debate forward? And were these large herbivores in general, or specific species such as deer, aurochs or domestic cattle? The answers to these questions may become clearer with further investigation into the general openness of the landscape during the Mesolithic/Neolithic transition period, especially with regard to whether any regional variations existed within what Rackham (1996) identifies as the 'wildwood'.

The Prehistoric Beginnings of the 'Wildwood'

Rackham (1996), in his book *Trees and Woodland in the British Landscape*, describes Britain's prehistoric forests by the general name of 'wildwood', suggesting that at the beginning of post-glacial Britain a series of different tree colonizations spread across the country from the south. Initially, at the end of the last (Devensian) ice age, climate change allowed birch (*betula*), aspen (*populus tremula*) and sallow (*salix*) to start replacing previous tundra grasses and heathers. Later, around 8500 BC, pine (*pinus*) and hazel (*corylus*) spread to replace the birch, which became uncommon until its recent resurgence. Pine in turn was generally replaced by oak (*quercus*) and alder (*alnus*), although certain areas, such as the Stonehenge landscape, appear to have retained their pine content for much longer (Branch 2014, personal communication).

Between 7300 BC and 4500 BC, this eventually gave way to lime (*tilia*) and elm (*ulmus*) and then holly, ash, beech (*fagus*) and maple (*acer*), although it should be noted that beech and lime penetrated little beyond the lowland zone of England. Huntley and Birks' (1983) use of pollen maps indicates that pine's early arrival into Britain is primarily due to its ability to migrate northwards at a faster rate (1,500 m per year) while elm, oak, lime and birch achieved a slower migration rate (up to 1,000 m per year). However, a tree's actual arrival time into Britain was further complicated by the various distances different species were required to migrate. During the last glacial period, oak and elm took refuge in the area of the Bay of Biscay while lime and beech had to commence their northward journey from Italy and the Balkans. Perhaps late tree arrivals were slow to establish due to the lack of vacant ground which they could occupy, having to wait for existing trees to die. Rackham (1996, p. 28) believes the arrival of lime began a period of relative stability where, to one degree or another, the whole of the British Isles became covered with forest until pollen profiles towards the end of this Atlantic period show a sudden collapse in the abundance of elm. Initial analysis of 'wildwood' pollen profiles around 4500 BC (Birks 1975) suggest the 'wildwood' consisted of a monotonous mixed oak forest with only a little lime and pine. However, Rackham (ibid., p. 28) later identifies that the reality would have been much more complex, suggesting the numbers for lime were greatly underestimated due to the fact that lime sheds significantly less pollen than oak. This is further complicated by the fact that tree density varies between species, according to the soil upon which they grow and the amount of water they are able to secure.

In the prehistoric 'wildwood', unlike present-day woodland, the areas around large trees were potentially full of saplings. When one of the giant trees fell, it would have created a gap, within which successors, often consisting of different species, would have been able to start to grow. At other times, whole areas of trees may have been destroyed by storm or fire, being replaced by an environment consisting of trees of a similar age.

Research carried out on Mesolithic sites at Woolaston on the Severn Estuary (Bell et al. 2005) and at Star Carr in Yorkshire (Milner et al. 2012) suggests that Mesolithic people repeatedly used woodland burning as a means of forestry management to improve their production of food such as hazel. Underwater excavations from the Mesolithic site at Tybrind Vig in Denmark, which dates between 5620–4040 BC (Andersen 2013, p. 213) have 'yielded the most important wood assemblage to date from the Late Mesolithic period', identifying that the local vegetation would have been characterized by human exploitation of the trees, especially the hazel bushes, which were used to furnish material for fish weirs and various types of hunting implements and tools.

At the Drayton Cursus, Barclay et al. (2003, pp. 102–08) identified seventy-eight examples of tree throw holes, covering around 30 per cent of the pre-alluvial ground surface. While this could be a result of wind throw, the fact that approximately 45 per cent of these holes show evidence of burning, either through charcoal flecks within them or soil scorching around them, strongly suggests some form of earlier human involvement. Worked flint discovered within the tree throw holes has led Barclay et al. (2003, p. 13) to suggest that pre-cursus tree clearance started in the fifth millennium and intensified during the early fourth millennium, while Grooved ware recovered from some of the holes potentially indicates that some form of secondary clearance also occurred. Human manipulation of the landscape had earlier been investigated by Ellis and Newsome (1991, p. 69) who suggested that the chalk downland in the Yorkshire Wolds went back to the Mesolithic period. These examples would enhance Zvelbil's theory (1994, p. 62) that, in a sense, Mesolithic people had started farming the forests and were indeed manipulating the local woodland area in order to produce the abundance of raw materials they required.

Whether the changes were due to natural elements or human intervention, the overall structure of the prehistoric 'wildwood' would have depended upon how particular trees fell. For example, a large beech, uprooted when still alive, crashes down on to its neighbours, creating a large gap in which other species may grow. However, a large lime rots at the base, therefore when it crashes down it leaves a stump from which its own sprouts may shoot, thereby continuing to occupy the original spot. A large oak usually dies standing, taking about thirty years for the root system to rot away, by which time the gap may have been filled by the expansion of neighbouring trees. Elm falls to pieces branch by branch, creating little surrounding damage, where any resultant gap tends to be filled by its own suckers.

Whitehouse's (1998) use of fossil beetle data identifies that a potential further range of natural disturbances and catastrophes, such as forest fires and storm damage, played equally important roles in the creation of open areas. This has led Whitehouse (Whitehouse and Smith 2004, p. 204) to suggest forest fires in particular, appear to have played an important role in maintaining the open character of at least some of these early Holocene forests, especially where they are dominated by flammable pine (*pinus*). This is supported by Wikars and Schimmel (2001, p. 197), who identified substantial numbers of the boreal insect fauna were fire-favoured, suggesting (ibid., p. 199) local vegetation patches and islands of openness may have emerged quite frequently, creating semi-permanent open spaces and the opening up of the woodland canopy. However, total catastrophes probably played only a minor part in the shaping of the 'wildwood' for, with the exception of pine, no British woodland can be totally destroyed by fire. And although storms can uproot trees, the hurricane in southern Britain during 1987 highlighted that the majority of uprooted trees were from planted woodland rather than ancient woodland. This suggests that prehistoric woodland would potentially have suffered less from storm damage than modern-day woodland.

The presence of grass pollen, including occasional grains from plants such as buttercup, cuckooflower, ragged robin, bugle and devil's bit scabious which do not flower in shade, identify that the 'wildwood' was not continuous. Glades, where deer, aurochs and domestic cattle could have congregated to eat tree saplings, coexisted alongside areas of wood-pasture perhaps initially created by Mesolithic hunter-gatherers and then later by Neolithic pastoralists' periodic burning of the landscape.

However, one of the greatest influences within the 'wildwood' was the decline of elm. Within a couple of centuries around 4000 BC, half of the elm vanished from across Europe. While Ingrouille (1995, p. 205), suggests the decline of elm was perhaps an attractive marker in the study of prehistoric landscape development, it appears that some landscapes were already open by this period and that lime, also a potential provider of browse, did not show any marked decline.

The theory that elms were selectively felled by barking to open the woodland for agriculture appears to have been rejected due to the fact that elm pollen levels fell by over 50 per cent. This would have required the felling of between forty-seven and eighty million trees within a very short period. Neolithic industrial use of elm bark for purposes such as the manufacture of twine or rope and for weaving into clothes was also an unlikely cause, as the decline appears to have occurred throughout Europe, even in areas where there was marginal human habitation.

Rackham (1980, p. 34) suggests that the most likely explanation for elm decline, given experiences in Britain from the 1970s episode of Dutch elm disease, is that elm (*ulmus*) suffered an episode or a succession of episodes of some form of elm disease. This followed on from earlier work undertaken by Brazier (1979, pp. 78) who identified that in Dutch elm disease the fungus *Ceratocystis ulmi* invades the tree's wood tissue, interfering with the tree's hormones, being transported from tree to tree by the elm bark beetle, *Scotytus*, which allows for rapid infestation over large areas within an extremely short time span.

Crane (2016, p. 85) believes the decline in elm was probably due to various prehistoric practices which 'may have encouraged the spread of the disease'. Woodland margins that had remained constant for millennia appear to have been more resilient to the disease than margins created suddenly by felling. These older margins appear to have acted as a protective wall around the mature elms which continued to thrive within the wood. However, the practice of pollarding by both Mesolithic and Neolithic populations appears to have exposed the mature elms to invasion by the beetles, due to the fact that the trimming of trees, which encouraged them to grow faster, also made them more prone to disease. Although elm disease saved millions of hours of tree clearance, it came at a cost. Not only did it see the loss of a good source of building material but, as elm's interlocking grain did not split, it also produced by far the most effective bows.

There are a large number of potential explanations for the openness of the landscape during the Mesolithic/Neolithic transition period. Was it manipulated by the local indigenous or immigrant population (Bell et al. 2005; Andersen 2013), was it a consequence of the grazing of large herbivores (Vera 2000), was it due to disease (Ingrouille 1995) or was it a consequence of climate change? It is currently difficult in most cases to accurately establish what happened to cause tree decline throughout the Mesolithic and Neolithic periods. To answer these questions will require investigation into each geographic region upon which later cursus monuments were constructed to establish whether variation within the degree of landscape openness were due to natural causes or other factors?

Vegetation during the Mesolithic/Neolithic Transition Period

Returning to the question of what type of landscape environment existed at the point of the Mesolithic/Neolithic transition period, Thomas (1999, p. 31) suggests woodland clearance for this period was nothing new. The tradition of woodland clearance appears to start in the Late Mesolithic when areas were burnt off and opened out to manipulate the forest flora, thereby stimulating browse for ungulates (Mellars 1976; Moore 1996 & 2003; Simmons 1975). The primary difference between Mesolithic and Neolithic periods seems to have been the extent of woodland clearance, where more extensive clearance appears to have taken place at the start of the Neolithic (Bell et al. 2005). However, did the introduction of British Neolithic agricultural systems see a fundamental change in the way people lived and interacted within the landscape? Thomas (1999, p. 25) questions whether arable farming and permanent settlement indicates significant aspects of Neolithic life, while Barrett (1994, p. 359) suggests that during the Early Neolithic, the non-intensive agricultural practices of horticulture and animal pasturing were potentially added to activities already undertaken by Mesolithic groups.

Barrett (1994) and Tilley (1994) believe these aspects of Neolithic life indicate how people perceived the landscape around them, indicating they might centre round rights of access rather than ownership, where the concept of territory may not have been relevant to communities which were still essentially mobile. Clay (2001, p. 2) further suggests that the concept of space and access regarding any particular landscape was potentially based on historical knowledge of the area passing down through the generations, previous sequences or events within the landscape having an influence upon the group's response to the area. Good experiences might have led to an area being frequently re-utilized whereas bad experiences might have led to the area being avoided. This supports Edmonds' (1999, p. 4) earlier suggestion that although landscapes were initially created in step with the understanding that communities had of their worlds, future generations would have been affected by this process adding myths or historic pasts to serve their own particular interests.

Although it is difficult to specify the precise balance between hunting and gathering on the one hand and farming on the other, Ray and Thomas (2018, p. 54) believe there is evidence to support the notion that hunting and gathering remained integral to the subsistence of people living in Britain after 4000 BC. They (ibid., p. 83) base this on the importance of migratory Neolithic populations acquiring an understanding of their new homeland prior to transferring one's way of life into a new environment. This would include the need to understand the soil and the types of vegetation that could be collected or harvested from it; the prevailing winds, precipitation and any water sources and how these will affect the growing season; and how wild animals within the area could affect both the growing of crops and the keeping of domestic animals. While hunter-gatherers appear to have been good colonizers, flexible and adaptable in the ways they could accommodate their way of life into new conditions, less of a case can be made for farmers, for whom minor variations in ecological conditions could mean the difference between life and death. However, Ray and Thomas (2018, p. 54) believe that what remains undisputed is that there appears to have been a relatively rapid transformation in the outward appearance and seemingly also in the basic concerns of communities in Britain between 4100 and 3800 BC that manifested itself in the use of pottery, polished tools and the domestication of animals, especially cattle. However, they (ibid., p. 54) stress that it is important not to assume that the artefacts and practices that manifested themselves in Britain in the fifth to fourth millennia could have only travelled alongside closed groups of pioneer people as they could have also been traded between different social groups which led totally different kinds of lives.

On the chalklands of southern England, the prehistoric landscape's environment was initially broadly proven and then further defined through research undertaken by Evans (1971, 1972, 1975; Evans and Jones 1979). Further research by Scaife (1980) and Thomas (1982) appeared to reaffirm this earlier work, as did the molluscan research undertaken by Mike Allen at Cranborne Chase (2002, pp. 54–68; 2007, pp. 263–73), within the Avon Valley as part of the Stonehenge Riverside Project (Parker Pearson et al. 2003, pp. 6–8; 2004, pp. 218–48) and within the greater landscape of the southern English chalklands (2017, pp. 144–64). This differs significantly from previous thinking – that the end of the last (Devensian) ice age resulted in forests which blanketed the whole country (e.g. Godwin 1975; Iversen 1973), where Neolithic communities later cleared pastures and plains to provide grazing for herds of domestic cattle and for the construction of monuments.

However, the study of post-glacial woodland succession on the British chalkland areas appears more complicated when compared to other regions of the British Isles. This is mainly due to difficulties in reconstructing the chalkland palaeo-environment, a fact recognized by archaeologists since the 1950s. Piggott (1954, p. 5) initially suggested that this was due to the lack of long pollen sequences presenting problems within this special environmental resolution. This means we need to differentiate more between the different types of woodland. Open woodland, with its herbaceous sub-canopy, would offer different opportunities and resources from those offered within closed woodland where the sub-canopy was strewn with fallen and rotten trunks. Evans' (1967 & 1972) examinations from a series of Neolithic soils buried beneath long barrows – such as Beckhamton Road, Horslip, South Street, West Kennet, Ascott-under-Wytchwood and Wayland's Smithy – provided examples of the type of landscape, as did Neolithic soils buried beneath the Windmill Hill causewayed enclosure, Silbury Hill and Avebury. This followed Evans' earlier work (1966), which had suggested that interpretations from the specific locations he had examined could provide representative examples for the wider landscape. However, a series of papers in Evans' memory has led Mike Allen and Julie Gardiner (2009, pp. 49–66) to suggest the assumption that these areas once supported forest is now challenged, for there appears to be significant variation regarding the presence of Neolithic woodland at a number of these locations. They summarize (ibid., p. 62) by indicating that the main landscape environment of the chalkland regions of Wiltshire and Dorset consisted of open post-glacial woodland, only developing into denser woodland in a few places.

Evans' (1967 & 1972) original *Age and Origin of British Chalk Grassland* has also been challenged due to analysis of pollen taken from mires on the chalklands of the Yorkshire Wolds (Bush 1988 & 1989; Bush and Flenley 1987). Evidence from Willow Garth Bog suggests an environment of open grassland had existed from early post-glacial times and still persisted into later prehistory. Their identification of significant variation occurring across the different buried soils led Rouse and Evans (1993) to suggest that it was at these precise ecological boundaries between areas of grassland and woodland that later Neolithic monuments were placed.

A theory often put forward by ecologists is that vegetation mosaics occur within most landscape environments which include natural gaps within the woodland cover – these glades encouraging both the growth of hazel, which requires open canopies and light to flower, and the growth of berries at the woodland margins. This combination of graze, browse and berries would potentially have allowed these areas to be exploited by grazing and browsing herbivores, making them prime Mesolithic and Early Neolithic locations, with a combination of human activities and herd movement further maintaining the scrubby downland glades. It can be no coincidence that these potentially open areas, such as Dorchester, Cranborne Chase and Stonehenge, attracted some of the first large human populations and, as a consequence, went on to become some of the areas of more concentrated Neolithic settlement and monument construction. However, this has been further complicated as Bell et al.'s (2005) research at Woolaston, Conneller et al.'s (2012) research at Star Carr and Andersen's (2013) research at Tybrind Vig all identify that a large degree of forestry management also occurred at major Mesolithic sites which did not go on to develop into Neolithic monument sites.

The anthropogenic perspective potentially underestimates the contribution played by natural disturbances. Fire and wind throw could have played an important ecological role within the creation of some forests, where the expansion of plants and animals appears to have been associated with cleared landscapes and pasture. Therefore, a mixture of both natural and human activity probably had a major impact on Holocene forested ecosystems. However, entomologists argue that the north European primary forest was potentially similar in structure to pasture woodland. This idea, supported by the conservation biologist Frans Vera (2000, pp. 52–55), indicates that the role of large herbivores maintaining open forests has been seriously underestimated. Nicki Whitehouse (Whitehouse and Smith 2004, pp. 203–12), enters this debate from the perspective of early Holocene insect assemblages. She identifies that dung beetles, usually associated with the dung of grazing animals, showed a persistent presence within many early woodland fauna. She believes this fossil insect evidence has fundamental implications in the interpretation of the palaeoecological record, suggesting environmental archaeologists and palaeoecologists appear more interested in the interaction between humans and the environment rather than identifying the ecological evidence which may have been preserved within palaeoecological records (ibid., p. 210).

Allen and Gardiner's (2009, pp. 60–62) research into various areas of the Wessex and Sussex downlands adds support to the vegetation mosaic theory. They identify evidence for generally dense closed-canopy woodland occurring in the north of Wiltshire, within which only small areas of more open-canopy woodland appear to have been represented. In contrast, the immediate environment of the Stonehenge area appears to have consisted of open-canopy woodland with the wider Stonehenge environment supporting a patchy course-grained vegetation, where any closed-canopy woodland potentially occurred on the sides of the Avon valley. Allen and Gardiner (2009, p. 61) also suggest that, an open park landscape was present in the Cranborne Chase area. This leads to the impression that during the Mesolithic/Neolithic transition period the South Wessex chalk downland consisted of a coarse-grained mosaic of closed-canopy woodland which contained significant degrees of open park landscape.

By way of contrast, Allen and Gardiner (2009, p. 61) identify that throughout the South Downs a closed-canopy woodland was present throughout the Mesolithic/Neolithic transition period. This would have resulted in early monument construction occurring within woodland clearings which were perhaps specifically created for them. Identifying a fundamental difference between the characters of two geographically close chalk landscapes suggests that the human response to these landscapes could have been affected by the difference in vegetation, where the South Downs chalk ridge appears to contain an unexplained absence of henges or cursus monuments. To gain a better understanding of these different environments, there follow a number of case studies across the various regions of the English chalkland belt.

Case Study for the Vegetation Environment of the Yorkshire Wolds

Willow Garth is a 5-hectare wet woodland site that lies in the Great Wold Valley approximately 4 km east-southeast of the Rudston Cursus complex. It is bounded by the winterbourne Gypsey Race chalk stream to the north, which in prehistoric times appears to have been frequently affected by several days' flooding each year. The woodland is flanked by arable fields which reveal a patchwork of peaty and sandy deposits. These enabled Bush (1988, pp. 453–62) to subdivide the site into five individual time periods (WG 1–WG 5), which allowed him (ibid., pp. 457–58) to identify that, unlike throughout most of lowland Britain where clear succession in most Holocene pollen diagrams shows birch invading but then being out-competed by later species such as oak, elm or hazel, at Willow Garth these later species appear to have failed to gain any kind of dominance. High levels of dry grassland pollen were also identified in the area, where use of general prehistoric British pollen diagrams would normally predict only around 5 per cent.

Insect taxa, characteristic of open grassland, provided further evidence of potential early landscape disturbance. This led Bush (1988, p. 458) to suggest that man was disturbing the forest around Willow Garth from the beginning of the recovery of birch (*betula*), which occurred around 6900 BC and potentially as early as 7200 BC. However, as palaeoecological evidence is not currently available from the Willow Garth site between the dates of 5980 BC and 2400 BC, it is not possible to state whether these grasslands remain throughout this period. Yet when pollen records for Willow Garth resume around 2400 BC, there is ample evidence for a mature grassland system.

We are able to fill in a lot of the missing data through earlier work carried out by Evans and Dimbleby (1976), who established Holocene palaeoecological records for the landscape of the Yorkshire Wolds during Manby's excavation of the Kilham long barrow. This lies approximately 2.5 km from the western end of the Rudston Cursus C and approximately 6 km from the Willow Garth site. While pollen preservation from the site was fairly poor, there was evidence for a period of chalk grassland prior to the construction of the barrow. This led Evans and Dimbleby (1976, p. 156) to suggest that it is possible that some break in the natural forest canopy existed since earlier

Mesolithic occupation. Earlier excavations undertaken by Manby (1963, p. 201) at the Willerby Wold long barrow, approximately 10 km to the north-west of the Rudston complex, allowed him to use molluscan data to identify that the long barrow had been constructed upon a grassland environment with woodland nearby. This led Manby (1976, p. 145) to conclude that much of the larger Rudston site complex was initially situated within an open grassland environment which had existed since the Early Mesolithic occupation.

Case Study for the Vegetation Environment of Cambridgeshire

An important element for any case study regarding the openness of the Cambridgeshire landscape during the Mesolithic/Neolithic transition period is the area around the fen edge. Earlier flooding of Doggerland (6500 BC) had interrupted the chalkland belt between Lincolnshire and Norfolk, to become what was later to be known as the Wash and the Fens, which resulted in the area remaining one of constant change.

Stuart (2006) identifies that, after the Doggerland flood, the sea encroached up the Witham Valley as far north as Lincoln, while the general coastline extended as far inland as Peterborough and approximately 5 km north of Cambridge. The rising water levels which flooded the fenland basin would have caused migrating herds to move inland, where, by the time of the Mesolithic/Neolithic transition period, they would have been required to cross the region in the very area where the later cursus monuments at Maxey, Etton, Barnack, Godmanchester and Eynesbury were constructed.

Alan Smith et al.'s (1989, pp. 207–49) investigations into the environmental impact of both Mesolithic and Neolithic activities upon the Cambridgeshire fen edge identify that a substantial opening of forest cover occurred around 6250 BC. This persists for some 700 to 1,500 years before the forest cover once again tries to re-establish itself around the time of the Mesolithic/Neolithic transition period.

Later investigations by French et al. (1992) in the lower Welland Valley allow them to identify the late glacial to early Holocene river development within the Maxey – Etton area. Here a low-sinuosity, braided river channel system was replaced with a higher-sinuosity, meandering river. Examination of twenty-five soil profiles from beneath the prehistoric earthworks of the Neolithic causwayed enclosure and cursus monument at Etton together with the cursus monument at Maxey enabled French (French and Pryor 2005, pp. 166–67) to suggest that, while the immediate river floodplain appears to be relatively open, much of the monument landscape was cleared of woodland just prior to the construction of the Etton causewayed enclosure. This would suggest that, prior to cursus monument construction, the majority of large herbivore movement would have been along the floodplain. However, during the life of the Etton causewayed enclosure, both the site and the river floodplain appear to have remained relatively open. This could perhaps suggest that the monument is increasing the ease of movement across the landscape, either to entice wildlife into the area, perhaps to make hunting easier or to ease the movement of domestic cattle across the area. However, as the monument reaches the end of its life, the monument landscape becomes increasingly dominated by thorn scrubland.

Case Study for the Vegetation Environment of the Chiltern Hills

The Chilterns dominate a landscape incised by valleys which rise to just over 900 ft. The high ground principally consists of a clay-with-flint cap which has resulted from the dissolution of the underlying chalk-with-flints deposits. This is unresponsive to cropmark production.

In the south of the county, pollen records (Wilkinson et al. 2000) show oak, elm, lime and hazel establishing in the area, while alder spreads westward down the Thames floodplain. This suggests that by 6900 BC, the period often referred to as climate optimum, the Thames floodplain was becoming closed woodland. This is supported by discoveries of Late Mesolithic sites at Chesham, radiocarbon dated from an aurochs bone 5006–4504 BC, and at Misbourne Valley (Farley 1983), radiocarbon dated from the bottom of tufa deposits to 5000–4000 BC, both indicating a closed-canopy landscape, of oak and ash, respectively.

Allen (Allen and Gardiner 2004) suggests that, along the Thames and other rivers in south-eastern Britain, there appears to be particular importance placed upon river islands and confluences which could have been used for early arable farming. However, charred emmer grains together with numerous hazelnut shells discovered in a midden at Lake End Road, Berkshire (3900 to 3500 BC), potentially shows that a gathering culture continued alongside any small-scale arable use of the landscape. This potentially suggests that the Thames Valley area to the south of the Chiltern Hills primarily consisted of a closed woodland environment well into the period when cursus monument construction was occurring.

To the north of the Chiltern Ridge, the Wolverton Cursus Complex nestled between the low-lying limestone hills of the River Great Ouse. Here the floodplain was transformed by the construction of a Neolithic mortuary enclosure, four cursus monuments and a hengiform pit circle at the point where two major braided palaeochannel systems bracket the monuments to both the north and south. Tree throw holes peppered the entire exposed gravel surface, suggesting a formerly densely wooded landscape. Significantly, the ditches of the cursus monuments consistently cut the tree throw hollows, implying that the monuments were situated in a cleared or at least partially cleared landscape (Hogan 2013b, p. 7).

In the Chiltern Hills, molluscan data extracted by Stafford during excavations by Hey et al. (2007) at Whiteleaf Hill, a prominent chalk ridge upon the Chiltern escarpment containing a Neolithic barrow, identified no evidence for open-country faunas within the pre-barrow deposits. Although some level of woodland clearance appears to have occurred prior to construction of the barrow, the small number of open-country colonists suggest this was not long beforehand. This potentially indicates a localized clearance in an otherwise wooded area, perhaps suggesting the clearance was temporary, possibly for the sole purpose of construction. Other molluscan evidence from the region (Pitstone), also suggests a mixed deciduous environment was present in the area between the sixth and the fourth millennia BC.

Holgate (1995, p. 3) believes that 'woodland clearance does not start within the valleys of the Chiltern Hills until the early second millennium BC'. This is supported by an English Heritage earthwork survey report carried out by Brown and Field (2000, p. 5) which suggests 'Bronze Age/Iron Age society of the Chiltern Hills farmed the slopes of the hills avoiding the heavily forested valley bottoms'. This suggests the Chiltern scarp consisted of open landscape while the valleys remained forested until at least the Bronze Age. This could have been the reason why, unlike in other areas where cursus monuments tend to be sited upon the floodplains, the potential cursus monument at Ivinghoe Beacon (Gover 2000) was constructed as an upland monument upon the chalk hilltop.

Case Study for the Vegetation Environment of the Thames Valley

The area of the Thames Valley can be split into three bands. To the south, the claylands and upland gravels of the Thame and Kennet river corridors cut through the topographic zones, providing both boundaries and corridors. In the centre, the Berkshire and Marlborough chalkland downs comprise a scarp edge bisected by a series of valleys. To the north, claylands and upland gravels once again dominate the large expanses of undulating ground along the vales of central and north Oxfordshire.

While French, Allen and Scaife (2007) each suggest that certain areas of the chalkland belt, such as Cranborne Chase and the Yorkshire Wolds, may never have been totally covered by forest, Hey (Hey and Robinson 2011, p. 80) by contrast suggests there was little existence of comparable environments within the Berkshire and Marlborough Downs. Similar to the Sussex Downs, a closed-canopy woodland appears to have been present throughout the Mesolithic/Neolithic transition period.

A woodland environment also appears to remain the dominant feature in the Late Mesolithic landscape of the Upper Thames Valley region, where there is little direct evidence for woodland clearance. However, at Ascott-under-Wychwood (Benson and Whittle 2007) and at Goring (Brown 1995), there were indications that some clearings in the woodland were used, perhaps opportunistically, by the local community. The dominance of woodland environment within the Upper Thames Valley region is supported by the number of tree throw holes across the site of the Drayton Cursus. Although these tree throw holes could have been a result of wind throw, the fact that about 45 per cent of these holes show evidence of burning, either through charcoal flecks within them or soil scorching around them, suggests some form of human involvement. This would indicate that tree clearance occurs just prior to construction (Barclay et al. 2003, p. 65). The dominance of woodland environment within the Upper Thames Valley region is further supported by Hey's investigations at the Dorchester-on-Thames Cursus, where she indicates that, with the exception of the area around the earlier long barrows, the cursus monument also appears to have been constructed within a closed-canopy woodland (Hey 2016, personal communication, February).

Case Study for the Vegetation Environment of South Wessex and Dorset

One of the most important results from the Stonehenge Riverside Project (Parker-Pearson et al. 2012) was the work produced by Michael Allen, Charles French and Rob Scaife, where new sequences of Holocene landscape change were discovered through their investigation of sediment sequences and their analysis of pollen and molluscan data. This has led Allen (Allen and Gardiner 2012, pp. 42–43) to suggest that previous interpretations that the prehistoric Stonehenge chalkland landscape consisted of a uniform post-glacial closed deciduous woodland are being brought into question, as evidence increasingly emerges that the woodland was in fact more open than previously suggested. French et al. (2012, pp. 2–4) support this theory, suggesting the notion of extensive forest clearance by the Late Neolithic for agriculture was too generalized for the Stonehenge landscape, believing the environment consisted of a mosaic of both woodland and grassland.

Earlier research undertaken by Scaife (1995, pp. 51–52) also supported this theory that the Stonehenge landscape was potentially more open than previously thought. He suggests that Pit 9580 had been dug in an open pine (*pinus*) and hazel (*corylus*) woodland cleared for the erection of the four post-pits. However, the fact that a new mollusc fauna established itself indicates there was open country in the vicinity, from which this fauna originated. Nick Branch (2014, personal communication), who undertook boreholing within the wider Stonehenge landscape as part of the Blick Mead project, suggests stable calcareous grassland may have greater antiquity than previously thought.

To the east of the Stonehenge landscape, Wessex Archaeology (2015) discovered a Mesolithic posthole (A-148) on Boscombe Down, which was confirmed through radiocarbon dating to have been erected between 8460 BC and 8250 BC. Analysis of land snail data from within the posthole identifies that the post was also originally situated on the edge of open countryside next to woodland.

To the west of the Stonehenge plain, the Wylye is a typical shallow chalkland river that meanders along a broad, fertile floodplain. It is flanked by the steep, rolling scarp slopes of Salisbury Plain to the north and the Great Ridge to the south. Thirteen Early Neolithic long barrows are located within the Wylye Valley. All appear to be intervisible with at least one other, their viewsheds potentially influenced by major topographical features. Allen and Gardiner (2009) indicate that the positioning could only have been achieved in a meaningful manner if the valley sides were largely unwooded. Pollen and molluscan analysis of the land surface beneath the surviving Corton long barrow demonstrates woodland clearance and the establishment of open grassland occur long before the construction of the monument.

Kennard (1936) was the first to identify that the Thickthorn long barrow at the southern end of the Dorset Cursus exhibited predominantly open-country faunas associated with rendzina grassland soils directly beneath the long mounds. Allen (2007) discovers similar results when investigating two long barrows on Gussage Cow Down in the upper Allen Valley. From this he (ibid.) is able to suggest that, not only had the barrow been built within a grassy clearing, but that the area had been clear for some time prior to any monument construction. Allen's (2007) investigations further discovered that other areas along the Dorset Cursus, such as those associated with the Wyke and those at Bottlebush Down, appeared open, while pollen data recovered from the Easton Down long barrow (Cruise 1993) indicates a mosaic of open-canopy landscape including grassland, bracken and hazel with woodland nearby.

Summary of the Openness of the Chalkland Belt

Using pollen, molluscan and fossil data, it has been possible to identify that large sections of the English chalkland belt consisted of open landscape during the Mesolithic/Neolithic transition period. Through a predominance of open-country snail shells around both Stonehenge cursus monuments, Allen et al. (2012) identified that the monuments were constructed upon open landscape. This is supported by research undertaken as part of *Stonehenge: A Landscape Through Time. An assessment of the evidence for large herbivore movement and hunting strategies within the Stonehenge landscape during the Mesolithic* (Saunders 2019, pp. 173–200), which also identifies that large sections of the Stonehenge landscape consisted of open landscape.

Regarding other regions, a combination of Manby's (1976) earlier research and their own enabled Bush and Flenley (1987) to identify that large areas around the Rudston Cursus in Yorkshire were open. Prior's (1982) identification of sandy silt within the Maxey Cursus ditches enabled him to indicate that construction occurred upon open landscape. The discovery of humus, derived from leaf litter, allows Hey (Hey and Hind 2014) to identify that the Dorchester-upon-Thames Cursus was constructed within a linear opening in the forest cover. A similar picture emerges from excavations along the length of the Dorset Cursus (Kennard 1936), while at the Drayton Cursus, Barclay et al.'s (2003) investigation of charcoal found at the site points to an oak-alder woodland that appears to have been cleared only at the time of monument construction. It is becoming increasingly clear that, while woodland may have been present on a handful of sites, most cursus monuments appear to have been constructed upon areas which have probably consisted of open countryside for generations.

However, differences in the landscape environment did occur across the various regions.

- The Yorkshire Wolds environment appears to have consisted of open grassland from early post-glacial times, continuing into later prehistory.
- The fen-edge environment appears to have consisted of woodland which was cleared around 6500 BC. This clearance lasted between 700 to 1,500 years, at which point the forest cover re-established itself. Further clearances occurred at cursus monument sites immediately prior to monument construction.
- The upper slopes of the Chiltern Hills appear to have consisted of open landscape while the valleys remain forested until the Bronze Age.

- This pattern appears to continue into the Thames Valley region.
- The Wessex region around the Stonehenge Cursus and the Dorset Cursus appear to have consisted of open-canopy woodland together with established dry grassland or open grassland conditions.
- The Sussex Downs appear to have consisted of dense closed-canopy woodland.

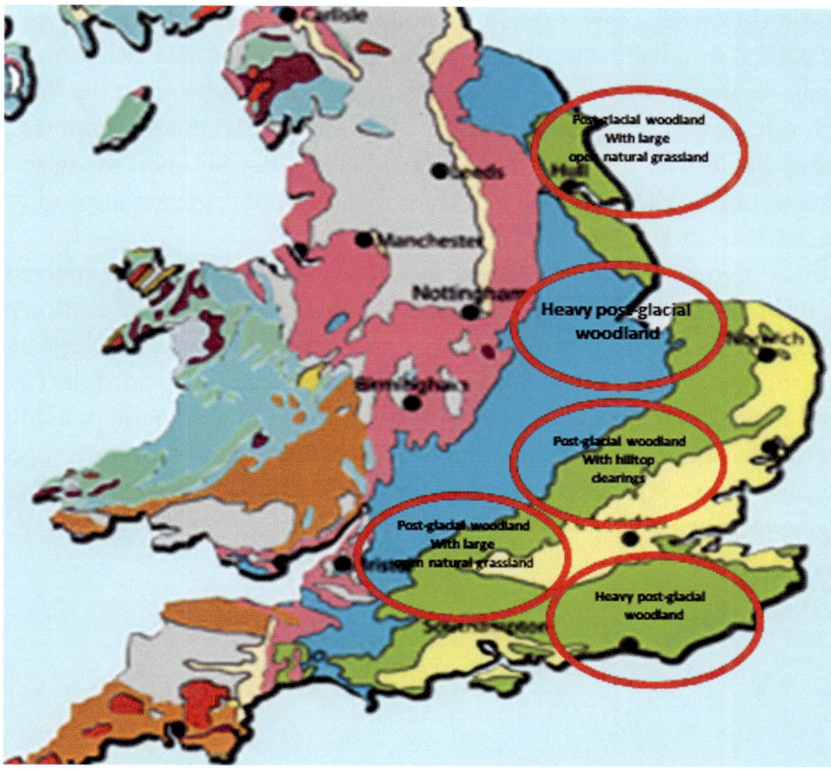

Map 1: Summary of vegetation environment regions. Reproduced with the permission of the British Geological Survey ©UKRI. All rights reserved.

An increase in palaeo-environmental data is finally enabling archaeologists to move away from site-specific reconstruction modelling to an overall environmental reconstruction. This has allowed the contrasting opportunities that broad-scale patterns of woodland gave to human communities to be modelled. Use of this technique has enabled Spikins (2000, pp. 219–34) to identify that, as earlier birch and hazel woodland was replaced by less productive woodland, Mesolithic communities shifted away from the lowlands to the more productive uplands. It becomes clear that various disturbance factors contributed to the Mesolithic/Neolithic transition period's patchwork style of woodland. While the greater section of landscape remained generally wooded, large sections of open landscape persisted, particularly in places that were kept open by a mixture of grazing animals and human intervention.

Vera (2000, pp. 52–55) highlighted the need to understand the various factors which disturb woodland. However, Vera's theory could have been placing too great an emphasis upon grazing herbivores, rather than considering all factors within the disturbance. While factors such as fire, floods and storms would have created openings within the woodland, once created, any opening constrained by topography was likely to be perpetuated by grazing animals. Research by Bell and Nobel (2012, pp. 80–92) has identified a seven-fold increase in herbivore abundance in woodland that had been manipulated by Mesolithic communities.

Palaeoecologists continue to establish the regional variability of woodland history across the British Isles. Allen and Gardiner's (2009, pp. 93–107) evidence from the Sussex and Wessex Downs regions identifies that Neolithic societies living on the Sussex Downs followed a different trajectory from those in Wessex. They suggest this was potentially because of pre-existing open areas in Wessex, which allowed increases in local population centres. This led to the adoption of significantly different types of material culture, depositional practice and monument construction from those found within the South Downs region, where there appears to be no evidence for large openings or obvious retardations to the woodland development.

However, did the Sussex Downs retain its woodland cover due to these differing practices, or was it that the natural woodland cover led to differing practices? Martin Bell (2016, personal communication) believes any answer may lie in the differing soil depths between the Wessex and Sussex Downs. 'The South Downs Integrated Landscape Character Assessment' supports Bell's assumptions, identifying that heavier soils within the 10 m of clay with flint upon the surface of the chalk throughout this section of the chalkland belt were able to support greater areas of woodland than chalkland areas without these features.

The fact that different regional Neolithic societies underwent different trajectories, and that this appears to be linked to the landscape, could perhaps indicate that a combination of factors such as the landscape's natural topography and the openness of the landscape may have been important factors in cursus monument placement and alignment. However, it is also possible that other reasons could lie behind their placement, such as the geology of the local area, perhaps holding some form of symbolic or ritual association, or the natural concentration of herd movement occurring at river crossings or passages between steep hills. Each of these aspects will be looked at in greater detail over the following chapters.

CHAPTER 3

Geology, Geography and Topography

We have seen that the differing soil depths between the Wessex and South Downs could have led to their various Neolithic communities following different trajectories. Is it also possible that different geologies at the local level likewise played a major part in a community's selection of the landscape upon which cursus monument construction would take place?

The geology upon which the cursus monuments within this data set are located includes a fairly even spread between chalk (34 per cent), clay (32 per cent), greensand (20 per cent) and muddy limestone (14 per cent). This potentially suggests that the bedrock geology upon which cursus monuments were constructed was not perhaps a primary factor in their location. But could specific local areas of geology hold the key for cursus monument placement? This idea came to me during excavations of the southern bank on the mini-henge at Marden in the summer of 2017, as part of the University of Reading Archaeological Field School, where I overheard a conversation between the dig director, Jim Leary, and a group of visitors to the site. This led to a long conversation between Leary and myself about his personal thoughts regarding the placement and alignment of Neolithic monuments in general, and of cursus monuments in particular.

Similar to Jacques' (Jacques et al. 2014, p. 29) suggestion that one explanation for the Stonehenge Greater Cursus could have been that it memorialized the special hunting grounds used by local communities for thousands of years, Leary (2017, personal communication, July) believes that the natural world was, in some way, being memorialized within many of these monuments and that the people who constructed them were perhaps celebrating, not just the monument itself, but the actual topography, geology and geography of the landscape upon which they were built.

This significance of geology appears to be supported by both Clive Waddington's (1999, p. 171) research into the Mesolithic/Neolithic transition period within the Milfield Basin, Northumberland, which highlights that the Mesolithic settlement had been located on the raised gravel terraces of the valley floor, providing easy access to a wide diversity of ecological zones and the identification by Philip Aikens, Member of the Council for the Suffolk Institute of Archaeology (2016, personal communication, November) that the Fornham All Saints Cursus had also been constructed within an area of significant drift geology. While the first raised terrace upon which the monument was constructed appears to consist of a narrow band of sandy gravel, within a few metres the landscape between the river and the cursus monument changes to one consisting mainly of peat. And immediately to the south of the monument, the geology of the landscape changes once more, this time to boulder clay.

Although it was constructed approximately 1,500 years later than the cursus monuments discussed within this book, the southern bank of the mini-henge at Marden appears to have been deliberately constructed at a point where significant deposits of yellow gravel would have been visible on top of the greensand bedrock. Leary et al. (2013) believe that a similar feature occurred during the construction of the nearby Silbury Hill. Leary et al.'s (2013) Silbury Hill Project, which developed as a response to the sudden dangerous collapse of a forgotten antiquarian excavation shaft that had not been correctly backfilled, enabled investigation of the previous excavation tunnel to be carried out between 2007 and 2008, which highlighted the prehistoric sequences for the mound's various construction phases.

Figure 2: Yellow gravel deposit on southern mini-henge bank – Marden

It became apparent (Leary et al. 2013) that Silbury Hill was not a construction project formed from a single or even a few phases of construction, but that it had become the focus of an array of disparate activities, where the mound appears to have developed, mutated and evolved through time. Ingold (2000, p. 179) has suggested that although the form of monuments were probably conceptualized before their construction, and therefore appear to have had an ultimate purpose, they could not be disentangled from their surroundings, their materials and the conditions of their construction. This has led Leary et al. (2013, p. 204) to suggest that 'there appears to have been no contrast between the process of bringing the monument into being and the construction of the monument itself'. This is supported by Pryor's (1998, p. 364) excavation of the Maxey Cursus near Peterborough, which has led to the suggestion that the Maxey Cursus appears more like a project in progress rather than a structure built to a predetermined plan. He (ibid., p. 364) believes that people may have visited the site on a seasonal basis, adding new bits to the structure year after year. This would indicate that the construction activity was more important than the actual cursus architecture. If, as Pryor suggests, the most important thing was the activity of disturbing the landscape, what was the motive for doing this? It could perhaps have been, as suggested by Jacques, to memorialize previous uses of the landscape (see Jacques and Phillips, 2014, pp. 7–27 and Jacques et al. 2014, p. 29), or perhaps, as suggested by others, to put an end to any previous uses (Harding 1999, p. 32; Pearson and Field 2011, pp. 37–39).

Conneller (2011, p. 24) feels the type of materials used within monument construction require to be taken more seriously, contending that the matter is not inert to the final imposed form, but that they act together, the form being defined by the limits of the material (ibid., p. 24). This appears to support Boivin and Owoc (2004, p. 10) who suggest that in many instances, non-western societies recognize particular types of soil as being animate, where strong symbolic and ritual association such as colour and texture are given to the soils and stones resulting in people travelling great distances to acquire certain materials from particular sources. This could possibly be the reason the Fornham All Saints Cursus was constructed at its precise location upon an area of significantly changing geology. Although people may not have had to travel great distances to find the materials incorporated within most monuments, much of which could have been found within the general locality, Leary et al. (2013, p. 214) believe that the construction of Silbury Hill identifies that 'there does seem to have been some degree of deliberate selection of materials, suggesting perhaps that some symbolic meanings underpinned their selection'.

Similar examples of deliberate selection of materials appear also to have occurred at Duggleby Howe in Yorkshire (Gibson et al. 2009), the potential location of another of the cursus monuments within this study group. At Avebury, Watson (2001, p. 301) identifies that riverine clay had been used as a packing material within the stone holes despite other clays being more immediately available. This leads Watson (ibid., p. 301) to suggest that the clay may have been selected due to its colour, the dark brown riverine clay contrasting entirely against the pale bedrock clay. Leary et al. (2013, p. 215) believe that the materials used within Neolithic construction appear to have been just one element of a monument in progress where the various phases of surrounding ditches appear to have been more than just quarries for raw material, appearing to have greater significance than is usually ascribed. The continuous recutting appears to indicate that, as with the mound construction sequence, the cutting of the ditch and its apparently deliberate backfilling were important elements in the construction and reworking of the site, perhaps designed to bring about continual environmental change, perhaps resulting in a change to the wildlife in the area.

Leary et al. (2013, p. 218) believe that alongside the actual construction of the mound and ditch, there is one further significant aspect to consider, that being the reason behind the precise location of the mound. Although Silbury Hill was set in a landscape overlooking earlier Neolithic monuments such as Windmill Hill and the West Kennet Long Barrow, it appears to have deliberately occupied a liminal zone, as it sits on the edge of a chalk spur. This zone represents the edge of the dry chalk upland as it penetrates the wetter lowland area. Leary et al. (2013, p. 218) believe the geology of the area at the footprint of Silbury Hill is unusual due to the chalk having been overlain by a mantle of clay-with-flint material which is normally only found on the tops of upland areas. However, within this area, it appears to have eroded downslope, a fact that may not have gone unnoticed in the Neolithic period.

Leary et al. (2013, p. 218) also question whether, similar to many of the cursus monuments within this study group, perhaps the lowland setting for Silbury Hill, and the fact that it was adjacent to streams and springs are potentially the most important factors. While there is current uncertainty whether water flowed perennially or intermittently during its construction, standing pools would have developed once the porous chalk had been saturated and the stream reached maximum capacity, so the importance of the drainage system to the construction and meaning of Silbury Hill would have been paramount.

Ashmore and Knapp (1999, p. 137) had earlier suggested that similar to the focus of ritual and ceremony placed on water elsewhere, it was possible that the ditch extension to Silbury Hill had symbolic implications. Loveday's (2006) and Brophy's (2016) lifetime investigations into the potential reasoning behind the placement and alignment of cursus monuments note an association between cursus monuments and rivers and streams, while my research appears to identify an association between cursus monuments and riparian zones. However, Leary et al. (2013, p. 219) believe that 'what is less well known is that Neolithic round mounds appear to have also had a similar focus'. The Hatfield Barrow within the Marden Henge appears to have been adjacent to the River Avon, Conquer Barrow within the henge enclosure at Mount Pleasant lay adjacent to the River Stour, the Great Barrow at Knowlton lay close to the River Allen and Duggleby Howe was situated at the source of the Gypsey Race.

Leary et al. (2013, p. 220) also note that the broader area to the east of Silbury Hill was notable for its extensive drifts of sarsen stone boulders. Lyons and Machen (2001) have identified, through fitting range cattle with GPS collars, that the cattle tended to prefer some range sites over others because of the terrain, this preference appearing to be related to the presence of loose and embedded rock on some sites. This has led Leary et al. (2013, p. 220) to put forward the suggestion that 'perhaps there may have been a significance in the fact that these sarsen streams appear to flow towards the Swallowhead springs where the river emerged'. These weathered stones, remnants of a sandstone layer formed approximately fifty million years ago that once overlaid the chalk of the downland regions, littered the prehistoric landscape, forming significant restrictions to potential cattle movement along the western edge of the Marlborough Down ridge and in the Avebury valley to the east of the River Winterbourne.

Looking closely at the topography of the actual landscapes upon which cursus monuments have been constructed, we see that most monuments associated with a chalkland geology have been constructed upon the rolling downland. The exceptions are the Ivinghoe Cursus, which appears to have been constructed as an inverse cursus monument along the ridgeline of the Chiltern Hills, and the Kirby Underdale Cursus, which appears to have been constructed along the route of animal movement from the lowlands to the uplands of the Yorkshire Wolds. However, cursus monuments located upon other types of geology, such as muddy limestone, clay and greensand, appear to have been constructed upon raised gravel terraces. Although the majority of cursus monuments that have been constructed upon a geology other than chalk appear to have been situated upon the first raised gravel terrace, it is noticeable that, of the eleven cursus monuments constructed within the Thames Valley region, approximately 73 per cent were situated upon the second gravel terrace.

CHAPTER 4

The Link with Hunting or Herding

When attempting to identify a link between the placement of Neolithic cursus monuments and large herbivore movement, we first need to establish what type of large herbivore movement was involved. Are we talking primarily about the way deer moved around the landscape, or the way cattle moved around the landscape? And if cattle, were they wild or domesticated?

We have already identified the possibility that Mesolithic hunters managed their game to such an extent that they were effectively livestock farmers (Pryor 2015b, p. 78), deliberately intervening in the landscape, improving wildlife grazing by creating or enhancing clearings. We have also identified that, during the first few centuries of the Neolithic, the scale of domestic animal herds appears quite modest, which could initially make it difficult to specify what balance there might have been between hunting and gathering on the one hand and farming on the other.

Is it possible to establish how, thousands of years ago, Late Mesolithic or Early Neolithic communities went about hunting across landscapes within the vicinity of later cursus monument construction? This chapter aims to establish whether it is still possible to identify aspects of prehistoric landscape that could have helped ensure a successful hunt, to establish whether patterns of lithic scatters are able to identify the types of animal potentially hunted and to establish how others have studied prehistoric hunting habitats, identifying some of the difficulties they have encountered and potentially identifying how certain aspects of their methodologies may be used to overcome similar problems when analysing hunting-associated data across cursus monument sites. We will attempt to specify what the landscape can tell us with regard to the hunting of large herbivores in general, before concentrating on the individual species of deer and aurochs, and then looking at the way pastoralists moving with their domesticated cattle would have interacted with the landscape.

General Hunting of Large Herbivores

Although often referred to within prehistoric papers and books, hunting during the British Mesolithic/Neolithic transition period is seldom discussed in any great detail. When it is, the focus tends to be solely upon either the flint assemblages used or the faunal assemblages found, as in the work undertaken by Legge and Rowley-Conwy (1988, p. 94) on the Star Carr assemblage to identify the season within which the animals were killed. But could an in-depth study into the various hunting strategies and techniques used throughout prehistory enable the hunting sites themselves to be identified?

When archaeologists talk about the mobility of hunter-gatherer communities, the support for mobility within the archaeological record tends to consist only of the flint or chert flakes and the broken animal bones that have been found. This has led Speth (2010, p. 9) to suggest that there currently isn't any hunting in the archaeological record, as what we recover as archaeologists are the broken animal bones and stones potentially used to dispatch them. However, Speth (2010, p. 1) believes that archaeologists are able to use a uniformitarian framework where previous geological factors still operate and are therefore observable today. This would support the theory that it is still possible to survey previous prehistoric landscapes across later cursus monument sites, establishing features that may have assisted with the hunting of large herbivores. Yet Speth (2010, p. 1) warns that to avoid merely projecting the present into the past, it is important to identify any relevant variables associated with the problem under investigation, identifying how they are linked or interrelated to each other as this would allow for the generation of a hypothesis that can be tested using the archaeological data, and perhaps avoid the pitfalls associated with the ethnographic use of modern hunter-gatherer data.

Waddington (1999, p. 36) had earlier suggested that any attempt to understand the genuine patterning of an archaeological process needs to appreciate any processes that may have affected the residues since their initial discard. This suggests that a full understanding of any taphonomic context of the archaeological remains would be an important part of any landscape study. Zvelebil et al. (1992, pp. 193–226) identify that it is only when these taphonomic distorting effects have been taken into account that rigorous and meaningful interpretations can be made from the archaeological data. Walters (1992, p. 102) believes that an archaeological record does not accurately reflect the complete pattern of the site that once existed, but instead reflects only the biases of geological preservation. We will need to take this into account when considering any potential features in the vicinity of cursus monuments that could have assisted large herbivores as they passed through the area and whether these would have aided any overall hunting strategy.

Since the early 1980s, a far greater level of attention appears to have been given to the geoarchaeological setting and the taphonomy of the landscape in relation to any lithic scatter. For example, Waddington (1999, p. 37) attempts to overcome these difficulties by recording individual point co-ordinates for each individual find, rather than previous methodologies of using coarser resolution of area data. Waddington (ibid., p. 37) believes this means that subsequent analyses of the results are not constrained by coarse spatial referencing. Consequently, the unit of analysis becomes the artefact rather that an imposed geometric grid.

However, when investigating surface artefact scatters, it is important to determine any geomorphic processes that may have affected their modern-day distribution. Allen (1991, p. 39) suggests that prior to making any interpretation from the artefact distribution of surface collection it is necessary to understand both the nature and the history of the land surface. Allen (1991, p. 44) argues that the density of artefacts tends to be over-represented on the downslope due to the erosion of soil on hill crests and the subsequent deposition of soil on the foot slope. Although he was able to identify that even steep slopes remain relatively stable until they are disturbed and destabilized, Allen (1991, pp. 45–47) has identified that, once disturbed, downslope movement occurs within just a few years. In his experiments, in excess of 50 m of movement occurred within four years. This has led Allen (1991, p. 49) to suggest that hillslope deposits within southern England are directly or indirectly solely the result of human interference within the environment rather than a wholly natural phenomenon. Allen (1991, p. 54) sums up by stating that 'although colluvium may aid the preservation of ancient landscapes, it also creates archaeological blanks'.

McOmish et al. (2002, p. 11) identify that the increased efficacy of cultivation techniques has led to an increased rate of monument and feature destruction, where at a very basic level the landscape appears to be able to be divided into two broad zones: one of monument and feature survival, the other of monument and feature destruction. Investigation of the landscape at this local level is not new. But is everything seen during a landscape survey useful? Although any lump, bump or hollow could potentially have something to tell us, even if it appears to be of little significance, without further research it is usually impossible to tell. What may look homogeneous today may have originated at widely different times for varying reasons. Waddington (1999, p. 36) recommends separating out natural features, such as subsoil and drift-deposited material from anything archaeological before interpreting what the material might mean. However, in the attempt to identify prehistoric hunting sites, it could be precisely these natural features, such as banks and ridgelines, which may have assisted hunter-gatherer communities, that are of the greatest importance.

This is supported by Frison's (2004) investigations of prehistoric bison hunting methods across thousands of years in North America. Frison (2004, p. 11) suggests that animal and human behaviour along with the evidence from taphonomic analysis reveal information that is significant for prehistoric studies, an approach that has been vindicated in large part through his study of bison bone beds and bison jump sites. Undertaking investigation at the local level, Frison (2004, p. 73) identifies the use of arroyos, where prehistoric hunters either drove bison attracted by the lush grasses at the entrance into the arroyo trapping them against the head cut, or as at the Olsen-Chubbuck site stampeded the bison over the edge of a deep narrow sided arroyo, resulting in the death of an estimated 190 animals, the resultant bone pattern leaving no doubt from which direction the stampede occurred. Frison (2004, p. 74) also identifies several shallow variants to the arroyo trap which appear to rely on longer, steeper slopes. In these examples, erosion has quickly altered the original prehistoric profile, making them less visible to archaeological investigation. Although much of Frison's (2004) research occurs on the North American Plains in areas that have had little or no modern cultivation, could this be directly transferable to areas of the British Isles that have potentially suffered from some form of cultivation for the past three and a half millennia?

Similar geological processes do appear to be relevant when undertaking prehistoric field survey investigation in the British Isles. Although many of the slighter original features have been considerably altered both by erosion and deposition, significant features such as hillsides, valley slopes and ridgelines do appear to retain their prehistoric profile. This is supported by Taylor (1983, p. 11), who proposes that with a few notable exceptions, archaeologists have tended to be myopic when studying the remaining features of the prehistoric landscape, perhaps due to their understandable concentration on the human activity details that are their basic source of information.

The history of archaeology shows that the amount and location of materials discovered are probably not solely dependent upon the original landscape profile or upon the original occupiers of the landscape but on the events and changes that have taken place within the landscape since its original use. Taylor (1983, p. 20) also suggests that:

> The lack of prehistoric sites in some places, such as in narrow steep-sided valleys, or on the wide flood-plains of major rivers, is the result of the massive amount of erosion and deposition of soil and rocks that has taken place over the last ten thousand years, the result being that many, perhaps most, indications of use of the prehistoric landscape are in fact buried under many metres of silt and soil and are therefore, only found by chance when deep modern digging takes place.

The arguments put forward by Taylor (1983) and Frison (2004) tend to suggest that many of these original smaller prehistoric landscape features will no longer be detectable when attempting to undertake field surveys upon the landscape where they originally featured, due to post-prehistorical damage. Very few examples for direct prehistoric use of the landscape remain and, where they have been discovered, such as the identification of hunting sites upon the Stonehenge Plain (Parker-Pearson 2012, p. 230), their discovery has tended to be solely due to the minute flint flakes found within the region. However, in other instances, although the features may have significantly reduced, it appears to be still possible to identify prominent prehistoric features directly from the modern landscape itself.

Tilley (2010, p. 51) appears to be one of the few researchers to attempt investigation of the use of the prehistoric landscape at the local level when he identifies that, exposed hollow ways, created by herds of domestic or wild cattle disturbing the ground would have introduced new sensory experiences. Tilley's (2010, pp. 293–347) phenomenology research within the Exmoor National Park regarding what he identifies as stone settings could be significant as they appear to be related only to the inner part of the moor, occurring on gently sloping ground near to the tops of ridges and hills. Tilley (2010, p. 310) identifies that:

> unlike with some other monuments, from a stone setting there appears to have never been a panoramic view over the whole landscape. Therefore, their restricted view, limited to only one or two directions, appears to have had an intimate relationship to rivers and watercourses in general, and to the heads of coombes in particular. The high landscape location and the restricted view suggesting that these were places to look out from across the landscape rather than identifiable places to look or travel towards.

Tilley (2010, p. 338) believes that stone settings would have provided ideal locations in which to hide and observe, thereby marking places from which it was good to hunt at particular seasons and times of the year according to the wind direction. It appears that the ideal locations provided by these stone settings allowed for the observation of deer movement, where Tilley (2010, p. 339) believes that, 'perhaps the act of waiting for and sensing the deer in the surrounding landscape might have developed an intimate knowledge of the immediate landscape'. However, Tilley (2010, p. 340) stresses that 'any rituals associated with hunting would have been unlikely to have left depositional traces within the archaeological record'.

To identify the importance of any use of the landscape at the local level and the length of time this has been known, Tilley (2010, p. 340) cites Jefferies (1892, p. 37) who states:

> In front appears a coombe, overgrown with heather from summit to foot, and I stop suddenly. On the opposite slope are five hinds lying down, their heads visible above the heather, but too far for a good view. To stalk them it is necessary to go around the head or shallow upper end of the coombe and to get the wind to blow from them. Their scent is so quick that to approach down the wind is useless ... The hollow of the coombe carries the wind somewhat aslant just there from its general direction like a tube, else I think they would have scented me as it is.

Greater investigation into hunting techniques has been undertaken throughout Europe. Andersen (2013, pp. 497–502) investigated healed hunting wounds from the submerged Mesolithic site at Tybrind Vig to identify that red deer and wild boar appear to have been shot using a transverse-type arrowhead, perhaps to maximize blood loss. However, aurochs skeletons in the same region were found with the remains of multiple microlith arrowheads in the chest area, at both the Vig site, found in 1905 during turf digging (Hartz and Winge 1906), and the Prejlerup bog area (Sorensen and Petersen 1986, pp. 111–17). The fact that both aurochs had been shot using microlith arrowheads suggests different techniques were being used to hunt aurochs than for hunting red deer and wild boar. Perhaps the microlith arrowhead acted as a bodkin against the tougher hide and increased muscular nature of an aurochs. If we assume that bringing down a creature the size of an aurochs required the hunters to hit one of the vital organs, while the creation of a traumatic wound with the resultant blood loss appears to have been sufficient to dispatch a deer, this could enable us to identify the hunting of different species through lithic collections.

Tests conducted by Waguespack (Waguespack et al. 2009, p. 797) identify little difference between stone-tipped arrows and those having only sharpened wooden ends. While the arrows tipped with stone penetrated farther into simulated gel targets than those tipped with wood, the differences were relatively small (though statistically significant), with both attaining depths greater than 20cm. It may, therefore, have been of little consequence to hunters in most real-life situations whether they used stone-tipped arrows or not.

The submerged Mesolithic site at Tybrind Vig revealed other hunting implements, such as club-shaped wooden arrowheads, which Andersen (2013, p. 149) assumes were for use in the hunting of birds, as at least twelve different species were represented at the site. Other examples of this arrowhead, which would not have damaged the feathers during the kill, were also found at Ringkloster and at Ronaes Skov. Further finds from Tybrind Vig (Andersen 2013, pp. 123–30) show that the throwing spear continued to be used throughout the period, with more than thirty examples of what Andersen described as slender shafts being found. Experiments using slender shafts and atlati identify that the darts could have been hurled over 70 m, although accuracy at that range would have been extremely variable (Raymond 1986, p. 165). These shafts had a syringe feature at the pointed end, which Andersen (2013, p. 128) suggests was the functional part, 'being either the mount for a point of some other material such as bone or stone or possibly designed as a blood channel that allowed for easier withdrawal of the item'. A reconstruction of one of the earliest split-elm Mesolithic self-bows, found at the Holmegaard Moor site in Denmark, enabled Bergman (1993, p. 102) to establish the bow had a maximum range of between 150 and 200 m, enabling hunting to be carried out at a far greater range than had previously been thought. The bow had been carved from the tree so that its back was the flexible sapwood while the belly was the harder heartwood. These Mesolithic self-bows would have been extremely effective weapons in the hands of trained archers, where the impact of arrows shot from these bows would have proved fatal if striking critical points of the animal's body, such as the chest or the neck. Faunal assemblage evidence from the Tybrind Vig and Holmegaard sites identified that most injuries were indeed associated with the chest area. However, the type of arrow selected and the openness of the environment within which the animal lived would have affected the hunting techniques, while the maximum range of the weapon will have had a direct relationship to any potential hunting strategies that could have been used.

This leads to the question – what type of hunting strategies were likely to have been employed during the Mesolithic/Neolithic transition period? Myers (1989, p. 89) suggested that the warmer climate resulted in longer autumn and spring months with milder winters. This would potentially have made predicting the movement or timing of animal congregation more difficult. Myers (1989, p. 90) believes that a shift in hunting strategy occurred across the British Isles resulting in intercept hunting being largely replaced by encounter hunting. These involved hunters following or stalking prey, instead of waiting for the animals at predictable intercept points. However, encounter hunting would have been likely to result in other changes, such as to the settlement patterns and organizational structures previously used by the Mesolithic hunters.

Mithen (1990, p. 224) suggested that procuring smaller groups of animals in less predictable environments may have led to smaller hunting groups being dispersed across the landscape. This could have resulted in hunting strategies changing from a group activity to a more individual activity. Climate change throughout the Mesolithic would have resulted in hunter-gatherers being confronted with two choices – becoming increasingly mobile in order to exploit a greater area, or concentrating their efforts on particular resources within the original area. However, some degree of oscillation probably occurred between these strategies, with a mixture of strategies potentially prevailing during differing times and conditions. Jochim (1976) adds weight to this theory in his book *Hunter-Gatherer Subsistence and Settlement: A Predictive Model*, where he identifies that securing the appropriate nutritional levels became more difficult during the resultant change to encounter hunting, particularly at times when animal populations were lower. This could have resulted in hunting becoming a year-round activity, reducing the use of seasonal camps and increasing the risk of the population failing to meet its dietary requirements during the winter and early spring month – unless specific sections of the landscape drew animals to them at certain times of the year, perhaps allowing for a continuation of previous intercept hunting strategies.

It has long been realized that the easiest form of hunting is that of intercept and ambush. This technique is a fairly simple operation, where all the hunter has to do is get downwind and close to the line of travel. Could the natural topography of the open landscape where later cursus monuments were constructed have resulted in a continuation of this easiest form of hunting? Could the natural restriction to herd movement through the flooding of low-lying areas have resulted in predictable intercept points when the herd made use of spring meadow resources? Or could the evidence suggest that manipulation of the landscape due to monument construction actually drove wildlife away from the area, resulting in hunting ceasing to be an activity within these sections of the landscape where they started primarily to become areas used only by domestic cattle? To answer these questions will require further investigation into the landscape topography upon which cursus monuments were constructed, either to see how it would have interacted with the types of large herbivores that could potentially have been hunted at these locations or to establish that the landscape would have been used primarily for the grazing of domestic cattle. Therefore, in the next section, I intend to examine the different techniques required for the hunting of red deer and the hunting of aurochs, as well as any evidence that these areas were primarily used for the grazing of domestic cattle.

While the overall appearance of the landscape, as based on the geology and geomorphological processes within it, owes little to the intervention of man, his interference with vegetation and drainage over millennia will have had widespread local effects. Aspects of change and complexity are extremely important when studying the prehistoric landscape for, although the British Isles have been continuously occupied for around 12,000 years, only the last 3,000 to 4,000 years have seen organized settlements, field systems, trackways and religious sites which tend to appear prominently throughout the landscape. For this reason, I undertook a boots-on-the-ground field observation exercise across each individual cursus monument site that made up the study group of cursus monuments on or adjacent to the English chalkland belt. This has provided the largest representative sample to date for evaluating any potential landscape that could be associated with earlier hunting activities across English cursus monument sites.

Hunting Red Deer

Red deer (*cervids*) have been central to human cultures throughout the prehistoric period. Crane (2016, p. 12) suggests that just about all parts of a deer could be used for something. The hide could be used for lashings, containers, boat hulls, tents or clothing, while parts of the antler could be used for fabricating bows, projectile points, tools, pegs, toggles and gaming pieces. The guts were handy for containers, hooves for glue and sinew for reinforcing bows, bindings and sewing thread. However, in spite of their diverse, deep-rooted and longstanding relationship with human society, until the *11th International Conference of Archaeozoology in Paris in 2010*, no multi-disciplinary research on cervids had been produced. This conference highlighted again and again that consistent patterns regarding deer phylogeography were starting to emerge from the Iberian, Italian and Balkan peninsulas of the Mediterranean together with the Dordogne region of France, increasing understanding of overall prehistoric deer distribution patterns.

Closer to home, during Clark's (1954) excavations at Star Carr, Fraser and King (1954) used red deer antler to identify when the majority of red deer appear to have been killed. As red deer cast their antlers in April, only growing a new set by the following October, the fact that both cast and attached red deer antlers were present appeared to indicate that the site had been used between the winter and spring time. Using this data, Clark (1972) undertook a reconstruction of the land use, determining that Star Carr was a winter basecamp, from where people dispersed into the uplands for the summer. Although small flint scatters have been discovered in the uplands, the acidity of the peat means that no animal bone has survived, preventing any testing of Clark's theory. In the 1980s Legge and Rowley-Conwy (1988) used the cycle of animal teeth development and wear patterns, rather than shed or unshed antlers, to re-analyse the data from Star Carr in an attempt to increase understanding of seasonal animal movement and to identify when most Star Carr animals had been killed. Initially, they investigated roe deer jaws for the replacement of milk teeth with permanent premolars and the permanent eruption of M3s, both of which occur at twelve months. This, and the fact that roe deer are born in June, indicated that the Star Carr roe deer had been killed between May and June. However, the inclusion of red deer skulls with shed antler, which could only occur from the April onwards, led them (ibid., p. 94) to conclude that, 'most animals had probably been hunted during the late spring and early summer'.

Rowley-Conwy (2017, p. 205) indicates that as most other indicators from Star Carr strongly suggest a summer occupation, perhaps the antlers investigated by Fraser and King's had been traded from hand to hand throughout the various seasons. Perhaps being used as raw material for making points? Rowley-Conwy (2017, personal communication, December) believes that the importance of the birchbark canoe has been seriously underestimated. Weights that would have been daunting on foot would have been trivial when transported by boat. It is interesting to note that numerous rolls of birch bark were discovered during Clark's (1954) initial excavation, together with a resin cake which could have been mixed with clay and beeswax and used to seal the stitching.

Rowley-Conwy (2017, p. 204) suggests that at the last glacial maximum, around 18,000 BC, enormous amounts of ice flowed from the north into the Vale of York and from the east into the Vale of Pickering producing end moraines. These are irregular ridges of glacial sediments that formed at the front edge of the ice sheet, representing a standstill of the ice. By the late glacial period, around 12,000 BC, Rowley-Conwy (2017, personal communication, December) suggests that deposits from these end moraines resulted in two lakes being formed by melting ice-water, these being the lowland Lake Humber, which voided through the Humber Gap around 11,700 BC and Lake Pickering which voided through the Kirkham Gorge at a later, but as yet unknown, date. This created a lowland environment of early post-glacial washlands, consisting of shifting watercourses which produced unstable wetlands, lakes and temporary areas of dry land upon which Mesolithic flint scatters have been found. It therefore appears that communities from the early Mesolithic onwards could have used lightweight shallow-draught boats to fish, forage and hunt along the coasts, estuaries and river systems. It is Rowley-Conwy's (2017, personal communication, December) belief that hunter-gatherer communities used birchbark canoes to travel between the Vale of Pickering and the Vale of York washlands, where they set up their winter basecamps. However, indications by Legge and Rowley-Conwy (1988, p. 94) for the hunting season at Star Carr would obviously be closely associated with occupation of the site, which is not to say that hunting didn't occur in other areas throughout the year (Rowley-Conwy 2017, personal communication, December).

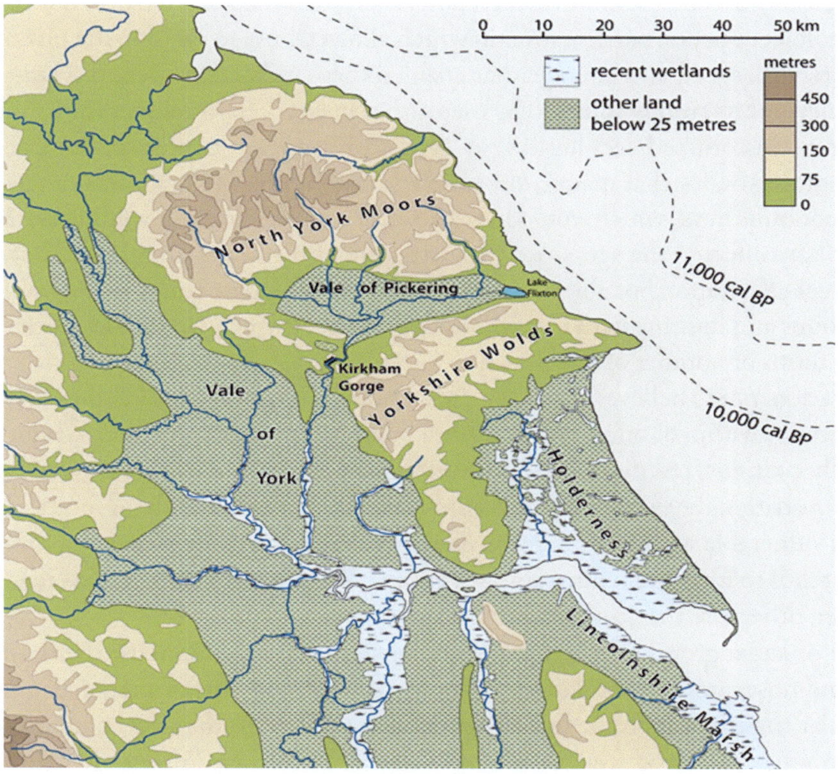

Map 2: Distribution of washlands (modified from Rowley-Conwy 2017 Fig. 23.4, courtesy of Peter Rowley-Conwy)

Andersen's (2013, p. 431) investigations at the Danish Mesolithic submerged Tybrind Vig site have identified that red deer calves appear to have been killed across all seasons, although there was little evidence for winter hunting. Andersen's data again highlights that the majority of red deer hunting occurred, as indicated by Legge and Rowley-Conwy (1988, p. 30), in the late spring and early summer. The data from Tybrind Vig suggests various forms of hunting were potentially used. The majority of hunting activity appears to have consisted of economic hunting for meat yield, the hunters targeting the smaller young males. This would maintain the deer population's carrying capacity and increase antler quality, while also providing sufficient food for the local community. Crane (2016, p. 12) identifies that 'an adult female red deer weighed between 70 and 150 kg, and an adult male between 100 and 250 kg. Some 50 to 60 per cent of a deer was edible meat'. So, the hunting of a young adult male would have produced around 35 to 45 kg of edible meat per successful hunt.

Trantalidou and Masseti (2015, p. 65) suggest that, antler traits only become useful if the adult antler have been grown under good nutritional conditions, the size, shape and growth being directly affected by the habitat's nutritional quality. This supports Andersen's (2013, p. 431) theory that younger animals were being culled as part of a herd management programme, which emphasizes the economic importance of red deer antler. The number of tines on an individual red deer's antler increases annually before stabilising between the age of six and twelve years. So killing a mature stag, while immediately providing two quality antlers, would remove the local supply of any future pairs of antlers from this animal during subsequent years, as it would take several years for a newborn calf to provide equivalent replacement antlers.

Discovery of the Star Carr (Clark 1954) antler headdresses initially suggests that some form of trophy hunting occurred during the rutting season, at least during the Early Mesolithic period. Red deer stags would be in their prime at this point, with magnificent antler sets and coats. However, as we have seen, being killed at this stage would immediately remove them from the economic equation. This has led Trantalidou and Masseti (2015, p. 72) to suggest that this factor, together with further zooarchaeological and iconographic evidence, potentially identified that red deer hunting was a rite of passage for adolescent males, participation in which allowed boys to develop the physical and moral qualities of strength, speed and courage that enabled them to become men, the hunter choosing a target male solely for its trophy value, to be used either in later ceremonies or for its status value.

However, prehistoric red deer hunting at Tybrind Vig does not appear to have been associated with trophy hunting or, for that matter, any form of random process. It appears to have been biased towards the economic hunt, which would have had clear consequences for wildlife management. Not only would it have affected the age, size or sex structure of the herd's population, but it would also have had a direct effect upon possible hunting methodologies. These could include solitary stalking, hunting in groups and hunting with or without dogs, each of which would have produced different efficiencies in terms of number of deer killed.

Rattray (2009, p. 82) believes that red deer's natural tendency to isolate themselves as individuals or small groups within or on the edge of forest cover had a major influence on hunting strategies. Being naturally cautious, red deer choose to feed under cover. So if the forest was capable of fully meeting their nutritional needs, red deer would have stayed permanently in forest cover, knowing they were less vulnerable within thick cover. Stags and hinds generally browse on different types of plants, which tends to force them into same-sex groups. Yet to sustain these groups requires sufficient food and water, otherwise the groups became fragmented.

A benefit of larger group sizes is that it gives maximum protection to the herd, since more sets of eyes, ears and noses are alert for danger. However, the whereabouts of a red deer group's territory at any particular time appears to be mainly dependent upon both the weather and the time of the year. Access to water and good feed are the most crucial factors. Red deer only tend to travel, feed and drink in the early morning and late evening, usually following covered corridors between feeding areas. Knowing where these covered corridors were would be the key to successful prehistoric hunting. Hoof marks, scuff marks, droppings and hair on trees and bushes would identify the trails, which tend to move horizontally around hillsides and slopes, keeping close to the treeline and thick bush.

Identifying the location of a deer trail would have allowed the hunters to hide themselves in positions to achieve the best possible shot as the deer used the trail to return to thicker bush. The hunting strategy would then become one of intercept, waiting for the deer to pass by and shooting it with an arrow selected to produce maximum blood loss. However, even with modern-day hunting using rifles, a deer often takes flight after being shot, to die of blood loss at a later stage. Therefore, the next stage of the deer hunt was likely one of following the blood trail until the remains of the deer were located.

Rowley-Conwy's (2017, personal communication, December) isotopic investigations of fauna discovered during earlier excavations of the Coneybury Anomaly (Richards 1990, pp. 40–61), appear to highlight that deer were inhabiting a different environment from the domestic cattle. The deer appear to have originated from forest cover while the cattle appear to have come from three distinct separate locations of cleared grazing ground around the Stonehenge area. This adds to the doubt over any potential correlation between the landscape upon which deer were hunted and the landscape where future cursus monuments were constructed.

While the antler headdresses discovered at Star Carr (Clark 1954) potentially identify some form of symbolic ritual involving red deer and the Early Mesolithic population, it is debatable whether this continues, as it has not been found within the context of any other Mesolithic site (Rowley-Conwy 2019, personal communication, January). Therefore, with the exception of some degree of stag trophy hunting, the fact that it appears mainly hinds and young males were being hunted during the Mesolithic/Neolithic transition period suggests that the hunting of red deer held a more economic than symbolic value. This appears to have included some form of sophisticated and potentially standardized system of antler collection, where the majority of antler requirement was met primarily by the organized seasonal collection of shed antler. This potentially suggests that by the Late Mesolithic/Early Neolithic period the hunting of red deer no longer held the same prestige that appears to have been attached to the earlier hunting of red deer or the hunting of other large herbivores, such as aurochs.

In conclusion, the fact that red deer hunting tends to have occurred in closed-canopy woodlands rather than upon open grasslands and the fact that by the Late Mesolithic period red deer hunting appears to have lost its prestige strongly suggest that there is not a correlation between the previous hunting sites of red deer and the landscapes where many later cursus monuments were to be constructed.

Hunting Aurochs

The hunting of red deer as they moved horizontally around the slopes and hillsides of closed-canopy woodlands raises serious doubt about there being any potential correlation between earlier red deer hunting sites and landscapes where later cursus monuments were to be constructed. Is it, however, possible to establish a potential correlation between sites upon which aurochs hunting occurred, especially the grassland sites which appear to have supported herds of aurochs throughout the spring and summer months (Rogers et al. 2018, p. 142), and the landscape where later cursus monuments were to be constructed? Carbon isotopic analysis carried out on two aurochs teeth from Blick Mead (Rogers et al. 2018, p. 142; supported by additional research by Legge and Rowley-Conwy (1988) and Van Vuure (2005)) suggest that aurochs herds appear to have had a tendency to make use of the first spring meadow resources and then remain on the grassland throughout the spring and summer months, only splitting into smaller groups, to head away from these grassland plains and into the forests to feed on winter browse. Is it also possible that the flooding of low-lying areas, adjacent to where the majority of later cursus monuments were constructed, led to natural restrictions regarding wild cattle herd movement, and could this have resulted in sufficient numbers of aurochs congregating at these locations to make them predictable intercept points?

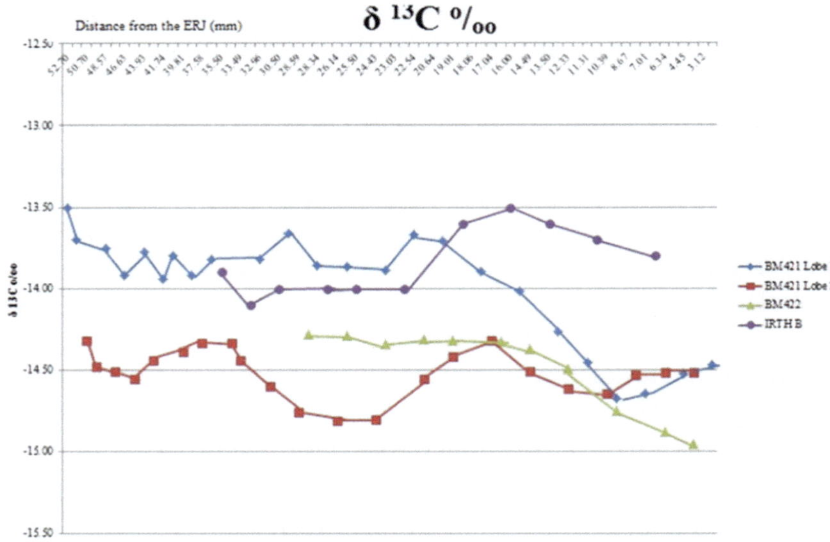

Figure 3: Results of $\delta^{13}C$ from Blick Mead (Rogers et al. 2018, p. 142) Courtesy of Peter Lang

Myers (1989, p. 89) believes that longer autumn and spring months together with milder winters during the Mesolithic/Neolithic transition period made the prediction of wild animal movement more difficult which resulted in a shift in hunting strategy. The possibility exists that the natural topography of open landscapes where later cursus monuments were constructed allowed for previous hunting strategies to continue there. However, if the hunting of aurochs did continue at these sites, would this have proved problematic when trying to keep domestic animals within the same landscape? Perhaps, as Crane (2016, p. 11) suggests, Early Neolithic communities killed aurochs due to the beast's preference for valleys and floodplains which would have put the herds in conflict with humans who wanted to use the same landscapes for their domestic cattle.

But would the numbers of aurochs within these herds have been sufficient to be problematic to the later Neolithic communities that constructed cursus monuments? Maroo and Yalden (2000) undertook studies to estimate the British aurochs population for the Late Mesolithic, by analysing pollen data to identify the type of vegetation cover across twenty-two British Mesolithic sites. To establish a plausible aurochs population density, they compared this to a balanced mammal population living in similar habitats as their identification for the British Mesolithic. It appears that the Biakowieza National Park in Poland met their criteria (Jedrzejewska and Jedrzejewski 1998). Assuming that a Neolithic aurochs population remained similar in numbers to previous Mesolithic populations and that the population of Mesolithic aurochs had a similar density to previous bison numbers in the Biakowieza area prior to the First World War, it could be estimated that there was an aurochs population of around 84,000 within the British Isles during this period. This would equate to roughly two aurochs per 5 sq km. However, if, as Crane (2016, p. 56) suggests, 'of the total land area of Britain, 60 per cent was wooded, and three quarters of that woodland was deciduous. Around 20 per cent of Britain was grassland, and the rest was occupied by fenland, heath and moor and various shrubs, herbs and ferns', this would mean that when aurochs grazed on the grasslands in the early spring (Rogers et al. 2018, p. 142) there would have been approximately one aurochs per square kilometre. To put this in perspective, this would equate to a herd of approximately twenty-five to thirty aurochs roaming the Stonehenge Plain during the summer months, which would appear to be a sufficient number to place them in conflict with Neolithic pastoralists.

The last living aurochs died in the forests of Jaktorow in Poland in 1627, the final specimens being heavily protected, corralled and even fed by gamekeepers during the winter months. Therefore, any data resulting from this period would potentially be unreliable in determining the general lifestyle of aurochs. In an attempt to overcome this problem, this study has used data from a mixture of data sets – from the wild Chillingham herd, which has been isolated within Chillingham Park in Northumberland for 750 years, although obviously any data from the wild Chillingham herd also suffers from the same problems as the last aurochs living in Jaktorow; from Shorthorn-Aberdeen Angus cross cattle which were abandoned when the local population deserted Swona in the Orkneys in 1974 and have now turned feral; from cattle introduced on to the Floodplain Forest Nature Reserve by The Parks Trust, Milton Keynes, to improve the wildlife value of the area; and from range cattle living in the American Mid-west – as these appear to be the closest available reference points for the potential lifestyle and lifespan of both ancient aurochs and potentially domestic cattle within the Mesolithic/Neolithic transition period.

Data from the Chillingham Park herd predicts an approximate maximum life expectancy for aurochs of about seventeen years for cows and thirteen years for bulls, where the cows matured around the age of four years. Although bulls matured earlier, between the ages of eighteen and twenty months, they appear to have taken longer to reach a dominant position within the herd.

The deserted feral herd at Swona is still going strong and currently consists of seventeen cows, three calves and three bulls, and is now classified as a new breed in the *World Dictionary of Livestock Breeds*. On average, two calves are born each year but, as the herd is self-selecting for hardiness, not all live to maturity. For most of the year the bulls, who appear to have a ranking structure, live apart from the main herd. Having been separated from the mainland for such a long time, their isolation has resulted in them reverting back to their wild behaviour. The fact that they consolidate their position around their calves when initially threatened would potentially make them vulnerable to a group hunting strategy, especially if they were ambushed at any point of maximum herd restriction.

The Floodplain Forest Nature Reserve, an area of around 48 hectares within the Ouse Valley Park, Wolverton, Milton Keynes, is the same area where excavation discovered a complex of five cursus monuments lying between ancient palaeochannels (Hogan 2013b). After excavation, the ancient palaeochannels were left open, becoming a series of pits and pools planted with self-set willow scrub together with weedy species such as willow herb and stinging nettle. However, to prevent the landform from becoming totally overgrown thereby impoverishing it for nesting birds and other wildlife, grazing cattle and horses were introduced. Martin Kincaid, Senior Biodiversity Officer for The Parks Trust (2018, personal communication, September) outlined that the original proposal was for the nature reserve to use solely Konik horses for the grazing management of the site. Koniks have been used for conservation management in the UK since the 1980s due to their grazing style, which creates a mosaic of differing vegetation heights superbly adapted to wetlands. Organizations such as National Trust, RSPB and Wildfowl and Wetlands Trust have used these animals on some of their larger wetland sites as they will happily wade up to chest height in water, swim well, they have self-trimming hooves. However, while Koniks do an excellent job, the Floodplain Forest Nature Reserve needed to supplement them with cattle during the spring and summer months, introducing a stocking density of about twenty cattle which currently graze seasonally through the drier months of spring and summer.

Although American range cattle are semi-domestic, their environmental requirements would potentially be similar to aurochs. As part of an investigation to reduce the impact of livestock on water quality, aquatic and riparian habitat, Melvin George et al. (2007, pp. 1–20) fitted GPS collars to a number of American range cattle to monitor their feeding, rest and watering positions. This shows that the distribution of the herd is at its most concentrated when feeding on the gentle slopes near to water and that the herd only increases its distribution area after it has depleted these feeding sites and needs to move to areas further from water or with steeper gradients. Aurochs, however, would have been able to move on to fresh feeding sites and would therefore have had less need to feed upon the steeper gradients.

Young (2017, p. 13) suggests that not only will cows eat the young green grass but given the opportunity they will return to the area to eat the grass again when it comes into seed. This would give the herd two fatty acids which they cannot make from any other source – linoleic acid and linolenic acid – both of which are needed to help the animals grow. Plentiful supplies of water are one of the primary focuses in a herd's decision about where to graze, ruminate, rest and drink. Riparian zones tend to receive increased use due to their provision of water, shade, thermal cover and a source of high-quality forage. These areas also influence the grazing distribution of the herd, although they are rarely used as a feeding site for more than a couple of days at a time.

George et al. (2007, p. 2) suggest that 'when grazing animals become familiar with a landscape, they retain information about the prime location focal points, such as where to find water, shade and areas to feed'. Perhaps the construction of cursus monuments could have been an early attempt to overcome that cattle memory, to stop herds of aurochs from returning to landscapes now required for the feeding of herds of domestic cattle. George et al. (2007, p. 4) suggest that 'cattle memory appears to provide the basis for ranking feeding sites when deciding where to graze next'. The cows appear to have a morning foraging bout lasting for around four hours followed by a second foraging bout of similar length at a different location in the late afternoon or evening. On average, cows spent one third of their day grazing, one third ruminating and one third resting. However, the time spent grazing appears to be highly dependent upon the forage quality and availability. The use of GPS collars has allowed George et al. (2007, p. 4) to establish that 'herds of American range cattle appear to have used a well-established trail system, making daily circuits of the high-quality seasonal riparian patches of grassland over a large rangeland pasture'.

George et al. (2007, p. 5) note that 'both the terrain over which cattle roam and their distance from water play extremely significant factors in their grazing capacity'. Gradient significantly reduces the distance a herd can travel without it starting to experience weight loss. He identifies that 'it only requires an 11-degree slope gradient to start to see a significant reduction in grazing capacity and therefore in the milk yield of lactating cows':

Slope gradient <10 = 0 per cent reduction in grazing capacity
Slope gradient 11–30 = 30 per cent reduction in grazing capacity
Slope gradient 31–60 = 60 per cent reduction in grazing capacity
Slope gradient >60 = 100 per cent reduction in grazing capacity

(George et al. 2007, p. 5)

The distance the herd travels from water appears to be another critical component. Sufficient water is required to take into account factors such as the climatic condition, the activity of the herd and the lactation status of individual animals. A lactating cow requires up to seventeen gallons of water per day to produce the four or five gallons of milk needed daily to feed a calf over the twelve-month period until it is weaned (Young 2017, p. 29). George et al. (2007, p. 5) suggest that 'the distance from water not only reduces a herd's grazing capacity but also has a direct effect upon the number of animals that can actually graze within the herd'.

Distance from water
0–1.5 km - 0 per cent reduction in grazing capacity
1.5–3 km = 50 per cent reduction in grazing capacity
>3 km = 100 per cent reduction in grazing capacity

(George et al. 2007, p. 5)

However, Crane (2016, p. 13) suggests that Britain's tilted topography, fed by rainfall producing 160,000 km of watercourses, would mean that cattle would seldom have been more than 1.5 km from running water. Could such factors potentially have resulted in the landscapes upon which later cursus monuments were constructed becoming prime routes for herds of aurochs as they moved to early spring meadow grasslands (Rogers et al. 2018, p. 142) or to fresh pasture, where ambushes were set by groups of hunter-gatherers at points of maximum restriction to the herd movement?

Rogers et al.'s (2018, p. 142) suggestion that 'Carbon isotopic ($\delta^{13}C$) levels show aurochs came out of forest cover in the early spring, perhaps feeding on the early grass growth of the spring meadows' together with Legge and Rowley-Conwy's (1988) theory that 'aurochs herds split into smaller groups and headed away from the plain and into the forest to winter on acorns and browse' appear to suggest a seasonality to aurochs movement where they spent eight months of the year feeding upon the grasslands before moving into the forest cover for winter. Pryor (2015b, p. 58) suggests that seasonality would not only have affected the movement of animals, the ripening of fruits, nuts and other wild foods but it would have also affected the availability of certain landscapes and other resources, which in wintertime could have been submerged in floodwaters. However, as Crane (2016, p. 14) suggests, it was not only the herds of aurochs that moved, but 'everything moved. In this climatically charged, geographically complex world, food sources were not distributed evenly, or throughout the year'. Therefore, each environment, whether it was grassland, woodland, glades, estuaries or coasts, would have had its own continually shifting species and seasons of abundance.

Rogers et al.'s (2018, p. 141) oxygen isotopic data ($\delta^{18}O$ values) for the aurochs teeth from Blick Mead (BM421 and BM422), while highlighting that both individuals came from either the same region or different regions with similar ($\delta^{18}O$) values, potentially indicates that the aurochs either originated in parts of Scotland or parts of eastern England, although there was some ambiguity over which actual region. However, consistent strontium results strongly indicate that the aurochs were either local to the Blick Mead area or from the chalklands along the Lincolnshire and Yorkshire east coasts.

Higgs (1961, pp. 144–54) had earlier identified that aurochs probably had some degree of seasonal movement in most open habitats, which he demonstrated by research undertaken on the fauna of the Haua Fteah cave in Cyrenaica, Libya, where he noted that a peak of aurochs bones at the end of the Pleistocene period equated to a woodland recession due to cooler conditions and reduced rainfall. It therefore appears that aurochs had some degree of seasonal movement, both moving to new riparian zones on the grassland and moving away from the grasslands to winter in forest cover. However, whether the aurochs whose teeth were found in the Blick Mead spring were roaming along the entire chalkland belt from as far as the Yorkshire Wolds is debatable. But even if the Blick Mead aurochs did not travel from as far as the Yorkshire Wolds (Rogers 2016, personal communication, October), it does appear that they travelled from outside the immediate Stonehenge area. This suggests that the aurochs whose remains have been found within the spring at Blick Mead may not have lived their entire lives within the Stonehenge area, but could potentially have travelled from afar, to be intercepted and killed within the Stonehenge area during the process of a much longer journey.

However, to identify any potential correlation between locations where the Mesolithic hunting of aurochs took place and the later open areas of landscape where cursus monuments were constructed will require three criteria to be met – firstly, that aurochs hunting occurred within areas of open landscape which were restricted by adjacent rivers or marshland; secondly, that these locations were the type of topography where later cursus monuments were constructed; and thirdly, that it is still possible to identify these hunting sites, either through recovery of faunal features or through lithic scatters.

Addressing the first criterion, that aurochs hunting occurred within areas of open landscape restricted by adjacent rivers or marshland, Holm (1991, pp. 89–100) supports the theory that hunting techniques have remained similar throughout millennia where even the invention and refinement of modern weapons, motor vehicles and aircraft appear to have had little effect on basic ambush patterns and techniques. The Canadian National Defence Force (2001, p. 1) has concluded, in line with Holm, that 'the basic elements of ambush tactics have remained constant over this period'. It therefore appears plausible to relate modern military tactics to those of prehistoric hunter-gatherers. A surprise attack upon slow-moving or temporarily halted animals by hunters lying in wait at well-laid ambush sites would have been an effective way in which to kill selected aurochs from within the herd. This would have offered a group of hunters the opportunity to concentrate a high degree of firepower by making maximum use of available resources. Selecting a site which prevented observation while being downwind of the herd would potentially have caught the animals by surprise. Suitable places would therefore include concentration points of known herd movement around watering points. In these areas, an instantaneous co-ordinated action against a surprised herd within range of the Mesolithic bow (Bergman 1993, p. 102), with members of the group covering all arcs of fire, was likely to lead to a successful hunt.

A skeleton found at Vig, in Odsherred, north-western Zealand. (Hartz and Winge 1906), which identifies an aurochs killed in the course of an unsuccessful hunting expedition, supports a group hunting theory. When found, the remains of numerous microlith arrowheads were recovered from the chest area. The fact that, after being shot, the beast had wandered into a bog, escaping the hunters to later die of its wounds, supports the notion that hunting occurred near water or marshland. Another aurochs skeleton, found at Prejlerup bog in north-western Zealand (Sorensen and Petersen 1986, pp. 111–17), shows that the aurochs had been shot by at least nine microlith arrowheads. Again, the beast had wandered into a bog to die after an unsuccessful hunt. This identifies the high degree of trauma required to kill an aurochs, which further supports the group hunt theory, while the fact that once again, after being shot the beast had wandered into a bog, escaping the hunters to succumb to its wounds, supports the notion that hunting occurred near water or marshland.

Prummel's (Prummel and Niekus 2011) discovery of a single female aurochs hunting and butchery site at Balkweg, in the valley of the River Tjonger, in Holland, included more than forty-nine aurochs' bones together with a flint blade found on the surface next to a recently dug ditch. Although most of the bones were spread over an area which measured approximately 2 sq m, two smaller concentrations of bone, consisting mainly of vertebrae, were found within a short distance alongside two Mesolithic blade fragments that clearly fit together. Approximately 40 km upstream, part of a burnt aurochs vertebra was also found. Prummel (Prummel and Niekus 2011, p. 1457) believed 'minimal displacement had occurred and since no skeletal element was represented more than once, they all probably belonged to one individual'. The most likely interpretation was that they were the remnants of a single episode of hunting and butchery, the animal having been skinned and butchered in the location of the kill, the burnt bones and the burnt flint blade indicating a fairly large group of hunters lit a fire to cook selected parts of the carcass, the ribcage having been opened and the meat removed from the vertebrae and ribs to be consumed locally.

All three of the above examples appear to establish that group hunting was a requirement to maximize firepower when hunting aurochs and that hunting occurred in locations adjacent to water or marshland. While environmental data was not available for these precise hunting locations, the fact that a group hunting strategy was used presumes some degree of open landscape, as this would have been a requirement to pass visible signals to one another.

Turning our attention to the second criterion, that aurochs hunting grounds consisted of the same type of topography where later cursus monuments were constructed, Taylor (1983) and Frison (2004) identify many difficulties when they suggest that many original prehistoric landscape features will no longer be detectable due to post-prehistorical damage. Very few examples for direct prehistoric use of the landscape remain and, where they have been discovered, such as the identification of hunting sites upon the Stonehenge Plain (Parker-Pearson 2012, p. 230), their discovery has tended to be solely due to the archaeology found within them. This has led both Gaddis (2002, p. 103) and Taylor (1983, p. 24) to indicate that as not all sources survive, maybe we can perhaps never expect to get the full story of what actually happened. In an attempt to overcome this problem, this study undertook a boots-on-the-ground field observation exercise across the landscape within the vicinity of each individual cursus monument site that made up the study group, while taking Taylor's (1983), Gaddis' (2002) and Frison's (2004) suggestions into account. This has provided the largest representative sample to date for evaluating the potential landscape associated with earlier activities across English cursus monument sites.

To assess the level of probability of each field observation being similar to any actual Neolithic landscape feature, the areas surveyed were subdivided into four groups – high probability, which includes features such as hillsides and high ridges, likely to have been similar during Neolithic times;

medium probability, which includes features such as excavated palaeochannels, where some degree of interpretation is required; low probability, which includes difficult-to-justify features; and landscapes destroyed since the Neolithic, which speaks for itself.

Sites with a high level of probability included the Rudston A Cursus, where field observation established a slight drop in terrain, still visible even after centuries of active ploughing, that ran east-west (OS Grid Ref TA 102668) approximately 1 km north of the southern terminal. As this is the only visible ridge, it is probably not a lynchet and was probably much higher during prehistoric times.

A natural ridgeline at Duggleby Howe, just to the north of the Duggleby Cursus (OS Grid Ref SE 881669) lies in an area just to the south of where maximum herd concentration would have occurred. Again, this is a natural ridgeline that would potentially have been similar during prehistoric times.

Any killing zone for the Fimber Cursus potentially operated from a high ridge that aligns in an east-west direction running parallel to the southern side of the cursus monument (OS Grid Ref SE 894609 to OS Grid Ref SE 896607), while the killing zone of the Kirby Underdale Cursus appears to be situated where two winterbourne watercourses concentrate. A natural high ridge between two becks, known locally as Eskhams, would potentially have been similar during prehistoric times and is the probable killing zone (OS Grid Ref SK 809588), offering a position to shoot down on the herd as it attempted to cross the beck.

And finally, at Stonehenge, Jacques (Jacques and Phillips 2014, p. 24) identified that palaeochannels produce a natural funnelling effect where the fording point was located within Stonehenge Bottom, while fieldwork (Saunders 2015, pp. 60–69) identified that the natural topography at this point would have resulted in the aurochs being within 45 m of hunters before they had any chance of catching sight of them, thus giving the hunters a massive advantage over their prey.

Sites with a medium level of probability would include the palaeochannels that could potentially have assisted prehistoric hunting at the Maxey and Etton cursus monuments, the Stanwell Cursus complex and the Wolverton Cursus complex; the ridgeline, where later populations make use of the lee of the hill to construct a Roman villa (OS Grid Ref TA 088668), about 750 m south-west of the point where the Rudston B Cursus meets the Gypsey Race winterbourne river; and the same ridgeline about 500 m south of Springdale Farm (OS Grid Ref TA 085682) which crosses the path of the Rudston C Cursus.

Sites with a low level of probability include both the Bag Enderby Neolithic pit alignment in Lincolnshire, where a difference in ground heights between two fields indicates a possible earlier ridgeline (OS Grid Ref TF 351726), and the Stenigot pit alignment, also in Lincolnshire (OS Grid Ref TF 244810), where once again a slight difference in ground height indicates an earlier ridgeline with extraction points to the south-west.

Of the further sites, over a quarter of them lie upon landscapes that have been destroyed by quarrying, urbanization or the creation of roads, railways or airports. This study also has to accept that there is currently little general archaeological consensus as to what modern landscape features could potentially constitute a typical prehistoric hunting site and that similar characteristics could apply equally across both archaeological and natural features (Bradley 2018, personal communication).

It therefore appears that the use of field observation investigation alone will not enable us to place any potential hunting-associated landscape within a specific Mesolithic or Neolithic period. It also appears to highlight difficulties with correlating cursus monuments to any individual landscape feature that could potentially have assisted with hunting activities. While initial results could appear to give the impression that, at the 64 per cent of cursus monument sites where it is still possible to use field observation, 83 per cent initially appear to have at least one feature that could be directly associated with the hunting of aurochs, when investigated further only 10 per cent of sites have a correlation that could be seen as being of a high probability. Therefore, when Pearson's correlation coefficient is applied, it identifies only a weak correlation which does not appear significant.

Looking at the third criterion, that it is possible to still identify these hunting sites through the recovery of fauna or lithic scatters, Barclay and Bayliss (1999, p. 17) suggest that 'one principle factor when excavating cursus monuments is that they are relatively clean of finds'. This is supported from data collected from the vicinity of fifty cursus monument sites on or adjacent to the English chalkland belt selected as part of this study's data group. Twenty-nine of the cursus monuments within the study group have been excavated, and aurochs bones have been discovered within the wider area around twenty-three of these monuments, which equates to around 80 per cent of the excavated cursus monument sites having some form of aurochs representation within the area. Although this obviously does not indicate any form of correlation between aurochs and cursus monuments, it does place aurochs within the various landscapes at the time of monument construction. However, at this time it is only possible to place aurochs directly within the immediate vicinity of ten cursus monuments (20 per cent), a figure significantly skewed since five of these monuments belong to the Wolverton Cursus complex.

At the Wolverton Cursus complex, Rajkovaca (In Hogan 2013b, pp. 38–39) identifies a near-complete aurochs first phalanx from the Wolverton 1 Cursus, while two tree throws (F.96/97 and F.98) produce fragmentary cattle remains which were provisionally identified as aurochs and further tree throws (F.106, F.111, F.112, F.114 and F.131) yield a number of cattle-sized limb fragments which were identified as probably belonging to aurochs. At the Drayton North Cursus, although most of the animal fauna appears to have been domesticated, Ayres and Powell (2003, pp. 159–63) were able to identify two aurochs bones. In the Thames Valley, thirty-five bones from a semi-articulated aurochs skeleton were discovered within the Mesolithic layer of a channel at the Thames Valley Park in Reading, next to the Sonning Cursus. At Stonehenge, Richards (1990, p. 83) suggests that thirty-eight fragments of unidentified large mammal found within the ditches of the Stonehenge Lesser Cursus were potentially from aurochs. At the Etton Cursus, Pryor (1998, p. 47) discovers two aurochs skulls, including their horns, buried upon an oak plank within ditch segment number 12 of the Etton causewayed enclosure. And at the Gussage section of the Dorset Cursus, Bradley (1991, p. 46) discovers two aurochs bones from the western ditch of the Dorset Cursus which dated to 4510–4950 BC.

As the use of field observation investigation alone does not appear to have been able to place potential hunting-associated landscapes within any specific Mesolithic or Neolithic period, investigation was also undertaken with regard to the distribution pattern of hunting-associated Mesolithic microliths and Neolithic arrowheads identified within the National Monument Number scheme. The scheme was able to recognize evidence with regard to the Springfield Cursus area that included seventy-seven unretouched blades and flakes discovered on the southern floodplain of the River Chelmer together with a further 400 Mesolithic flints found around the confluence of the River Can and the River Chelmer, which appear to form a funnel that would have aligned with potential large herbivore movement towards the later cursus monument. It was also able to recognize Parker-Pearson's (2012, pp. 235–36) excavations to the west of the Stonehenge monument that led to the discovery of a Mesolithic hunting camp, 400 yards south of the Mesolithic postholes. The important Mesolithic assemblages at Yatesbury, excavated by Kendall (1916) near to the River Winterbourne during the early part of the twentieth century, included a mixed scatter of Late Mesolithic and Early Neolithic flintwork, perhaps indicating a short-stay camp (Evans et al. 1993). In total, eight other Mesolithic find spots around Yatesbury suggest hunting activity potentially occurred in the region around the Galteemore Springs (Pollard and Reynolds 2002, p. 23), perhaps relating to the area where the Yatesbury (Avebury) Cursus was constructed. At the Gussage section of the Dorset Cursus, a striking distribution pattern for Mesolithic sites appears to emerge along the western ridge of Cranborne Chase, parallel to the later cursus monument, while in the area around Down Farm a further four sites that contain a high percentage of microliths were discovered (Green 2000, pp. 20–22). The scheme was also able to recognize potential Mesolithic hunting sites at the Harlaxton pit alignment, the Drayton North Cursus, the Drayton South Cursus and the South Stoke Cursus. However, this would only amount to 16 per cent of the total data set and therefore, when Pearson's correlation coefficient is applied, it again identifies only a weak correlation which does not appear significant.

The study also used the National Monument Number scheme to identify the location of Neolithic arrowheads, where arrowheads were identified at 50 per cent of cursus monuments within the study group. Although the majority of these finds appear to have been individual arrowheads scattered across the landscape, sixteen early Neolithic leaf-shaped arrowheads were found at North Burton, less than 500 m to the west of the Rudston D Cursus northern terminal, potentially suggesting the area was used for hunting during that period, and twenty-two Early Neolithic arrowheads were discovered at the Drayton South Cursus by Holgate (1986, p. 12), perhaps also suggesting that hunting continued at this site. However, it is not currently possible to identify whether these were used for the hunting of aurochs or, as suggested by Trantalidou and Masseti (2015, p. 72), were used within some form of trophy hunting that occurred post-construction of the cursus monument site.

Current evidence, therefore, suggests that any correlation between Mesolithic or Neolithic aurochs hunting sites and the later landscape where cursus monuments were constructed is not proven. Firstly, there is not the excavational evidence to conclude that aurochs hunting occurred within these areas since Pearson's correlation coefficient suggests only a weak, insignificant relationship. Secondly, investigation by field observation is unable to identify sufficient potential Mesolithic aurochs hunting grounds, again resulting in Pearson's correlation coefficient suggesting only a weak, insignificant relationship. And thirdly, investigation by National Monument Records is unable to identify sufficient Mesolithic lithic scatters, again resulting in Pearson's correlation coefficient suggesting only a weak, insignificant relationship. Although the recovery of Neolithic arrowheads moves the Pearson Correlation Coefficient to a medium level, this still appears to be insignificant and could potentially result from some form of post-monument trophy hunting (Trantalidou and Masseti 2015, p. 72).

It therefore appears that none of the three necessary criteria have been met that could identify any correlation between Mesolithic aurochs hunting grounds and the open areas of landscape where cursus monuments were later constructed. However, the fact that Pearson's correlation coefficient does increase to a moderate level, although still appearing insignificant, when Neolithic lithics are introduced into the equation potentially suggests a correlation between the placement and alignment of cursus monuments and the way Neolithic pastoralists roamed the landscape with their herds of domestic cattle.

It is interesting to note that 85 per cent of the locations within 5 km of a cursus monument site where aurochs bones have been discovered also contain the later bones of domestic cattle. Pearson's correlation coefficient would therefore suggest that any correlation with regard to aurochs was primarily to do with the landscape used by both them and domestic cattle. The series of warmer summers and cooler winters that began around 4100 BC (Crane 2016, p. 80) would have enabled Neolithic pastoralists to expand northwards on to pastures already being used by aurochs, with improved growing seasons allowing the pastures to support both species. It was only 300 years later, when the climate once again changed, reducing the growing season, that competition for these landscapes would have occurred. Investigation of British faunal assemblages from Early Neolithic small settlement sites (3900–3700 cal BC), from causewayed enclosure sites (starting around 3700 cal BC) and from excavated cursus monument sites (3600–3500 cal BC) show that these sites appear to be overwhelmingly dominated by the fauna of domestic cattle rather than aurochs, which appear to have become vanishingly rare. This would suggest that any correlation would be between the landscape where cursus monument construction occurred and some form of domestic cattle movement.

Domestic Cattle

A picture is starting to develop that suggests a potential correlation between the placement of cursus monuments and the way Neolithic pastoralists roamed the landscape with their herds of domestic cattle. Whether, as Ray and Thomas (2018, p. 84) propose, 'the British Neolithic was a co-creation achieved through contact and interaction between Continental Neolithic people and the indigenous British population' or through one group comprehensively replacing another, the fact that a new species of cattle enters the British Isles indicates that movement of people brought these animals into the country. And movement of people would lead eventually to the spread of cattle throughout the whole of Britain. As Crane (2016, p. 14) reveals, during this period everything moved due to continuously shifting ecosystems driven by a period of rapid global warming. Crane (2016, p. 18) suggests that in effect, the population lived through movement which resulted in the creation of paths. Yet as Pryor (2015b, p. 9) indicates, nomads lead highly structured lives following known routes covering the same areas of land from one year to the next.

Although initially created by repetitive footfall, the course of each path was probably constructed as the result of an innate negotiation between the way humans and their livestock needed to move across the landscape and the actual lie of the land. Lyons and Machen's (2001, p. 3), experimentation with cows fitted with GPS collars has demonstrated that cattle prefer some range sites over others due to the type of terrain, being discouraged through the presence of loose and imbedded rock. However, within a few generations these paths probably evolved into networks, which has led Crane (2016, p. 21) to suggest that 'the chalkland belt would potentially have been seen as a path leading across eastern and southern Britain'. The chalkland belt underpins a peripheral eastern band of hills that reach their greatest height in the Yorkshire Wolds, while another four ranges of linear chalk uplands stretch across the south of Britain, converging on Salisbury Plain. Of these four limbs, the longest and widest rolls diagonally across the south-east of Britain from Doggerland, by way of the Chiltern Hills and the Marlborough Downs to the plain. Two thinner but more distinctive bands of chalk, the North and the South Downs, approach as steep-flanked ridges from the east, while the stubbiest of the four limbs begins near the southern coastal promontory of Portland and reaches the plain by way of the Dorset Downs and Cranborne Chase.

Ray and Thomas (2018, p. 94) suggest 'it was the keeping of cattle that most tied these communities to their animals'. Therefore, although cattle would have provided a source of social and economic wealth that previous Mesolithic communities would not have known, the requirement to provide the valuable herd with plentiful supplies of water together with places to graze, ruminate and rest (George et al. 2007, p. 2) would potentially have resulted in continuous movement along the chalkland belt.

To succeed, and in fact to survive, these early pastoralists would have needed to know how to make themselves at home in the landscape. Alongside potentially using some of the same open areas as had been previously used by wild cattle, they would have needed to be adept at deliberately intervening with the landscape, creating or enhancing clearings to improve grazing for their domestic cattle. By this period, these initial pastoralists had successfully transported domestic cattle across the sea from the continent and across most parts of the British Isles, so it appears they were fully aware of the best methodologies for moving and looking after their most valuable asset, their cattle. They would have known instinctively that both the terrain over which cattle travelled and its distance from water were important factors in keeping the herd healthy (George 2007, p. 5). They would also have known where to graze the herd at various times of the year (Rogers et al. 2018, p. 142). Making use of the early spring meadows would have enabled a larger-size herd to be kept over winter, thereby increasing the wealth of the community, while the climate allowed for wintering of the herd in the closed-canopy forests, feeding on acorns and browse. However, for most of the year, herds of domestic cattle would have been grazed on the open grassland (Rogers et al. 2018, p. 142) or potentially alongside areas of wetland meadow (Kincaid 2018, personal communication, September; Van Vuure 2005).

Although working in the context of later Neolithic periods, Viner et al. (2010, p. 2812) have investigated the relative ease with which cattle could have been moved over long distances, where the requirement to provide ample pastureland has led to the conclusion that some form of prehistoric cattle movement occurred. Initial studies of the cattle from Durrington Walls (Viner et al. 2010, p. 2818) appears to suggest that 'most of the cattle originated in a geological domain adjacent to the chalkland. However, in some cases the minimum distances traversed by cattle is almost 100 km in a straight line'. If natural features such as hills, valleys and stretches of water were taken into account, the actual distance travelled would have been considerably greater. This identifies that the movement of cattle in the Neolithic, although difficult and time consuming, was not impossible. Later investigations from Durrington Walls (Greaney et al. 2018, pp. 26–31), where Richard Madgwick of Cardiff University was able to narrow down the potential origin of the livestock, indicate that 'one cow had travelled a minimum of 90 km, one cow had travelled a minimum of 145 km, four cows had travelled a minimum of 290 km and two cows had travelled a minimum of 320 km, where they appear to have averaged approximately 15 km per day'. This would result in a total travel time to move a herd from one end of the data set of this study group to the other being in the region of twenty-eight days, a journey that investigations at Durrington Walls (Viner et al. 2010 and Greaney et al. 2018) indicate was clearly possible.

Serjeantson's (2011, p. 15) review of excavated animal bones from the Neolithic and early Bronze Age of southern Britain, which focused on wild and domestic animals between 4000 and 1500 BC, identifies that the number of domestic cattle bones is greater than those of all other species in the Early Neolithic while the bones from aurochs appear to have become vanishingly rare. Cattle are the only domestic animal present in any number at the Coneybury Anomaly and the most numerous in the other assemblages. During the Early/Middle Neolithic, domestic cattle account for more than half of all remains across most sites, while in long barrows they account for nearly 70 per cent. This is supported by data presented by Peter Rowley-Conwy at the University of Buckingham in December 2018 which highlighted the animal fauna type from Early Neolithic sites such as Ascot-under-Wychwood, Hazleton North, Windmill Hill (pre-enclosure) and the Eton Rowing Course and from later causewayed enclosures at Abingdon, Hambledon Hill, Maiden Castle, Windmill Hill and Etton.

However, Serjeantson (2011, p. 25) suggests that although the relatively large number of cattle bones could potentially translate from the fact that people kept more cattle than other animals in the Early/Middle Neolithic, it could also be that people were deliberately depositing cattle bones in preference to the bones from smaller animals which would skew the data towards a predominance of cattle across all sites until the Late Neolithic.

Further investigation of domestic cattle fauna (Rowley-Conwy 2018, personal communication, December) identifies that in a milk herd most of the males are superfluous and, as such, many of the male calves are killed off soon after birth (Legge 1981b). This appears to be supported by the faunal data of male and female cattle from the causewayed enclosure sites at Etton, Hambledon Hill and Windmill Hill and appears to have been specifically the case if either grazing or manpower was limited. This would potentially suggest Neolithic herdsmen had mastered some of the strategies that can be used to encourage cows to let down their milk in the absence of the calf (Ryan 2005). However, if the management of the herd was not intensive, such as where spring meadow landscapes would have increased grazing capacity, allowing for larger-sized herds, male calves do not appear to have been killed until towards the end of the summer.

Serjeantson (2011, p. 68) further suggests that it is highly probable that cattle did not just provide food, but that they also played a major role in the ceremonial, ritual and religious life of the Neolithic community. This could be considered inevitable, given that the community depended upon their domestic herd for most of their daily food, that the care of cattle must have been the main activity in most people's lives and that social and economic prestige probably resulted from the owning of cattle. Serjeantson (2011, p. 71) believes this is supported by the fact that cattle skulls were placed within the ditches of numerous causewayed enclosures. Of the thirteen cattle skulls from the 1988 excavations undertaken at Windmill Hill, ten were in ditch terminals (Whittle et al. 1999, p. 359), as was a cattle skull in the enclosure at Corporation Farm, Abingdon (Barclay et al. 2003). Serjeantson (2011, p. 65) further suggests that, as a whole adult cow or bull provides more meat than can be eaten by a single family group, 'the killing of a cow or bull was probably only carried out on those occasions when firstly, numerous people were present to consume it and secondly, on occasions when feasting would have raised the social profile of the owner'.

However, although numerous excavations have occurred across cursus monument sites, it appears that very little cross-site compilation of domestic cattle remains has so far been carried out – perhaps due to cursus monuments' large size and more likely as a result of their reputation for having barren ditches and interiors which produce pitifully little by way of internal features, artefacts or dating evidence. This study has therefore assessed each excavated cursus monument site for the inclusion of cattle fauna.

Table 3: Cattle bone from cursus monument sites

Cursus monument site	Domestic Cattle Fauna	Excavation
Rudston Cursus A	(Harding, J. 2006, p. 119) identifies three pits at Low Caythorpe 1, along the course of the Caythorpe Gas Pipeline, which produced the partial remains of as many as nine domesticated head of cattle.	1877 W Greenwell 1958 C & E Grantham (Mentioned in Dymond D 1966) 1988 T G Manby
Rudston Cursus B	(Harding, J. 2006, p. 119) identifies three pits at Low Caythorpe 1, along the course of the Caythorpe Gas Pipeline, which produced the partial remains of as many as nine domesticated head of cattle.	
Rudston Cursus C	(Harding, J. 2006, p. 119) identifies three pits at Low Caythorpe 1, along the course of the Caythorpe Gas Pipeline, which produced the partial remains of as many as nine domesticated head of cattle.	
Rudston Cursus D	(Harding, J. 2006, p. 119) identifies three pits at Low Caythorpe 1, along the course of the Caythorpe Gas Pipeline, which produced the partial remains of as many as nine domesticated head of cattle.	
Duggleby Cursus	Not excavated	
Fimber Cursus	Not excavated	
Kirby Underdale Cursus	Not excavated	
Bag Enderby Pit Alignment	Not excavated	
Harlaxton	Not excavated	
Stenigot Pit Alignment	Not excavated	
Hanworth Cursus	Not excavated	
Fornham All Saints Cursus	Not excavated	
Stratford St Mary	Not excavated	
Barnack Cursus	Not excavated	
Eynesbury Cursus		Sections were cut across the Northern Cursus (Kemp 1998; Macaulay 1994)
Godmanchester Cursus		Excavations by Fachtna McAvoy and the Central Excavation Unit in 1988–91 at Rectory Farm, Godmanchester
Brampton Cursus		Mortuary enclosure at eastern end of Cursus excavated by Tim Malim in 1991

Table 3: *Continued*

Cursus monument site	Domestic Cattle Fauna	Excavation
Maxey Cursus	(Ainsley, C. In: Pryor et al. 2005, p. 84). The dominance of domestic cattle is remarkable at the Etton Landscapes site (485 relating to 64 per cent)	1962–63 WG Simpson 1979–81 F Pryor 1982–84 F Pryor
Etton Cursus	(Ainsley, C. In: Pryor et al. 2005, p. 84). The dominance of domestic cattle is remarkable at the Etton Landscapes site (485 relating to 64 per cent)	1982–87 F Pryor
Springfield Cursus	(Buckley, D. et al. 2001, p. 147) suggest fragments identifiable to cattle included four sesamoids, part of an ulna, two fragments of first phalanx, three fragments of vertebra and several fragments of unidentified long bone.	1979–84 Hedges and Buckley
Stanwell Cursus 1	Only two Neolithic features produced any animal bone: the HE1 enclosure and both ditches of the C1 Stanwell Cursus. However, the presence of domestic animals at the site during this period is noteworthy (Knight, S. In: Lewis et al. 2010). Two pieces were identified as cow (Bos Taurus), 34 fragments were identified as cow/red deer and 4 fragments were identified as large mammal.	1979–85 M O'Connell 2006–10 Framework Archaeology
Stanwell Cursus 2	Only two Neolithic features produced any animal bone: the HE1 enclosure and both ditches of the C1 Stanwell Cursus. However, the presence of domestic animals at the site during this period is noteworthy (Knight, S. In: Lewis et al. 2010). Two pieces were identified as cow (Bos Taurus), 34 fragments were identified as cow/red deer and 4 fragments were identified as large mammal.	1979–85 M O'Connell 2006–10 Framework Archaeology
Stanwell Cursus 3	Only two Neolithic features produced any animal bone: the HE1 enclosure and both ditches of the C1 Stanwell Cursus. However, the presence of domestic animals at the site during this period is noteworthy (Knight, S. In: Lewis et al. 2010). Two pieces were identified as cow (Bos Taurus), 34 fragments were identified as cow/red deer and 4 fragments were identified as large mammal.	1979–85 M O'Connell 2006–10 Framework Archaeology
Stanwell Cursus 4	Only two Neolithic features produced any animal bone: the HE1 enclosure and both ditches of the C1 Stanwell Cursus. However, the presence of domestic animals at the site during this period is noteworthy (Knight, S. In: Lewis et al. 2010). Two pieces were identified as cow (Bos Taurus), 34 fragments were identified as cow/red deer and 4 fragments were identified as large mammal.	1979–85 M O'Connell 2006–10 Framework Archaeology
Stanwell Cursus 5	Only two Neolithic features produced any animal bone: the HE1 enclosure and both ditches of the C1 Stanwell Cursus. However, the presence of domestic animals at the site during this period is noteworthy (Knight, S. In: Lewis et al. 2010). Two pieces were identified as cow (Bos Taurus), 34 fragments were identified as cow/red deer and 4 fragments were identified as large mammal.	1979–85 M O'Connell 2006–10 Framework Archaeology

(Continued)

Table 3: *Continued*

Cursus monument site	Domestic Cattle Fauna	Excavation
Biggleswade Cursus		2004 Albion Archaeology
Cardington Cursus	Not excavated	
Cople Cursus	Not excavated	
Ivinghoe Beacon Cursus	Not excavated	
Wolverton Cursus 1	(Rajkovaca, V. In: Hogan, S. 2013b, p. 38) The fieldwork at Manor Farm resulted in the recovery of 231 assessable fragments of animal bone of which 48 were identifiable to species of which 27 (56 per cent) were from domestic cattle. A further 47 were cattle sized but the specimen could not be further identified.	2008–11 Cambridge Archaeological Unit
Wolverton Cursus 2	(Rajkovaca, V. In: Hogan, S. 2013b, p. 38) The fieldwork at Manor Farm resulted in the recovery of 231 assessable fragments of animal bone of which 48 were identifiable to species of which 27 (56 per cent) were from domestic cattle. A further 47 were cattle sized but the specimen could not be further identified.	2008–11 Cambridge Archaeological Unit
Wolverton Cursus 3	(Rajkovaca, V. In: Hogan, S. 2013b, p. 38) The fieldwork at Manor Farm resulted in the recovery of 231 assessable fragments of animal bone of which 48 were identifiable to species of which 27 (56 per cent) were from domestic cattle. A further 47 were cattle sized but the specimen could not be further identified.	2008–11 Cambridge Archaeological Unit
Wolverton Cursus 4	(Rajkovaca, V. In: Hogan, S. 2013b, p. 38) The fieldwork at Manor Farm resulted in the recovery of 231 assessable fragments of animal bone of which 48 were identifiable to species of which 27 (56 per cent) were from domestic cattle. A further 47 were cattle sized but the specimen could not be further identified.	2008–11 Cambridge Archaeological Unit
Wolverton Cursus 5	(Rajkovaca, V. In: Hogan, S. 2013b, p. 38) The fieldwork at Manor Farm resulted in the recovery of 231 assessable fragments of animal bone of which 48 were identifiable to species of which 27 (56 per cent) were from domestic cattle. A further 47 were cattle sized but the specimen could not be further identified.	2008–11 Cambridge Archaeological Unit
Benson Cursus	Not excavated	
Dorchester Cursus		1947–52 Atkinson 1981 Chambers 1988 Bradley and Chambers 2010–17 Gill Hey
Drayton St Leonard	Not excavated	
Drayton North Cursus	(Ayres, K. and Powell, A. In: Barclay et al. (2003, p. 159) A total of 248 bone fragments were discovered from the east cursus ditch. Around half were identifiable, the majority being domestic cattle sized (100). The west cursus ditch provided similar results with a predominance of cattle bones in the identifiable material (20).	1977 Michael Parrington 1979–82 Ainslie and Wallis 1985–86 Oxford Archaeological Unit
Drayton South Cursus	(Wilson, R. In: Barclay et al. 2003, p. 29). Various features yielded scattered bone, but only those from the Neolithic pit 107 were recorded as soil samples from this feature were sieved for bones. Mainly domestic pig but one fragmentary domestic cattle tooth	1921–37 E T Leeds 1994 Oxford Archaeological Unit

Table 3: *Continued*

Cursus monument site	Domestic Cattle Fauna	Excavation
Buscot Cursus	Not excavated	
Lechlade Cursus	(Ayres, K. and Powell, A. In: Barclay et al. (2003, p. 207) The Lechlade Cursus produced a total of 148 fragments of animal bone, the majority unidentifiable. Most of the bone was recovered from the upper fills of the east cursus ditch. Of the identifiable fragments the majority were cattle sized (28).	1965 Vatcher & Vatcher 1985 Oxford Archaeological Unit
North Stoke Cursus		Rescue excavations were undertaken in the summer of 1950, again in the spring of 1951 and for a fortnight in the summer of 1952. Among those who worked on them were Hector Catling, Salvatore Puglisi and Isobel Smith. 1982 Case
South Stoke Cursus	Not excavated	
Stadhampton Cursus	Not excavated	
Sonning Cursus	Not excavated	
Stonehenge Greater Cursus	72 per cent of identifiable animal fauna recovered from the Stonehenge Greater Cursus comprised of domestic cattle (Serjeantson 2011, p. 16).	1947 J Stone 1963 Christie 1983 J Richards 2008 M Parker Pearson et al.
Stonehenge Lesser Cursus	(Richards, J. 1990, p. 83) A total of 178 fragments of animal bone were recovered. With the exception of red deer antlers very few bones were recovered from the bottom of the ditches. 38 fragments were identified as large mammal which Richards (1990, p. 83) suggests could possibly be aurochs, while 15 fragments were from domestic cattle.	1983 J Richards
Yatesbury Cursus (Avebury)	Not excavated	
Gussage Dorset Cursus	(Legge, A. In: companion volume of Barrett et al. 1991, p. 20) undertook the bone analysis which identified the Neolithic fauna were completely dominated by species suited to a woodland habitat, particularly cattle and pig.	1953 Atkinson 1986 Bradley 1991 Barrett et al. 1992 Green
Pentridge Dorset Cursus	(Legge, A. In: companion volume of Barrett et al. 1991, p. 20) undertook the bone analysis which identified the Neolithic fauna were completely dominated by species suited to a woodland habitat, particularly cattle and pig.	1953 Atkinson 1982 Barrett et al. 1984 Barrett et al.

Serjeantson (2011, p. 14) suggests that 'the improving climate of southern Britain between the fourth and third millennium BC allowed cattle to live outdoors all year round (Schulting 2008)', where it seems that Britain's milder winters would have enabled the wildwood to provide enough browse and shelter. However, Crane (2016, p. 96) identifies that from around 3800 BC climatic conditions became less favourable, dropping to a minimum by around 3500 BC, producing colder and stormier winters. Therefore, from around 3650–3600 BC, the time that northern communities started to develop an affinity with elongated cursus monuments, as cereal cultivation began to decline communities would have needed to forage for wild foods, keep more domestic animals and become less dependent upon permanently occupied sites, which would have resulted in people being forced to revert back to the more mobile lifestyle of their ancestors.

It therefore appears that a correlation could exist between the landscape where cursus monument construction occurred and the movement of domestic cattle. Pearson's correlation coefficient suggests a strong, although still insignificant, relationship exists between the Neolithic communal use of the landscape to herd domestic cattle and the use of the landscape as a cursus monument construction site. The movement of Neolithic domestic cattle would potentially be similar to that identified by Rogers et al. (2018, p. 142) through the carbon isotopic ($\delta^{13}C$) analysis for the movement of aurochs, where domestic cattle would be required to forage and browse in the forest cover to survive the winter months. The cattle would again require moving in the early spring, this time to specialized feeding grounds created by areas of floodplain, where spring meadows resulted in earlier grass growth. Throughout the rest of the year, the cattle would need to be regularly moved to riparian zones with plentiful supplies of water in line with the type of terrain identified by George et al. (2007, p. 5).

Pryor (2015b, p. 79) has long believed that 'the earliest British fields were actually laid out for the use of livestock', where the layout of these fields perhaps suggests that their primary function was to contain and manage domestic animals, as they are entered via corner entrances which are sometimes linked by apparent ditched or hedged droveways that appear to lead to handling areas close to the settlement. This appears to be supported within the Milfield Basin by the prehistoric construction of the Coupland complex (Waddington 1999, p. 136) which has been radiocarbon dated to around 3800 BC, just prior to the commencement of cursus monument construction. The complex is defined by Loveday (2006, p. 108) as an avenue, while Waddington (1999, p. 134) defines it as an enclosure and droveway. However, phosphate analysis of the feature by Emily Mercer of Bradford University (Waddington 1999, p. 136) has identified 'a high incidence of cattle faecal remains and trampling by stock, suggesting the droveway may have been used for the movement of stock in and out of the Coupland enclosure, possibly during the seasonal movement of herds thought to have taken place on these uplands at this time'. Waddington (1999, p. 136) has overlaid aerial photographs of the course of the structure on to detailed geomorphological maps to suggest that 'the structure used a deep naturally incised gully at its northern end, to lead cattle directly to the water's edge, while at the southern end, the structure led to the Galewood Depression, which in turn led to the River Glen'. Waddington (ibid., p. 136) further suggests that 'the southern course of the feature might have also been used for the daily requirement to water stock'.

It therefore appears that the use of a digital mapping software programme within a Geological Information System (GIS) highlighting the terrain's topography could potentially identify areas of the landscape used by Neolithic domestic cattle herds. If we plot the features that would aid pastoralists' movement of cattle, such as slope gradient and the openness of landscape, against items which would restrict herd movement such as river confluences, floodplains and spring locations, in conjunction with George et al.'s (2007, p. 5) identification that slope gradient and distance from water are directly proportional to cattle grazing capacity, could we potentially identify a correlation between the movement of Neolithic domestic cattle and the location and alignment of cursus monument sites? However, it is important to note that the identification of any potential cursus monument landscape cannot be achieved through the analytical use of a GIS system alone. It is vital that the researcher also puts boots on the ground, to physically investigate the site and its surrounding area.

CHAPTER 5

Field Research

Methodology

Existing evidence points to a potential correlation between how domestic cattle would have been moved through their spring and summer landscapes and the locations chosen for cursus monument construction. Viner et al.'s (2010) and Greaney et al.'s (2018) investigations at Durrington Walls appear to suggest that by this period, communities were moving domestic cattle across hundreds of kilometres. But it was not only across these vast regions that cattle were being moved. Movement of cattle would also have been required at a local level, as part of the annual management system. It is highly probable that domestic cattle would have required exactly the same environment as wild cattle, therefore Rogers et al.'s (2018, p. 142), Van Vuure's (2005) and Legge and Rowley-Conwy's (1988) suggestions that herds appear to have had a tendency to make use of the first spring meadow resources and then remain on the grassland throughout the spring and summer months, only splitting into smaller groups to head away from these grassland plains and into the forests to winter on acorns and other fruits, would hold true for domestic cattle. After foraging and browsing in forest cover to survive the winter months, the cattle would need moving in the early spring, this time to specialized feeding grounds created by areas of floodplain where spring meadows resulted in earlier grass growth. Throughout the rest of the year, the cattle would need to be moved regularly to fresh riparian zones with plentiful supplies of water, in line with the type of terrain identified by George et al. (2007, p. 5).

The fact that cursus monuments are currently seen as an enigma (Loveday 2006, p. 11) could suggest previous investigative methodologies have been unsuited to the task. Or the problem could be that, in many cases, explanations appear to have been based upon a few ideal examples and then extrapolated across the remaining monuments. Perhaps the reason cursus monuments appear enigmatic is that previous methodologies have not included sufficient data to provide accurate answers.

This raises further issues such as, were the correct questions being asked in the first place, were the correct methodologies being used and were researchers looking at the correct time span for the monument? Previous methodologies have tended to concentrate on the finished monument, thereby concentrating upon the post-construction period, which appears to have resulted in the development of theories that are potentially applicable to the use of the monument over many periods during its lifespan, rather than answering the question of why the monument was constructed where it was. Previous research carried out on cursus monuments within the British Isles has mainly been undertaken in isolation. The research outlined in this book is one of the first attempts to examine cursus monument topographical links alongside the pattern of movement of domestic cattle and could cast much-needed new light on Neolithic transitions within the British Isles.

Reasons for Choosing the Chalkland Belt

This study combines the use of various archaeological methodologies in relation to the understanding of cattle movement during the Neolithic period. It aims to establish a correlation between the style of domestic cattle movement that would have occurred in the spring and summer on the open grassland areas and the construction of cursus monuments. Cattle movement and its effects have been overlooked in the archaeological record for cursus monuments, and this needs to be addressed.

The primary reason for my selection of cursus monuments on or near to the chalkland belt is firstly Crane's (2016, p. 21) identification that five ranges of linear chalk uplands from the Yorkshire Wolds, the Chiltern Hills, the Marlborough Downs, the North and the South Downs and the southern coastal promontory of Portland stretch across Britain to converge on the 750-sq-km heartland of Salisbury Plain, the most extensive chalk plateau in Britain, and second Rogers et al.'s (2018, p. 141) oxygen isotopic data ($\delta^{18}O$) values for the aurochs teeth from Blick Mead (BM421 and BM422) which highlighted that both individuals potentially originated in parts of eastern England.

Crane (2016, p. 18) suggests that cattle movement resulted in the creation of paths where, imprinted by repetitive footfall, each path became the result of an innate negotiation between the pastoralist's desire, the lie of the land and the stubbornness of the cattle. Although most paths would have needed to follow George et al.'s (2007, p. 5) suggestions for cattle movement, their course would not necessarily have been expressly premeditated, nor their form constructed. For paths were, in effect, 'memories, retrodden year after year, refined with shortcuts and easements … Every turn had a reason: a boulder avoided; a bog bypassed; a gradient cheated; a ford sought'. Did these paths eventually become droveways along which people moved their animals between different grazing areas? And did some of these droveways obtain some form of ritual importance alongside their more practical function, to warrant the construction of cursus monuments?

Selected Cursus Monuments on or near the English Chalkland Belt

My research begins on the chalkland belt of the Yorkshire Wolds at the Rudston complex of cursus monuments (Rudston A, B, C & D), a site recognized around 150 years ago, although initially taken to be a series of prehistoric barrows (Greenwell 1877). It then moves east to the possible cursus monument site at Duggleby, adjacent to the source of the Gypsey Race chalk stream and close to the Duggleby Howe round barrow. Other cursus monuments within this area include the Fimber Cursus, interpreted as a cursus monument from aerial photographs, and the possible Kirby Underdale Cursus.

Although numerous long barrows were recorded within the chalkland belt of the Lincolnshire Wolds (Clay 2001), no cursus monuments are currently known to have been constructed in Lincolnshire. However, Jones (1998, p. 100) suggests 'this apparent absence of cursus monuments in the area may be due to linear post/pit alignment monuments having been adopted as an alternative to cursus monument construction'. The study therefore includes three examples of these alignments at Stenigot, Bag Enderby and Harlaxton.

South of the Lincolnshire Wolds, the chalkland belt was significantly affected by the flooding of Doggerland (Cunliffe 2012, pp. 47–57). Catastrophic effects due to the expanding North Sea had for millennia required cattle movement to divert inland on to the mudstone and muddy limestone geologies surrounding what was to become the Cambridgeshire Fens. This study, therefore, includes research of known cursus monument sites around the fenland edge at Barnack, Eynesbury, Godmanchester, Brampton, Maxey and Etton. Moving back to the chalkland belt, it investigates a probable cursus monument within the parish of Hanworth in Norfolk, the Fornham All Saints Cursus at Bury St Edmunds and the Stratford St Mary Cursus, in Suffolk.

The study then moves adjacent to the chalkland belt's southern edge to investigate the Springfield Cursus, near Chelmsford, and the Stanwell Cursus monument complex at Heathrow Airport Terminal 5, both of which are situated upon the silty and sandy clay of the Eocene Thames Group (Harwich and London Clay fms). It then moves on to the chalkland belt's northern edge, investigating the possible complex of cursus monuments near Cardington and the cursus monuments at Biggleswade and Cople, each of which are situated upon the mudstone of the Kellaways and Oxford Clay fms.

Although the Chiltern Chalk Ridge appears to be largely unresponsive to cropmark production, potentially due to its clay with flint capping overlying the chalk bedrock, the study investigates the possible Ivinghoe Beacon Cursus, discovered though geophysical investigation undertaken by Gover (2000), before moving north-west, on to the mudstone of the Kellaways and Oxford Clay fms, to the monument complex at Wolverton, near Milton Keynes, that contains five cursus monuments.

The study then moves on to a series of cursus monuments situated within the Thames Valley. These are the Benson Cursus, the Dorchester-on-Thames Cursus, the Drayton North Cursus, the Drayton South Cursus, the Drayton St Leonard Cursus, the North Stoke Cursus, the South Stoke Cursus, the Buscot Wick Cursus, the Lechlade Cursus and the Stadhampton Cursus, all of which were situated upon the mudstone and sandstone of the Gault and Upper Greensand formations, lying between the Thames and the earlier Mesolithic settlements of the Kennet Valley found at Thatcham near Newbury, before moving into the Kennet Valley itself to investigate the Sonning Cursus near Reading.

This then takes the study to what could be classified as the jewels in the crown, the cursus monuments of Wiltshire and Dorset including the Greater and Lesser Stonehenge Cursus monuments, the Yatesbury (Avebury) Cursus and finally the Gussage Cursus and the Pentridge Cursus which make up the Dorset Cursus.

This study of fifty cursus monuments either on or adjacent to the English chalkland belt is potentially the largest study of cursus monuments undertaken to date. It analysed both the various elements that make up the cursus monuments in the study group and the various explanations that have been put forward by others for their potential functions. It used univariate analysis to examine each variable, highlighting the distribution factors as a percentage of the total study group through the use of bar charts. It then used a frequency distribution bar chart to illustrate the importance and ranking of each factor. To support or reject the assumptions identified through use of the univariate analysis, it also undertook calculations using Pearson's correlation coefficient, thereby establishing any potential correlation between the various data sets and identifying a quantifiable measurement for their relationships. This study set the significance level (alpha) at 0.05 as this sets the confidence level at the conventional setting of 95 per cent to coincide with the 5 per cent convention of statistical significance in hypothesis testing.

The information obtained through the field research of this study was compared with information obtained through previous cursus monument studies in an attempt to clarify the current 'enigma' state of cursus monument research.

Table 4: Cursus monument location identification scheme

DJS Ident No	Cursus monument location	DJS Ident No	Cursus monument Location
1	Rudston Cursus A	26	Biggleswade Cursus
2	Rudston Cursus B	27	Cardington Cursus
3	Rudston Cursus C	28	Cople Cursus
4	Rudston Cursus D	29	Ivinghoe Beacon Cursus
5	Duggleby Cursus	30	Wolverton Cursus 1
6	Fimber Cursus	31	Wolverton Cursus 2
7	Kirby Underdale Cursus	32	Wolverton Cursus 3
8	Bag Enderby Pit Alignment	33	Wolverton Cursus 4
9	Harlaxton	34	Wolverton Cursus 5
10	Stenigot Pit Alignment	35	Benson Cursus
11	Hanworth Cursus	36	Dorchester Cursus
12	Fornham All Saints Cursus	37	Drayton St Leonard
13	Stratford St Mary	38	Drayton North Cursus
14	Barnack Cursus	39	Drayton South Cursus
15	Eynesbury Cursus	40	Buscot Cursus
16	Godmanchester Cursus	41	Lechlade Cursus
17	Brampton Cursus	42	North Stoke Cursus
18	Maxey Cursus	43	South Stoke Cursus
19	Etton Cursus	44	Stadhampton Cursus
20	Springfield Cursus	45	Sonning Cursus
21	Stanwell Cursus 1	46	Stonehenge Greater Cursus
22	Stanwell Cursus 2	47	Stonehenge Lesser Cursus
23	Stanwell Cursus 3	48	Yatesbury Cursus (Avebury)
24	Stanwell Cursus 4	49	Gussage Dorset Cursus
25	Stanwell Cursus 5	50	Pentridge Dorset Cursus

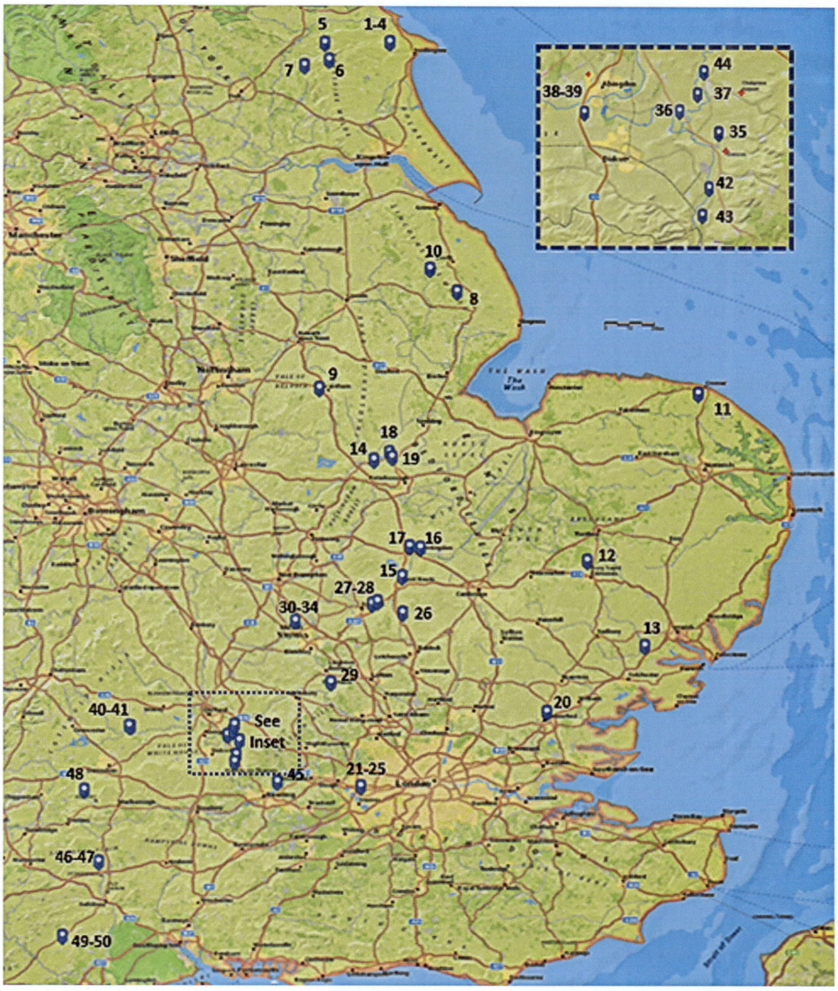

Map 3: Distribution of cursus monuments within the study group (Using DJS location identification scheme. Reproduced with the permission of the British Geological Survey ©UKRI. All rights Reserved)

Terrain Evaluation, Movement and Floodplains

Evaluation of the fieldwork data was undertaken using digital mapping software to analyse the slope elevation characteristics of the landscape surrounding each cursus monument, the important element being the identification of the slope gradient between 5 and 10 degrees. This is suggested by George et al.'s (2007, p. 5) research to be the significant level above which we start to see a reduction in grazing capacity of the herd. The routes suitable for use by Neolithic pastoral communities to move their cattle across the terrain were initially established using the terrain slope elevation within ArcGIS® software by Esri. For additional clarity, individual cursus sites were plotted using © OpenStreetMap and © OpenTopoMap, taking into account slope gradients in excess of 10 degrees which would have resulted in reductions of around a third in the size of the herd. This would have been an extremely important factor when we consider the social prestige and status that appears to have been placed upon the owning of cattle in the Neolithic. While this study examines potential Neolithic domestic cattle movement across each individual cursus monument, it can only indicate an optimum direction of travel, taking into account slope gradient (George et al. 2007, p. 5), wind direction (prominently from the south-west in the British Isles) and other geological features.

Geological map insets, reproduced with the kind permission of the British Geological Survey ©UKRI, are used to highlight areas that would potentially have been vulnerable to inland flooding, thereby potentially creating spring meadows. These are highlighted through the use of the Indicators of Flooding mapping series within the Geology Roam section of Digimap. These indicators are extremely useful for identifying potential prehistoric flooding, as the map is based solely upon observation of the types of geological deposit present. They do not consider any man-made influences such as house building or flood protection schemes, nor do they consider low-lying areas where flooding could occur but where there are no materials indicating flooding in the geological past. Flooding from rivers occurs when either the capacity of the river channel is exceeded and water overflows, known as fluvial flooding, or heavy intense rainfall is unable to be discharged by the river system, known as pluvial flooding. However, the map only highlights the inland fluvial flood deposits which are further divided into two zones. Those areas susceptible to the first influx of flood waters are highlighted in dark blue while those areas susceptible only during extreme flood events are marked in light blue. It is therefore the areas in dark blue that would have produced regular reliable water meadows.

Individual Cursus Monuments

Rudston Cursus Complex

A concentration of four cursus monuments are to be found around the village of Rudston, in the East Ridings of Yorkshire, lying on a bend towards the eastern end of the Gypsey Race winterbourne chalk stream, the main watercourse in the area, which flows from the high Wolds area, west through the Great Wold Valley to reach the North Sea at Bridlington.

Rudston A Cursus

The Rudston A Cursus (OS Grid Ref TA 099657 to TA 101680) was identified to the south of Rudston by aerial photographs (RAF 541/546/4066–7 and St Joseph 1961) of the Great Wold Valley. The northern terminal of the cursus monument appears to start at the western side of the Bridlington Gate plantation before descending south-south-west into the Great Wold Valley. It passes to the east of Rudston village to cross the Gypsey Race stream at an oblique angle, before travelling in a south-westerly uphill direction to cross a minor road at Burton Agnes Balk. Here it takes a sudden dog-leg turn to the south-south-east before swinging around to its original direction, where it ends on the eastern slope of Rudston Beacon.

The first person to excavate the site, Greenwell (1877), originally mistook it for a series of prehistoric long barrows. Yet when Grantham's of Driffield later undertook excavations in 1958, they found the long embankment continued much further than previously anticipated. Investigation of aerial photography by St Joseph (1961) finally established that the cursus monument consists of extensive parallel ditches that stretch for at least 2.5 km past the eastern side of Rudston village.

The southern end of the Rudston A Cursus monument, consisting of squared terminals with rounded corners, appears to start in a field adjacent to the Woldgate and Burton Agnes Balk roads. Constructed as a substantial ditch with inner bank, the ditch's southern arm has been largely destroyed by the quarrying of a chalk pit, although cropmarks of where the bank had been remained clear on aerial photographs. The lack of cropmarks past this point suggests that the cursus monument never extended any further to the south. As the cursus monument begins its northerly descent, it appears to lose its eastern ditch under the Burton Agnes Balk road, although this can still be traced intermittently on both sides. However, as it swings gently north-north-west, the banks appear to be almost entirely ploughed out, the only visible sign being a chalk spread to mark their potential position.

Medieval and modern ploughing appear to have obscured any trace of the cursus monument as it continues across the floor of the Great Wold Valley, however it is highly probable that the cursus monument crossed the stream of the Gypsey Race, which was potentially more braided during prehistoric times than the canalled example we see today. This is supported by the map insert which still identifies first influx fluvial flooding to a point east of where the Gypsey Race was canalled. Cropmarks, visible on aerial photography, identify two parallel ditches set approximately 60 m apart to the north of the Rudston to Bridlington road, which appear to represent a continuation of the cursus monument, although the northern end of the Rudston A Cursus monument has not been established.

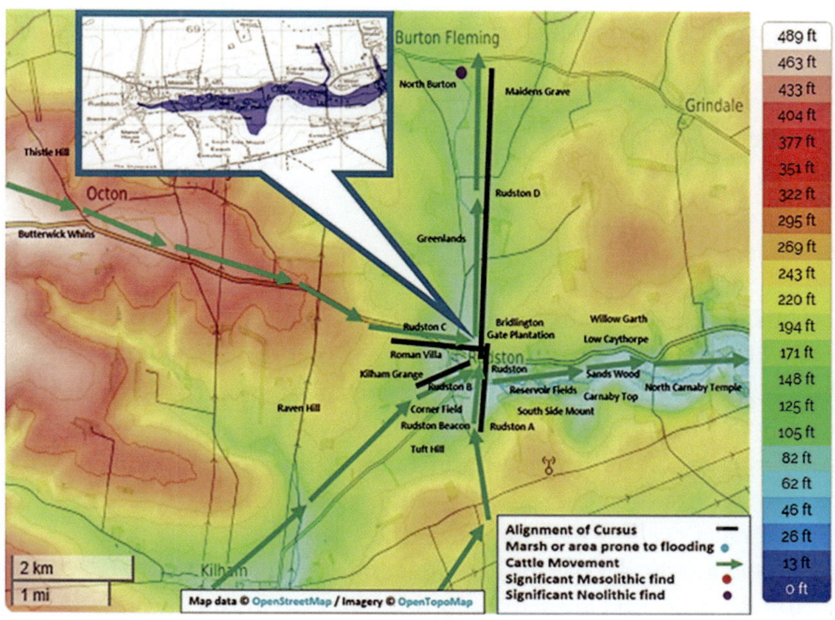

Map 4: Cattle movement across the Rudston Cursus Complex (Main map data created using © OpenStreetMap / Imagery © OpenTopoMap. Inset reproduced with the permission of the British Geological Survey ©UKRI. All rights reserved)

Harding's (2006, p. 119) investigations of excavations of Neolithic pit clusters during the laying of the Caythorpe Gas Pipeline appear to highlight deliberate deposition of animal bone. These differ significantly from the pit cluster fauna discovered earlier by Bramwell (in: Manby 1974) which suggested the pits usually included the remains of a single aurochs, an ox and one to three pigs. The three pits at Low Caythorpe 1 produced the partial remains from as many as nine domesticated cattle together with sixteen pigs, while the Carnaby Temple site produced the bones of eight pigs, and each of these major species appears to have been represented within individual pits by a small collection of fragmentary bone, leading Harding (2006, p. 119) to suggest that 'symbolic practices at North Carnaby Temple and Low Caythorpe were potentially linked to communal events like feasting, hence the deposition of the animal bone'.

Pastoralists travelling with domestic cattle from the south of the region towards what was later to become the location of the Rudston A Cursus would have been required to pass through the breaking ridge that occurs around the site of the southern terminal, due to the high ground of Tuft Hill to the west (OS Grid Ref TA 094656) and South Side Mount to the east (OS Grid Ref TA 110663). It appears that, as the herd dropped into the Great Wold Valley seeking the main watercourse of the region, the natural topography of the landscape would have forced it to move in a north-north-easterly direction where it would have crossed the winterbourne course of the Gypsey Race and the springlines associated with it to the east of Rudston, before continuing either in a northerly direction alongside where the Rudston D Cursus was constructed or in an easterly direction along the Gypsey Race. The fact that the herd appears to travel along the valley floor in accordance with George et al.'s (2007, p. 5) recommendations for cattle movement across a landscape could potentially suggest that the Rudston A Cursus monument commenced life as a droveway, perhaps identifying an initial practical function of the landscape prior to any possible later ritual importance. Harding (1999, p. 31) identifies two causeways in both the eastern and western ditches, which he uses to support his theory (Harding 1999, p. 31) that the monument was acting as some form of 'barrier', implying that some form of control was being asserted to sideways movement across either the landscape or the monument. However, field observation fails to identify any major reason for segregating the landscape. Perhaps this was solely to manage the pasture upon which the herd fed. It seems more probable that the function of any droveway, prior to the construction of Rudston A Cursus, would have been either to lead the herd to water or, if the monument continues into the Bridlington Gate Plantation, to assist with the herd's crossing of the Gypsey Race.

Rudston B Cursus

The Rudston B Cursus lies between Kilham Grange (OS Grid Ref TA 081669) on the south-western edge of Rudston village and heads in a north-easterly direction, where its eastern terminal disappears under the village (OS Grid Ref TA 094675). Aerial photography by St Joseph (1961) identifies that the Rudston B Cursus appears to be part of a larger complex of cropmarks that can be traced for approximately 700 m in a north-easterly direction. The cursus monument appears to have a squared tapering western terminal but, due to extensive ploughing, no surface remains appear to have survived, although a swelling under a hedge-line within the south-eastern bank (OS Grid Ref TA 083670) could potentially identify the last remnants of the monument. The cognitive archaeologist Paul Devereux (Pennick and Devereux 1989, p. 51) notes in his survey of cursus monuments that 'typical of these fascinating antiquities, nothing of any worth has been found along the length of the Rudston B Cursus to give any clues for its nature or function'.

Pastoralists travelling with domestic cattle from the south-west towards the main watercourse of the region, the Gypsey Race, would have needed to move their herd in a direct alignment with the Rudston B Cursus, due to the fact that any movement through the valley appears to have been initially restricted between the high ground of Raven Hill (OS Grid Ref TA 055659) and Tuft Hill (OS Grid Ref TA 094656). This would have required the herd to follow the north-easterly direction of the current Kilham to Rudston road. However, as the herd continued its way into the Great Wold Valley, the natural topography of the landscape would keep it moving in this direction until it met the winterbourne Gypsey Race, the springlines associated with it and any floodplains that resulted from excessive inundation where the flow of water exceeded the capacity of the stream – a factor that is supported by the later canalling of the Gypsey Race in an attempt to prevent this occurring and the first influx fluvial flooding identified by the map insert. The herd would therefore have reached the Gypsey Race just to the west of the village of Rudston where cropmarks of the later cursus monument disappear.

The fact that the herd appears to travel in accordance with George et al.'s (2007, p. 5) recommendations for cattle movement could potentially suggest that the Rudston B Cursus also commenced life as a droveway, perhaps again identifying an initial practical function of the landscape prior to any possible later ritual importance. The fact that Harding (1999, p. 31) also identifies a causeway in the southern ditch of this monument could again support his theory (Harding 1999, p. 31) that the monument was acting as some form of 'barrier', implying that some form of control was being asserted to sideways movement across either the landscape or the monument. Again, field observation fails to identify any major reason for the Neolithic community's desire to segregate the region, other than perhaps to manage the pasture upon which the herd fed. As the eastern terminal disappears under the village of Rudston, it is not currently possible to determine whether the monument actually crossed the Gypsey Race. Therefore, while it is probable that the droveway, prior to the construction of the Rudston B Cursus, led the herd to water, it is not possible to determine whether it also acted as a crossing point.

Rudston C Cursus

The Rudston C Cursus (OS Grid Ref TA 089680 to TA 099680) was initially identified through cropmarks on aerial photographs by St Joseph (1961). Of the four cursus monuments that make up the Rudston Cursus Complex, it is the monument that has the least known about it, which appears to be due to it having received the least amount of archaeological attention within the region. The alignment of the Rudston C Cursus runs east-west, where it appears to have cut across the Rudston D Cursus immediately north of Rudston village. The monument's two parallel ditches, about 60 m apart, are still visible as cropmarks for about 1 km. However, the western terminal appears to fade out near to the York road, while at the eastern terminal, after crossing the Gypsey Race, the ditches disappear upon entering the Bridlington Gate Plantation. Since Dymond's (1966) initial description, the Rudston C Cursus has been found to be much longer than his initial estimation of 1,000 m, due to a further 500 m having been discovered. Although this has increased the known length of the monument, its total length remains unknown as the eastern terminal has yet to be discovered.

Pastoralists travelling with domestic cattle from the west-north-west would have been required to enter this section of the Great Wold Valley in the area around Butterwick (OS Grid Ref TA SE 995712). To travel towards the main watercourse of the region, the Gypsey Race, in accordance with the landscape's natural topography recommended by George et al. (2007, p. 5) would require the herd to pass between the Butterwick Whins and Thistle Hill (OS Grid Ref TA 022706), moving along the spine of the hill where the current B1253 is located. Although the Rudston C Cursus does not appear to have a direct alignment with this route, being off by approximately 10 degrees, it does appear to align with the route of the Gypsey Race as it moves eastwards after passing Rudston.

The fact that the herd appears to travel in accordance with both George et al.'s (2007, p. 5) recommendations for the movement of cattle and the later positioning of the current B1253 could potentially suggest that the Rudston C Cursus is also following the optimum route down the valley and therefore also potentially commenced life as a droveway. The fact that Harding (1999, p. 31) also identifies two possible causeways in the southern ditch and another possible causeway in the northern ditch could again support his earlier theory (Harding 1999, p. 31) that the monument was acting as some form of 'barrier', implying that some form of control was being asserted to sideways movement across either the landscape or the monument. However, once again field observation fails to identify any major reason for segregating the landscape. Perhaps this was solely to manage the pasture upon which the herd fed. It seems more probable that the function of the droveway, prior to the construction of Rudston C Cursus, would have been either to lead the herd to water or, as the monument appears to cross the Gypsey Race, to assist the herd's crossing of the winterbourne.

Rudston D Cursus

To the north of Rudston village, running north-south roughly parallel to the winterbourne Gypsey Race, is the largest cursus monument within the Rudston Cursus Complex, the Rudston D Cursus (OS Grid Ref TA 099717 to TA 096679). This appears to have been over twice the length of any of the other three cursus monuments within the complex. The construction design of the Rudston D Cursus appears different to the other cursus monuments within the complex, its northern terminal appearing flattened, while other identified terminals within the complex are all rounded. The monument's northern terminal is situated due east of the village of Burton Fleming, from where it begins a southerly trajectory, running straight for several hundred metres until changing on to a slightly more secure southern alignment. From here, the cursus monument maintains a dead straight course for about 1.5 km, stopping just short of the later Rudston monolith in the centre of Rudston village. Here, archaeologists have found that a section of the Rudston C Cursus appears to have cut across it. Although the Rudston D Cursus is more than 4 km long, only three, possibly four, small causeways appear to have cut the entire length of its continuous ditch and inner bank.

A significant factor with regard to the Rudston D Cursus could be the recovery of sixteen early Neolithic leaf-shaped arrowheads, found at North Burton, 500 m to the west of the Rudston D Cursus northern terminal. As Trantalidou and Masseti (2015, p. 72) suggest, 'this could potentially indicate that some form of trophy hunting occurred post-monument construction at the site', and Harding's (2006, p. 119) investigations of excavations of Neolithic pit clusters during the laying of the Caythorpe Gas Pipeline potentially support this, due to his suggestion that 'symbolic practices, potentially linked to communal events like feasting, were occurring in the immediate area'.

Pastoralists travelling with domestic cattle from the south of the region would have been required to move parallel to the main watercourse, the Gypsey Race. This would appear to have the herd travelling in accordance with George et al.'s (2007, p. 5) recommendations, which would have resulted in the cattle movement having a direct alignment with the later Rudston D Cursus. This would have given herders control of their cattle as they moved along the widest section of the Great Wold Valley, the point where the Gypsey Race ran between the high ground of High Caythorpe Hill to the east (OS Grid Ref TA 123694) and Greenlands to the west (OS Grid Ref TA 086695). However, it would not have been until the herd reached the restricted section of the valley between Maidens Grave (OS Grid Ref TA 102714) and the Gypsey Race adjacent to Maidens Grave Farm (OS Grid Ref TA 097715) that the natural topography of the landscape would potentially have exerted significant control over this movement.

The fact that the herd appears to travel in accordance with George et al.'s (2007, p. 5) recommendations for cattle movement potentially suggests that the Rudston D cursus monument is also following the optimum route down the valley and therefore also potentially commenced life as a droveway. The three possible causeways in the western ditch and another possible causeway in the eastern ditch, as identified by Harding (1999, p. 31), together with the fact that the monument runs parallel to the Gypsey Race, could imply that the monument was acting as some form of barrier, perhaps controlling access to this section of the landscape. Prior to the canalling of the Gypsey Race, first influx fluvial flooding would have occurred alongside the Rudston D Cursus, potentially creating spring meadows with their early grass growth. The causeways could imply that some form of control was being asserted over access to these spring meadows, enabling Neolithic pastoralists to move cattle on to the floodplain in the early spring, which would have enabled herd sizes to be increased.

Duggleby Cursus

At Duggleby, in the Great Wolds Valley in Yorkshire, a pair of irregular parallel ditch lines identified through cropmarks discovered during aerial photography (RCHME/EH/HE Aerial Photographers comment, MacLeod, 1997. Duggleby Howe Causewayed Enclosure Project) have been interpreted as a possible cursus monument (OS Grid Ref SE 879669 to 892670). These lie along the first gravel terrace of the sloping valley side to the east of Duggleby, within 300 m of the source of the winterbourne Gypsey Race. The ditch lines appear to be 160 m apart and run for approximately 1.2 km in an east-west direction. Hurst (1983, p. 78) believed 'their width, their irregular, misaligned incorporation within earlier monuments, their extension beyond their apparent terminal and the fact that they appear to be aligned around the earlier outer causewayed enclosure ditch at Duggleby Howe are atypical to those of a cursus monument'.

Riley (1980, pp. 174–78) had earlier recognized that the later Neolithic Great Duggleby Howe round barrow, which is 38 m in diameter and stands 6 m high, was constructed within the centre of an earlier Neolithic causewayed enclosure, a feature that included interrupted ditches and which was more than 370 m in diameter, covering about 10.5 hectares, making it larger than the causewayed enclosures around Avebury or Mount Pleasant in Wessex. The Duggleby Cursus would have run between this causewayed enclosure and the source of the Gypsey Race, suggesting that the Duggleby area was as important a ritual centre for the western Wold area as Rudston appears to have been for the eastern Wolds.

Manby (1976, p. 145) had earlier identified that 'two phases of woodland clearance probably occurred in the region, possibly separated by a period of woodland regeneration'. However, the fossil soils preserved beneath the Kilham and other long barrow sites on the eastern Wolds, together with the evidence for forest clearance provided by the large number of imported Langdale stone axe finds in the parish of Wharram le Street and the possible tree holes that were filled with characteristic brown forest soil at Wharram Percy all suggest that this section of the Yorkshire Wolds consisted of open grasslands during the Neolithic cursus monument construction period.

The Duggleby Cursus lies parallel to a raised terrace, south-east of a springline at the source of the Gypsey Race (OS Grid Ref SE 876672). Pastoralists travelling with domestic cattle in a northerly direction towards the village of Duggleby would have been significantly restricted on their western side by both the escarpment that follows the current line of the Yorkshire Wolds Way and by Grimston Hill (OS Grid Ref SE 862674) and on their eastern side by the Stonepit Slack (OS Grid Ref SE 877648). The prominent springline (OS Grid Ref SE 879672) occurring just to the north of this location would have further restricted cattle movement, especially during the winter and early spring months, forcing the herd to divert to the north-east along the valley of the Duggleby and Mowthorpe Wolds. This would appear to have the herd travelling in accordance with George et al.'s (2007, p. 5) recommendations, which would have resulted in the cattle movement having a direct alignment with the Duggleby Cursus as it moved parallel to the Gypsey Race.

The fact that the herd appears to travel in accordance with George et al.'s (2007, p. 5) recommendations for the movement of cattle potentially suggests that the Duggleby Cursus is also following the optimum route down the valley and therefore also potentially began life as a droveway, perhaps again identifying an initial practical function of the landscape prior to any possible later ritual importance. Lack of excavation and difficulty in identification of aerial photographs have made it difficult to identify any causeways in the side ditches of the monument. However, field observation has identified that the monument runs parallel to the marshy ground at the start of the Gypsey Race. This could potentially imply that the monument was also acting as some form of barrier, asserting control over access to the spring areas and the resultant water meadows. This would have enabled Neolithic pastoralists to move cattle on to the floodplain in the early spring, which would have enabled herd sizes to increase.

Fimber Cursus

Aerial photographs (Stoertz 1997 Ancient landscapes of the Yorkshire Wolds, aerial photographic transcription and analysis pp. 27–29) of the valley to the north-east of Fimber showed cropmarks of two almost parallel ditches that have been interpreted as a Neolithic cursus monument. They run along the valley floor in an east-west direction where the northern ditch appears to have been constructed between OS Grid Ref SE 893610 and OS Grid Ref SE 907 610 while the southern ditch appears to have been constructed between OS Grid Ref SE 893610 and OS Grid Ref SE 906610 (Ramm 1974).

The total length of the cursus monument appears to be around 1.3 km, although the terminals have yet to be discovered so its true length remains unclear. The western end appears to start somewhere close to a minor road leading towards Towthorpe and extends along the valley floor before passing under the buildings of Fimber Grange. Here it becomes untraceable at a point where a natural crossroads has been formed by the meeting of four valleys that appear to have acted as a natural focus for the cursus monument. The western end of the monument currently appears to be open, as there is no visible terminal, however it is not clear if the ditches terminated here or if they continued further and are simply not visible as cropmarks past this point. The monument does not appear to have had parallel sides, as the distance between the ditches varies between 18 and 27 m at a point west of Fimber Grange and between 30 and 37 m at a point east of the Grange.

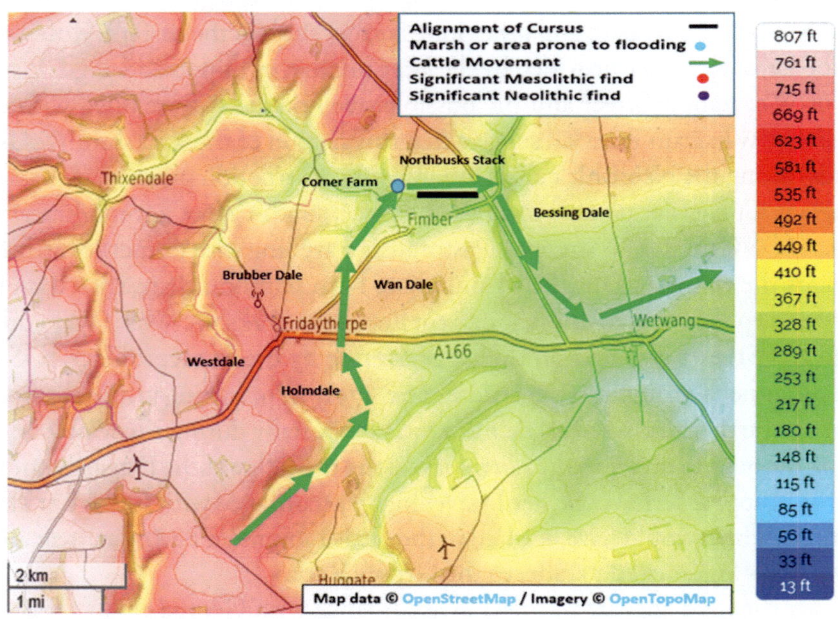

Map 5: Cattle movement across the Fimber Cursus (Main map data created using © OpenStreetMap / Imagery © OpenTopoMap. Inset reproduced with the permission of the British Geological Survey ©UKRI. All rights reserved)

The Fimber Cursus lies on the first raised terrace of a valley floor, overlooked by a steep ridge on the southern side of the monument. Pastoralists travelling with domestic cattle in a northerly direction towards Fimber would have been initially restricted on their western side by the Westdale escarpment (OS Grid Ref SE 868588) and on their eastern side by the Holmdale escarpment (OS Grid Ref SE 875584). For the herd to travel in accordance with George et al.'s (2007, p. 5) recommendations, it would have had to pass through the restrictions between Brubber Dale (OS Grid Ref SE 868609) and Wan Dale (OS Grid Ref SE 879605), which would have forced the herd to turn north-east before once more being required to change direction to move in an easterly direction along the path of the cursus monument, passing between Corner Farm (OS Grid Ref SE 894609) and Northbusks Slack (OS Grid Ref SE 896613). Although there is no current evidence for rivers or springs within this area, it is notable that, at the point of maximum herd concentration, a natural dew pond forms due to water running off the surrounding hillsides (OS Grid Ref SE 893609). After passing parallel to the cursus monument, the herd would have been required to turn south-south-east through Bessing Dale before continuing its journey in a north-easterly direction.

The fact that the herd appears to travel along the valley floor in accordance with George et al.'s (2007, p. 5) recommendations for the movement of cattle potentially suggests that the Fimber cursus monument is also following the optimum route down the valley and therefore also potentially began life as a droveway, perhaps again identifying an initial practical function of the landscape prior to any possible later ritual importance. Lack of excavation and difficulty in identification of aerial photographs have made it difficult to identify any causeways in the side ditches of the monument. However, two dew ponds on Chanctonbury Hill, in Sussex, have potentially been dated to the Neolithic period (Hubbard and Hubbard 1907). And while there is currently no evidence that the dew pond at the western terminal of the Fimber Cursus dates to this period, the fact that it is sited in a similar location potentially highlights that the valley had the ability to act in a similar way to the Chanctonbury Hill landscape, thereby potentially indicating that the Fimber Cursus monument was constructed on a landscape that was initially part of a wider cattle management system.

Kirby Underdale Cursus

Aerial photographs analysed by Stoertz (1997) identified a linear cropmark feature at Kirby Underdale, in Yorkshire, that she interpreted as a possible cursus monument. The monument appears to consist of two parallel ditches about 30 m apart. They run uphill in a westerly to easterly direction for some 800 m (OS Grid Ref SK 823594 to SK 807586) from Kirby Underdale, along the ridge of Uncleby Hill to the top of Uncleby Wold. Identification of the monument's terminals is not possible because the cropmarks at the western end disappear at a road that runs uphill from Kirby Underdale to Uncleby, presumably terminating at this road, while the cropmarks at the eastern end disappear at a disused quarry after passing close to three possible round barrows.

For the herd to travel in accordance with George et al.'s (2007, p. 5) recommendations, pastoralists travelling with domestic cattle towards the upland zones of the Yorkshire Wolds would have been initially restricted as they moved in a north-easterly direction along the Awnhams valley passing the North Cliff escarpment (OS Grid Ref SK 793563) on the eastern side and Barf Hill (OS Grid Ref SK781569) on the western side of the valley. As the herd reached Howe Hill (OS Grid Ref SK 806587), it would have been forced to turn north-east before crossing both the Waterloo and Uncleby Becks. This means it would have been moving in a direction parallel to the Kirby Underdale Cursus as it travelled along the ridgeline of Uncleby Hill, the difference in elevation between the westerly and easterly terminals being around 60 m, with the sharp escarpments of North Dale and Painsthorpe Dale further restricting movement to both the north and the south of the ridgeline.

The Hawk Hill springs feed the entire length of the Uncleby Beck to the south of the cursus monument while the Jenny Wren's springs feed the beck running down North Dale. This would have resulted in winterbourne watercourses concentrating at the point where the Uncleby Beck meets the Waterloo Beck. At this point, a high ridge between the two becks, known locally as Eskhams, accessed the start of the potential droveway which, running parallel to the cursus monument, appears to have enabled access into the high Wolds.

The fact that the herd appears to travel along the most accessible ridgeline in accordance with George et al.'s (2007, p. 5) recommendations for the movement of cattle potentially suggests that the Kirby Underdale Cursus is also following the optimum route down the valley and therefore also potentially began life as a droveway, perhaps again identifying an initial practical function of the landscape prior to any possible later ritual importance. The requirement to keep the herd away from the edge of the extremely steep escarpments of both North Dale and Painsthorpe Dale again potentially suggests that the cursus monument started life as a droveway which was part of a wider cattle management system allowing access to the high Wolds for Neolithic pastoralists and their herds.

Bag Enderby Pit Alignment

To the north of Bag Enderby (OS Grid Ref TF 351725), in the Lymn Valley in Lincolnshire, an undated pit alignment has been identified through the use of aerial photography (RCHME 1992–1996. National Mapping Programme, Lincolnshire. TF3572:LI.17.8.1). The cropmark of a short alignment of paired opposed pits (126 m by 46 m) has been interpreted as a prehistoric ceremonial monument, possibly a timber cursus monument, that potentially dates from the Early Neolithic. The pairs of pits, which are widely spaced at regular intervals, do not appear to form a linear boundary but appear to fall into the same category as Brophy's (2016, pp. 61–69) classification for Scottish Neolithic pit alignments. Similar examples of pit-defined avenues have been identified at Wilford and South Muskham on the Nottinghamshire side of the Trent Valley and from the Valley of the Gwash at Ryhall, in Rutland.

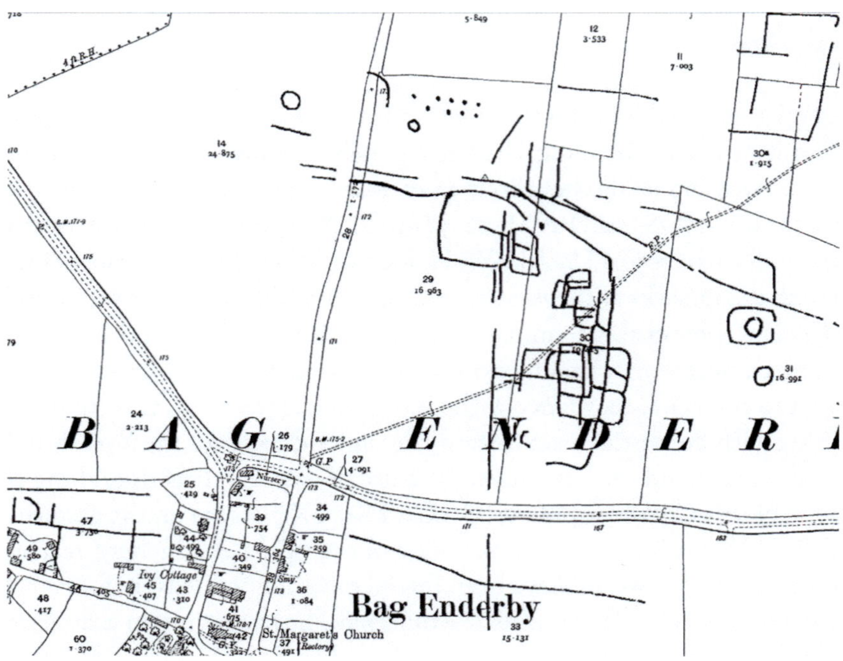

Map 6: Bag Enderby pit alignment HER Number PRN 45338 – Crown copyright 1905 (Reproduced with kind permission of Ordnance Survey)

The Bag Enderby pit alignment appears to have consisted of at least five pairs of parallel pits, which Clay (2001, p. 9) believes 'could possibly be a timber cursus monument'. The pit alignment appears to have been constructed on a flat plateau approximately 200 m north of the hamlet of Bag Enderby, with the steep slopes of Warden Hill (OS Grid Ref TF 345735) and Cloven Hill (OS Grid Ref TF 360738) to the north. To the south, a ridgeline (OS Grid Ref TF 348719) drops steeply to the River Lymn. Field observation has identified that, between this ridgeline and the river, the ground shows signs of multiple springs.

It appears that, for the herd to travel in accordance with George et al.'s (2007, p. 5) recommendations, pastoralists travelling with their domestic cattle would have needed to travel in a north-westerly direction, along the River Lymn. Their movement would have been initially restricted by Cinder Hill, which lies in the river valley south of Harrington (OS Grid Ref TF367701). This would have caused the herd to move on to the wider plateau to the south-east of Bag Enderby. At Bag Enderby, herd movement would once again have been restricted by Warden Hill and the springline that runs along the southern ridgeline as it drops down to the River Lymn.

The fact that the herd appears to have been required to travel along the valley terrace in accordance with George et al.'s (2007, p. 5) recommendations for the movement of cattle potentially suggests that it was not only cursus monuments that appear to follow the optimum route down the valley. Clay (2001, p. 9) suggests 'Lincolnshire constructed timber cursus monuments, now identified as pit alignments, as an alternative to ditched cursus monuments'. In all other aspects, this pit alignment appears to follow the same criteria as its ditched cousins. It is therefore highly probable that this pit alignment also potentially commenced life as a droveway. The fact this monument is a pit alignment, potentially a timber cursus monument, makes it impossible to establish whether any features existed that would have had a similar function to causeways, giving potential control over sideways movement and control over the landscape, as suggested by Harding (1999, p. 31). However, marshland running between the River Lymn and the monument would suggest control of the spring meadow grassland was extremely likely.

Harlaxton Pit Alignment

At Harlaxton, near Grantham in Lincolnshire, a number of linear and pit-defined monuments have been discovered (OS Grid Ref SK 891339) through the use of aerial photography (RCHM, 1995, National Mapping Programme SK 83NE aerial photo overlay). These include both single pit alignment features and a possible pit avenue which may have been a timber cursus monument. Together these make up a multiple alignment of four lines of pits and a ditch.

The complex extends for about 250 m in a north-westerly direction by 45 m in width. Within Lincolnshire, full cursus monuments appear to have been absent, but linear pit-alignment monuments seem to have been adopted as a regional variation to the typical cursus monument building that occurred in other parts of the British Isles. Harding (1995, pp. 117–36) indicated that this use of pit alignments may have reflected the changes taking place within the Neolithic society in this part of the East Midlands. Bradley (1984a, p. 65) suggests that 'the established order may have felt under threat' and that this linear form of monument that appears to have been allied to cursus monuments, may be regarded as 'representing an attempt to re-establish the unity of society through an explicit link with traditions of the past' (ibid., p. 65). There is earlier evidence of human activity in the region along a ridge to the north-west of the pit alignment that lay between North Lodge Farm and Harlaxton Clays Wood. Here, prehistoric implements including scrapers, broken Neolithic axes, the re-used flakes of polished axes and transverse, leaf-shaped and barbed and tanged arrowheads have been found in a spring area (OS Grid Ref SK 873348).

The pit alignment complex at Harlaxton was situated on a prehistoric route across the Jurassic limestone at a point where minor river valleys provided easy passage through the limestone massif that link the Welland Valley with the Vale of Belvoir. The Harlaxton pit alignment lay in a south-easterly to north-westerly direction upon a raised terrace between the later Grantham Canal and Harlaxton Lower Lodge Farm. However, the construction of the canal in the 1790s has resulted in spoil significantly altering this section of the lower part of the valley.

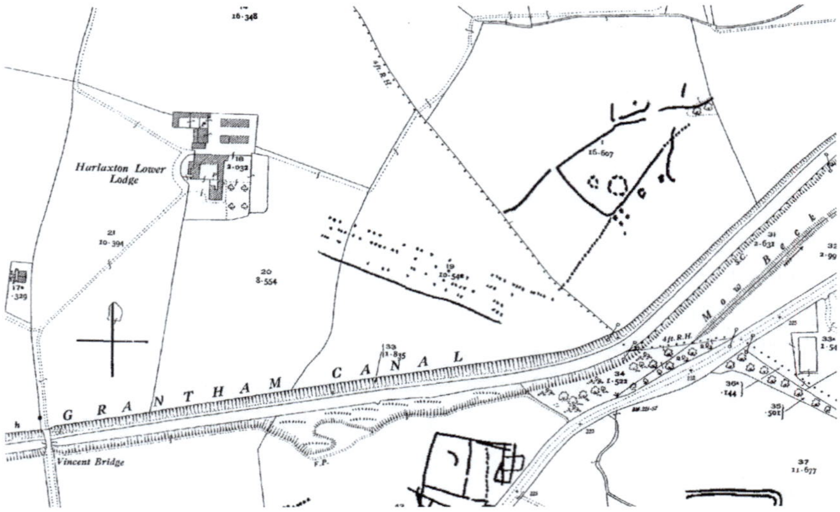

Map 7: Harlaxton pit alignment HER Number PRN 33382 – Crown copyright 1905 (Produced with kind permission of Ordnance Survey)

For the herd to travel in accordance with George et al.'s (2007, p. 5) recommendations, pastoralists with their domestic cattle would appear to have been required to travel from the south-west towards the Harlaxton pit alignment. This movement would initially have been restricted between the hill mound to the north of Denton (OS Grid Ref SK 863330) and Roland Hill (OS Grid Ref SK 865317). To the north of Harlaxton, Four Winds Hill (OS Grid Ref SK 876328) would have forced the herd to change direction to the north-east prior to reverting to its original easterly direction due to the ridgeline between North Lodge Farm and Harlaxton Clays Wood. A series of springlines around Echo Farm (OS Grid Ref SK 898338) would have further restricted cattle movement to the south-eastern edge of the pit alignment. However, this would mean that the pit alignment was constructed at an acute angle of 20 degrees to that of animal movement, a feature that has not been observed elsewhere throughout the study group.

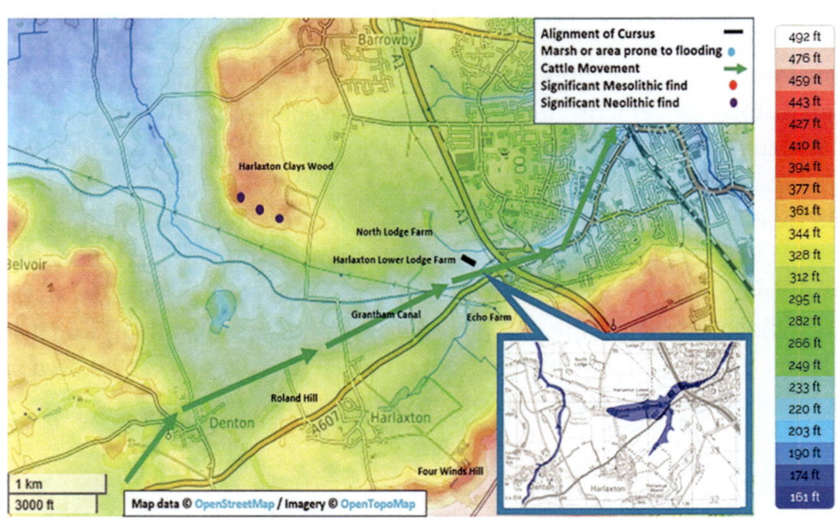

Map 8: Cattle movement across Harlaxton Pit Alignment (Main map data created using © OpenStreetMap / Imagery © OpenTopoMap. Inset reproduced with the permission of the British Geological Survey ©UKRI. All rights reserved)

The Harlaxton pit alignment appears extremely like the multiple pit or post holes that define the Inchbare North and Inchbare South timber cursus monuments in Scotland. Known only as cropmarks, between them they consist of at least eleven parallel pit alignments. This has led Brophy (2016, p. 124) to cast doubt on the exact nature of these monuments, questioning whether 'they were ever actually enclosures'.

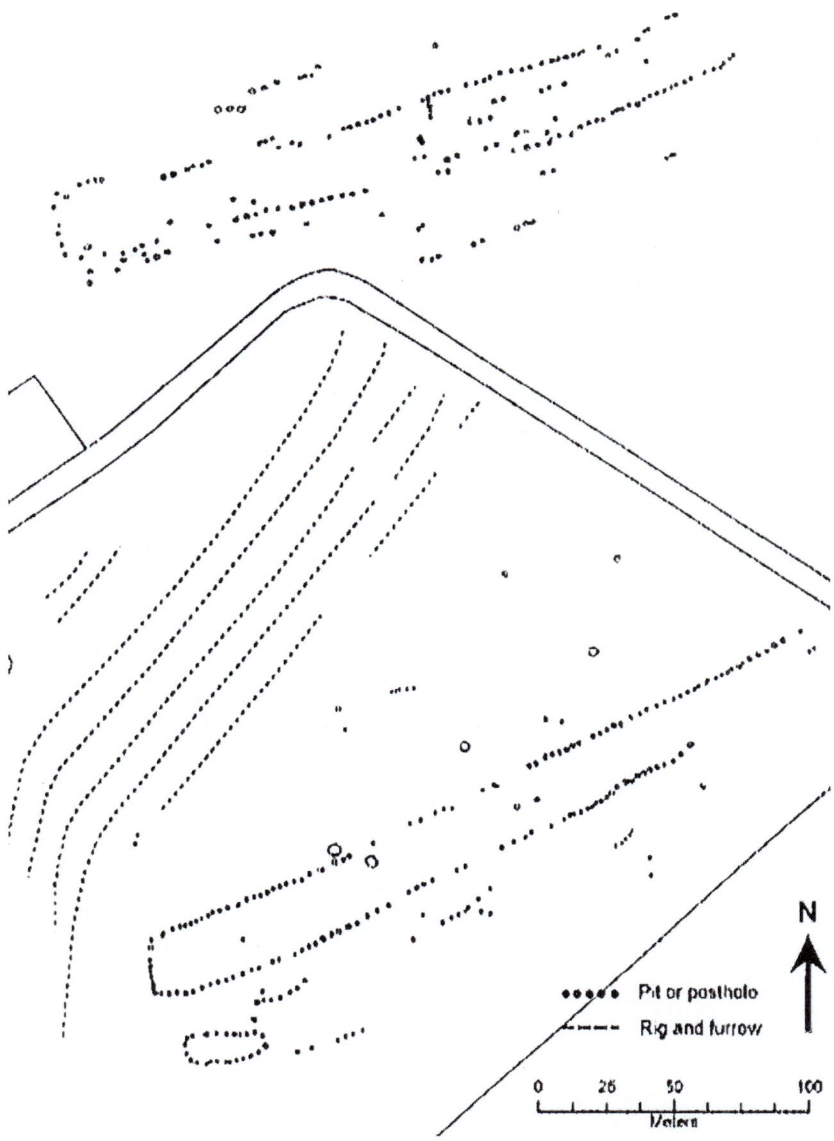

Map 9: Inchbare North and South pit alignments (After Millican 2016. Reproduced with kind permission of BAR Publishing, <www.barpublishing.com>)

Brophy (2016, p. 122) believes that 'the construction of some cursus monuments appear to have been a drawn-out affair, with monuments growing, shrinking or being reworked in one or both directions, where the re-establishment of the monuments sometime after their initial construction identifies secondary construction took place'. This would support Thomas's (2007, p. 244) earlier investigations at Holywood North and at Holm, where he identified 'posts being replaced time and time again'. Brophy (2016, p. 124) suggests that, in line with Thomas's work, 'the Inchbare cursus monuments could be viewed as a large-scale Holm, where multiple lines of posts were erected following the same orientation but perhaps not all standing at the same time'.

Brophy (ibid., p. 124) further believes that 'such multiple alignments would seem to embody a strong social memory and need to mark these locations in this way, perhaps indicating a significant endurance to the alignments'. Discussing the Inchbare alignments, Brophy (2017, personal communication, November) believes that 'the earliest construction phase may be a couple of centuries earlier than the presumed period for general Scottish timber cursus monument construction'. Regarding the Harlaxton pit alignment, it is interesting to note that the Inchbare North and South pit alignments also run at an acute angle (30 degrees) to possible cattle movement through the landscape.

Stenigot Pit Alignment

To the west of Stenigot (OS Grid Ref TF 245811 to TF 244812), in the Bain Valley in Lincolnshire, an undated pit alignment has been identified to the east of Fox Covert through the use of aerial photography (RCHME. 1992–1996. National Mapping Programme. Lincolnshire. TF2481:LI.216.8.1). The cropmark, consisting of a double alignment of pits measuring 95 m by 15 m, has been interpreted as a prehistoric ceremonial monument, possibly a timber cursus monument, potentially dating from the Early Neolithic. The dozen pairs of parallel pits, which are widely spaced at regular intervals, do not appear to form a linear boundary but, as with other Lincolnshire monuments, appears to fall into the same category as Brophy and Millican's (2015) classification for Scottish timber cursus monuments.

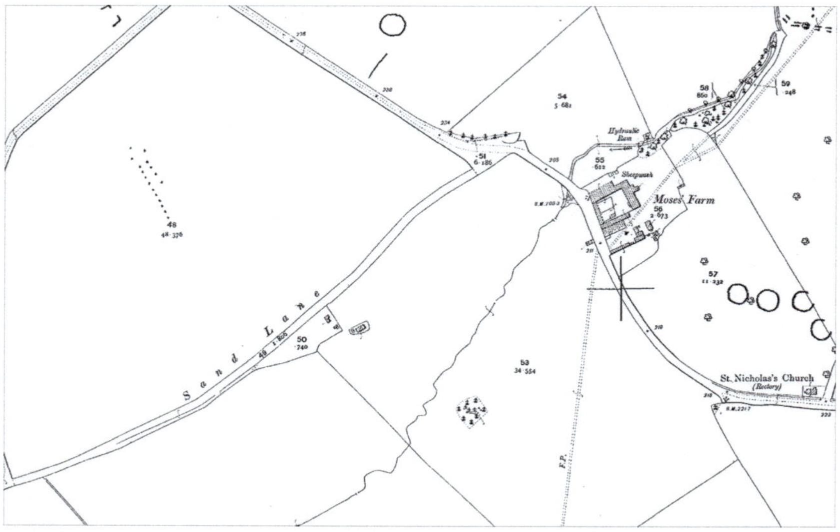

Map 10: Stenigot pit alignment HER Number PRN 44857 – Crown copyright 1905 (Produced with kind permission of Ordnance Survey)

The only other evidence for early human activity within the Bain Valley comes from the cropmark of a possible prehistoric long barrow (OS Grid Ref TF 269815) approximately 3 km east of the Stenigot pit alignment, from where it would have been visible on the hillside.

The Stenigot pit alignment lies in a south-easterly to north-westerly direction upon a raised terrace just north of a significant springline of the River Bain (OS Grid Ref TF 244805) and to the south of Nob Hill (OS Grid Ref 246821). Pastoralists travelling with domestic cattle from the south-east of the region would have been required to move parallel to the main watercourse, along the northern bank of the River Bain. This would appear to have the herd travelling in accordance with George et al.'s (2007, p. 5) recommendations, which would result in the cattle movement having a direct alignment with the Stenigot pit alignment. Cattle travelling in a north-westerly direction along this section of the Bain valley would have been initially restricted between Colley Hill (OS Grid Ref TF 251802) and Red Hill (OS Grid Ref TF 264808). This would have caused the herd to move in a parallel direction to the pit alignment. The springline at Moses Farm (OS Grid Ref TF 251810) appears to further restrict any potential cattle movement just before the southern terminal of the pit alignment.

The fact that the herd appears to travel along the valley terrace in accordance with George et al.'s (2007, p. 5) recommendations for the movement of cattle potentially suggests that, as with the Bag Enderby pit alignment, the Stenigot pit alignment, which Clay (2001, p. 9) proposes 'is a timber cursus monument', acts in a similar manner to ditched cursus monuments and also potentially commenced life as a droveway, perhaps again identifying an initial practical function of the landscape prior to any possible later ritual importance. Again, however, the fact this monument is a pit alignment, potentially a timber cursus monument, makes it impossible to establish whether any features existed that acted in a similar function to causeways, giving potential control over sideways movement and control over the landscape as suggested by Harding (1999, p. 31). However, marshland running parallel to the River Bain and the monument, especially that which was concentrated at Fox Covert, would suggest control of the spring meadow grassland would have been extremely likely.

Hanworth

At Hanworth, in the county of Norfolk, the Hanworth Cursus lies along the uppermost ridge of high ground, midway between the Scarrow Beck and the Hagon Beck, both of which are tributaries of the River Bure. A pair of diverging ditch lines that run to the south-west of a possible causewayed enclosure have been identified by cropmarks on aerial photographs (Edwards – NAU air photography 1986) as a probable cursus monument (OS Grid Ref TG 207362). The cursus monument, which aligns south-west to north-east, appears rectangular, having parallel sides that are about 400 m in length and up to 55 m wide. The south-western squared terminal has rounded corners which meet the side ditch at an angle of approximately 90 degrees. However, the angle between the terminal and side ditch at the monument's south-eastern corner appears slightly greater.

The Hanworth Cursus appears to have been associated with numerous nearby features. These include two round barrows and a double concentric ring ditch towards the cursus monument's south-western terminal and a ring ditch and small curvilinear enclosure which lie within the monument's banks. It also lies 1.5 km north-west of the Roughton causewayed enclosure, which has also been identified through the use of aerial photography (Dyer/18-DEC-1996/RCHME: Roughton Causewayed Enclosure Project). The area could potentially, therefore, have formed part of a larger prehistoric funerary and ceremonial landscape, examples of which have been found at Cardington in Bedfordshire and at Springfield in Essex, which both have cursus monuments located similar distances from causewayed enclosures. Although the causewayed enclosure appears to be genuine, breaks in the side ditches may not necessarily be archaeological, but could be a product of the cropmarks having been masked by natural geology.

The cursus monument's north-eastern terminal is unclear, however faint linear cropmarks are visible on several of the aerial photographs (National Mapping Programme, The Archaeology of Norfolk's Coastal Zone, EH Project No: 2913, by Albone, Massey and Tremlett), which has led the National Mapping Team to suggest that it may continue beyond the ring ditch. But it was not until investigation of the area affected by first influx fluvial flooding was examined through field observation that the landscape started to make sense. Although the uppermost ridge of high ground is slight, it appears that pastoralists travelling with domestic cattle in a north-westerly direction would have been required to cross the landscape between the Scarrow Beck and the Hagon Beck in accordance with George et al.'s (2007, p. 5) recommendations. This would have initially restricted the herd in the area that lies between Alby Hill (OS Grid Ref TG 194343) and Gallows Hill (OS Grid Ref TG 236349). However, as they passed to the north-east of Hanworth, a spur valley that runs parallel to the cursus monument in the area of Meadow Farm (OS Grid Ref TG 197361) would have forced them to change direction again to the north-east, resulting in the herd moving parallel to where the later cursus monument would be constructed.

The fact that the herd appears to travel in accordance with George et al.'s (2007, p. 5) recommendations for cattle movement could potentially suggest that the Hanworth Cursus again commenced life as a droveway. And the narrowing of the landscape through first influx fluvial flooding which occurs on both sides of the cursus monument, together with the fact that the Historic Environment Record 18190 appears to identify a possible three causeways in the side ditches of the monument, could support Harding's (1999, p. 31) theory that the monument was acting as some form of 'barrier', implying that some form of control was being asserted to sideways movement across the landscape. First influx fluvial flooding appears to significantly reduce the area of landscape between the Scarrow Beck and the Hagon Beck, creating extensive areas of spring meadow grasslands. By controlling access to these spring meadows, Neolithic pastoralists would have been able to move their herds on to grassland earlier in the year, thereby increasing the size of the herd that could be kept over winter and increasing the social and economic wealth of the community.

Fornham All Saints Cursus

At Fornham All Saints, in the Lark Valley near Bury St Edmunds in Suffolk, the cropmarks of a dog-legged cursus monument (OS Grid Ref TL 829687 to TL 842672) were discovered using aerial photography (RCHME Air Photography Unit 1995), where the overall length of the cursus monument appears to have extended for just under 2 km, with the ditches approximately 42 m apart. The monument is aligned roughly parallel to the River Lark, running in a north-westerly to south-easterly direction comprising of three straight lengths, each of which are on slightly different orientations, giving the monument a dog-legged appearance. The northern terminal has not yet been located, seeming to disappear before the road leading to Mill Farm. However, the southern terminal is still visible from the air, where it appears to end next to a circular cropmark.

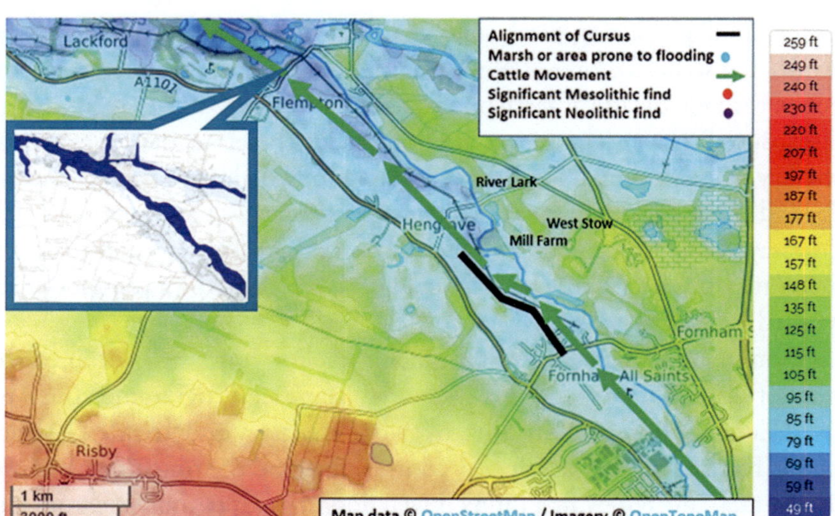

Map 11: Cattle movement across the Fornham All Saints Cursus (Main map data created using © OpenStreetMap / Imagery © OpenTopoMap. Inset reproduced with the permission of the British Geological Survey ©UKRI. All rights reserved)

The Fornham All Saints Cursus appears to have been constructed on a narrow band of sandy gravel on the first raised terrace to the south-west of the floodplain of the River Lark, an area of significantly changing geology. The landscape between the river and the cursus monument appears to consist mainly of peat, while immediately to the south of the monument the geology changes to a boulder clay (Philip Aitkens, Member of the Council for the Suffolk Institute of Archaeology, 2016, personal communication, November). Although numerous drainage ditches, dating from the medieval period, and later canalling of the River Lark have significantly drained the area between the cursus monument and the river, field observation still provides evidence for this landscape having been marshland during earlier periods. This is further supported by the fact that planning permission is currently being sought to recreate a natural wetland nature reserve in the area to the north of Mill Farm. This will lie between the northern terminal of the cursus monument and the River Lark (Philip Aitkens, Member of the Council for the Suffolk Institute of Archaeology, 2016, personal communication, November). This is supported by the map insert which identifies that this appears to be the first monument that would appear to have been affected by both the first influx fluvial flooding areas as well as flooding to areas that would have only been susceptible during extreme events.

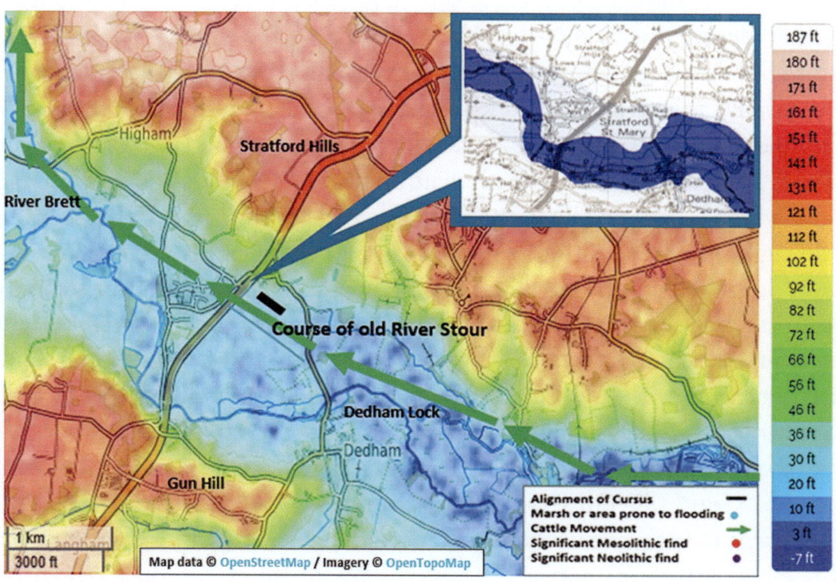

Map 12: Cattle movement across the Stratford St Mary Cursus (Main map data created using © OpenStreetMap / Imagery © OpenTopoMap. Inset reproduced with the permission of the British Geological Survey ©UKRI. All rights reserved)

Pastoralists travelling with domestic cattle in a north-westerly direction as they move along the Lark Valley would have been significantly restricted between the floodplain of the River Lark and the higher ridges that lie to the south of the monument. This would have forced the herd to move parallel to the later cursus monument. The fact that the herd appears to travel along the valley floor in accordance with George et al.'s (2007, p. 5) recommendations for the movement of cattle could again potentially suggest that the Fornham All Saints Cursus commenced life as a droveway, perhaps yet again identifying an initial practical function of the landscape prior to any possible later ritual importance.

The narrowing of the landscape through both extreme event flooding and through first influx fluvial flooding, together with the fact that Loveday (2006, p. 32) appears to identify possible causeways in the side ditches of the eastern terminal of the monument, could also support Harding's (1999, p. 31) theory that the monument was acting as some form of 'barrier', implying that some form of control was being asserted to sideways movement across the landscape. Field observation suggests that this would have enabled Neolithic pastoralists to assert some form of control over access to the spring meadows in the landscape between the monument and the River Lark, enabling them to move their cattle on to the floodplain in the early spring, which would have enabled herd sizes to increase.

Stratford St Mary

At Stratford St Mary, in the Stour Valley in Suffolk, cropmarks of a probable cursus monument (OS Grid Ref TM 048343 to TM 0463345) were discovered using aerial photography (Aerial photographers' comment RCHME/EH/HE 1996, Andrew Millar). The cursus monument appears to form part of a monument complex that was situated within a loop of the River Stour which also included a hengiform circle at the south-eastern terminal. Orientated north-west to south-east, the cropmarks of an elongated enclosure are visible, running approximately 300 m in length and 60 to 70 m in width. While the rounded corners of both ends are visible, much of its course has been destroyed under urban development.

The Stratford St Mary Cursus lies on a gently sloping gravel terrace to the east of a river confluence where the River Stour is joined by the River Brett. However, it should be noted that the River Stour Navigation Company undertook significant canal work of the section of river between Stratford St Mary and Dedham Lock during the early eighteenth century (1705–13). Therefore, the section of river we see today, and which is depicted by John Constable's oil painting, is not the river that would have been present throughout the Neolithic. The old course of the river ran some 100 m to the north of the current River Stour. Alluvium soils along the old course of the River Stour strongly suggest that floodplains occurred on both the northern and southern banks of this old river. These would have significantly restricted animal movement between the river and the cursus monument.

Pastoralists travelling with domestic cattle in a north-westerly direction towards Stratford St Mary would appear to have been required to travel along the steep hillsides of the Stour Valley adjacent to the northern bank of the old course of the River Stour. Cattle movement would have been further restricted by Gun Hill (OS Grid Ref TM 0403320) and the Stratford Hills (OS Grid Ref TM 048354), adjacent to each side of the cursus monument, forcing the herd to move parallel to the monument. The fact that the herd appears to travel along the valley floor in accordance with George et al.'s (2007, p. 5) recommendations for the movement of cattle could again potentially suggest that the Stratford St Mary Cursus commenced life as a droveway. Significant narrowing of the original landscape through both extreme event flooding and first influx fluvial flooding could support Harding's (1999, p. 31) theory that the monument was acting as some form of 'barrier', implying that some form of control was being asserted to sideways movement across the landscape, thereby gaining some form of control over access to the spring meadows. Field observation suggests this would have enabled Neolithic pastoralists to assert some form of control over access to the spring meadows in the landscape between the monument and the River Stour, enabling them to move their cattle on to the floodplain in the early spring, which would have enabled herd sizes to increase.

Barnack Cursus

Aerial photographs (RCHME/EH/HE Aerial photographers comment 1995, Winston) of the area that lies about 1.2 km north-north-east of the village of Barnack, in Cambridgeshire, appear to identify the cropmark of a cursus monument that is orientated roughly north-east to south-west (OS Grid Ref TF 083066 – TF 084067). It appears to have been rectangular in shape, comprising a single ditched enclosure that measures 115 m by 20 m. Further cropmarks are also visible on these aerial photographs, identifying a series of hengiform and ring ditches in the immediate vicinity. One of these, at the south-western end of the Barnack Cursus, is approximately 20 m in diameter, having a single west-facing entrance (Harding and Lee 1987).

The Barnack Cursus lies on a raised terrace 17 m above ordnance datum. This is about 7 m above the ordnance datum for both the Maxey Cursus and the Etton Cursus, both of which lie within palaeochannels some 5 km to the north-east. While this may initially suggest that the Barnack Cursus was not directly affected by these palaeochannels, which Pryor (2005) discovered when excavating these latter monuments, a range of hills that surround the cursus monument to both the south and the west, together with these palaeochannels, could potentially have been a significant factor in narrowing the entrance to the area. This would have resulted in a natural concentration of herd movement at a point which appears to have been on an ancient pathway, due to the Neolithic causewayed enclosure to the south at Upton. It is notable that this was the same area through which later generations were to build the Roman road of Ermine Street and the Stamford to Oundle railway, which closed in 1929. This potentially suggests that this was the optimum route for cattle to move across the area.

Therefore, investigation of the wider landscape appears to suggest that pastoralists travelling with their domestic cattle would have been required to travel in accordance with George et al.'s (2007, p. 5) recommendations. Although without major excavation of the palaeochannels it is now impossible to identify the actual route, it could again potentially suggest that the Barnack Cursus commenced life as a droveway. However, the development of the Fens means that, without major excavation, it is no longer possible to identify any significant narrowing of the landscape either through extreme event flooding or through first influx fluvial flooding.

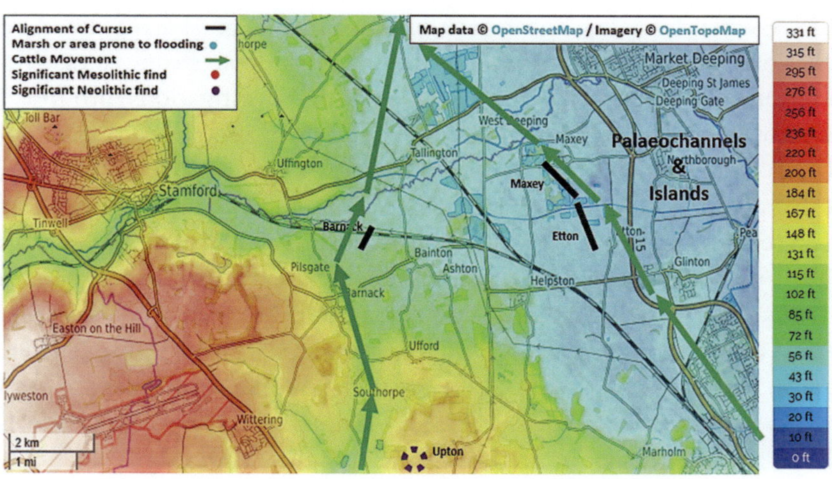

Map 13: Cattle movement across the Barnack, Maxey and Etton Cursus monuments (Main map data created using © OpenStreetMap / Imagery © OpenTopoMap. Inset reproduced with the permission of the British Geological Survey ©UKRI. All rights reserved)

Maxey Cursus

Aerial photographs (St Joseph 1956) of the area between the village of Maxey and the River Welland identified the location of a cursus monument (OS Grid Ref TF 125078 To TF139063) that is orientated roughly north-west by south-east and appears to have had a marked change in its direction around the mid-point. Cropmarks indicated that the north-western end of the cursus monument began near the banks of the current path of the River Welland, although its terminals are no longer visible, possibly being concealed by alluvium. The cursus monument extends for about 1.7 km, although the two principal ditches appear to be discontinuous. The southern ditch continues as far as the Maxey Cut to the south, while the northern ditch appears to have ended some distance before this point. However, further investigation of aerial photographs (RCHME 1997) revealed that a 30 m section of ditch appears to have continued approximately 140 m to the south of the Etton causewayed enclosure.

The destruction of large sections of the cursus monument through gravel extraction resulted in extensive archaeological excavations between 1957–58, 1962–63, 1979–81 and 1982–84, mainly around the central area and the southern end of the monument, which identified that the cursus ditches were about 58 m apart and 2 m wide. Both appeared very shallow, having no evidence of either a bank being constructed or any episodes of recutting. A discontinuous layer of comminuted charcoal was noted near the base of both ditches, which could potentially be associated with land clearance.

Later investigations by Pryor (1998, p. 110) suggested that 'the north-western length of the cursus monument was constructed long after the south-eastern portion, after the latter's ditches had silted up'. This led French (Pryor 1998, p. 7) to suggest that 'the early Holocene river development within the Maxey area of the lower Welland valley witnessed the replacement of a low sinuosity, braided channel system by a higher sinuosity meandering river'. By the Neolithic period, a combination of rising post-glacial sea levels and increased alluvium had resulted in these monuments being sited close to the floodplain. This would therefore appear to have placed the cursus monument in direct alignment with any potential cattle movement across the area.

The Maxey Cursus lies on a raised terrace parallel to a series of palaeochannel systems. It runs in a north-westerly direction before slightly changing alignment at its mid-point. It was sited upon Maxey Island, an oblong-shaped piece of high ground in the middle of the floodplain. Investigation by French (Pryor 1998, p. 7) established that the palaeochannel systems appear to have been slow moving, containing extensive marshland vegetation along their edges. He suggests the cursus's damp, unkempt ditches had been constructed in a meadow environment prone to occasional flooding. However, Robinson's (Pryor and French 2005, pp. 153–62) investigations into insect assemblages found to the west of the cursus monument identified that the area potentially consisted of extensive tree and shrub cover with only a degree of open grassland.

It appears that pastoralists travelling with their domestic cattle on to the Maxey Cursus floodplain would have been required to negotiate the various braided river channels among which the Maxey Cursus lay. Fenland drainage and modern farming methods have resulted in such significant changes to the landscape that field observations are no longer able accurately to determine any probable direction of travel. However, the creation of the Floodplain Forest Nature Reserve within the Ouse Valley Park, Wolverton, Milton Keynes (Kincaid 2018 – personal communication) upon the landscape where five cursus monuments were discovered does indicate the riparian nature of these palaeochannels where the probable direction of travel appears to be in parallel to the channels.

Pryor's (1998, p. 2) indication that a possible causeway existed in the northern ditch of the Maxey Cursus would appear to support Harding's (1999, p. 31) theory that the monument was acting as some form of 'barrier', implying that some form of control was being asserted to sideways movement across the landscape, thereby asserting some form of control over access to the spring meadows, enabling Neolithic pastoralists to move cattle on to the floodplain in the early spring, which would have enabled herd sizes to be increased.

Etton Cursus

Investigations undertaken by Upex, of aerial photographs taken during the extremely dry summer of 1976 between the villages of Maxey and Etton, identified the location of both a causewayed enclosure and a cursus monument. The monument appears to have been orientated north-west by south-east (OS Grid Ref TF 138074), being roughly parallel to the Maxey Cursus. However, the visible cropmarks from this section of the lower Welland valley potentially show only a portion of a far larger ancient landscape. The fact that large areas of the monument are potentially hidden from aerial photography by the extensive spreads of alluvium led French (Pryor 1998, p. 3) to suggest that understanding of this ancient landscape is currently far from complete, and that it is entirely possible that other sites similar to Etton still lie beneath the alluvial clays that cover the countryside between Maxey and the Fen edge.

Close examination of the aerial photographs reveals a faint ditch that runs parallel to the Maxey Cursus for several hundred metres to the east, which appears to end within the Etton causewayed enclosure. Another parallel ditch seems to run to the east, both ditches appearing in section on the freshly machined sides of the Maxey Cut. These ditches were further observed approximately 400 m south-east of the causewayed enclosure during excavations that took place prior to construction of the A15 Glinton bypass (French and Pryor 2005).

The original thoughts were that the cursus monument was an off-centred southerly extension of the Maxey Cursus. However, later excavations by Pryor between 1982 and 1987 (Pryor 1998) show it to be a separate monument, the Etton Cursus. The site lay in the crook of a palaeochannel, which apparently resulted in the location resembling a low-lying floodplain island immediately south of the higher floodplain of Maxey Island. This led Loveday (2006, p. 146) to suggest that, 'of all the cursus monument complexes throughout southern Britain, it was the Maxey/Etton complex that most closely reflected the multiple monument pattern witnessed at Rudston'. Loveday (2006) also indicated that it appears to lie along one of the principle entry routes for Group VI axes.

Another interesting fact from the monument complex is that, lying between the parallel ditches of the Etton Cursus, Pryor (1998, p. 47) discovered two aurochs' skulls, including their horns, buried on an oak plank within ditch segment number 12 of the causewayed enclosure, potentially identifying that the landscape had been suitable for supporting cattle, whether wild or domestic, for millennia. Ainsley's investigations (in: Pryor 2005, p. 84) give the impression that the dominance of domestic cattle at the Etton Landscapes site was remarkable, where bones from 485 identified cattle were recovered, relating to 64 per cent of the assemblage.

It appears that pastoralists travelling with their domestic cattle on to the Etton Cursus floodplain would have been required to negotiate the various braided river channels among which the Etton Cursus lay. Fenland drainage and modern farming methods have resulted in such significant changes to the landscape that field observations are no longer able accurately to determine any probable direction of travel. Pryor's (1998, p. 4) indications that at least three possible causeways existed in the northern ditch and a further three possible causeways existed in the southern ditch of the Etton Cursus would appear to support Harding's (1999, p. 31) theory that the monument was acting as some form of 'barrier', implying that some form of control was being asserted to sideways movement across the landscape, thereby asserting some form of control over access to the spring meadows, enabling Neolithic pastoralists to move cattle on to the floodplain in the early spring, which would have enabled herd sizes to be increased.

Eynesbury Cursus Complex

At Eynesbury, to the south of St Neots in Cambridgeshire, excavations undertaken by Macaulay (1994) and Kemp (1998) revealed the remains of a large Neolithic complex. There are, specifically, three potential Early Neolithic cursus monuments (OS Grid Ref TL 181582 to TL 184584), for which OSL (optically stimulated luminescence) gave a date range of between 4860 BC and 3450 BC. This would make the Eynesbury Cursus complex one of the earliest in the British Isles. However, caution should be used when dating by the OSL method, as it typically contains a degree of uncertainty of up to 10 per cent of the age of the sample.

Discovered in 1959 as cropmarks on aerial photographs by St Joseph (1961), the southernmost cursus monument of this group of three appears to have been orientated east-west, although only the eastern end appears to have been visible in the aerial photographs. The western end was presumably represented by the current course of the River Ouse, which flows roughly south-west to north-east at this point in the landscape. Unlike the other two cursus monuments, the southern cursus has not been excavated. The middle cursus monument is aligned north-south, its south-eastern corner slightly overlapping the northern side of the southern cursus monument towards its eastern end. An undated double enclosure to the north of this monument was also identified, perhaps representing another cursus monument, however the stratigraphical and chronological relationship between these three monuments is currently unresolved.

The Eynesbury Cursus complex lies within the Ouse Valley on a gentle sloping gravel terrace to the east of a river confluence, where the Hen Brook joins the River Great Ouse. Floodplains lie adjacent to both the eastern and western banks of the river as it moves in a northerly direction from Little Barford through St Neots. Pastoralists travelling with their domestic cattle would appear to have been required to travel along the valley floor in accordance with George et al.'s (2007, p. 5) recommendations, being restricted within the gap between the eastern side of this floodplain and Alington Hill (OS Grid Ref TL 187559). This would have required the herd to follow the direction of both the later Roman road of Ermine Street and the current East Coast railway line as it passed parallel to the cursus complex, again suggesting that the Eynesbury Cursus complex potentially commenced life as some form of droveway, where the middle and northern cursus monuments appear to follow a similar direction to the palaeochannels, once again perhaps identifying an initial practical function of the landscape prior to any possible later ritual importance.

The fact that significant narrowing of the landscape occurs during both extreme event and first influx fluvial flooding and the fact that the southern cursus appears to complete a corralling of the spring meadows between the three monuments, the palaeochannels and the River Great Ouse itself, together with Malim's (1999, p. 79) indication that the middle cursus monument had at least two causeways in the western ditch and a further possible causeway in the eastern ditch, would appear to support Harding's (1999, p. 31) theory that the monument was acting as some form of 'barrier'. This implies some form of control being asserted to sideways movement across the landscape, which would have given some form of control over access to the spring meadows. This seems to be further supported by the pit alignment enclosure which would appear to enclose the three cursus monuments, the palaeochannels and the River Great Ouse. The dimensions of the enclosure do not appear dissimilar from those of the horned enclosure at Godmanchester, which was 336 m long by 168 m wide at its narrower, cursus end.

Godmanchester Cursus

The Godmanchester Cursus lies in the Ouse Valley to the north-east of Godmanchester, in Cambridgeshire. (OS Grid Ref TL 255709) but cropmarks from aerial photography (Oblique aerial photograph reference number TL 2571 30-JUL-1984 NMR 2173/1299–1316) have only been able to identify the north-eastern terminal. The cursus monument appears to have been aligned north-east to south-west, although the south-western terminal appears to have been lost beneath the urban development of Godmanchester.

The cursus monument appears to have been about 90 m wide at its north-eastern terminal and seems to have terminated against the south-western side of a Neolithic trapezoidal enclosure. Excavations undertaken in 1988–91 suggested that the cursus monument was the later of these two monuments, having been radiocarbon dated to the early fourth millennium BC. These excavations identified that sections of the cursus ditch showed extensive variation regarding both the width and the depth of the ditch, while causeways have been identified along the north-western side of the ditch, close to the terminal. The ditch also appears to have been accompanied by an internal bank. Cropmark evidence confirmed that the cursus monument extended for at least 500 m before disappearing beneath the urban development of Godmanchester. This led Malim (1999) to suggest that it may have terminated at a tributary of the Ouse, which would suggest the total length of the monument to have been around 1 km. Like the Eynesbury Cursus, the Godmanchester Cursus also appears to have included a pit alignment enclosure as part of the complex. This could suggest they performed similar functions, perhaps indicating that the Eynesbury and the Godmanchester enclosures had local parallels.

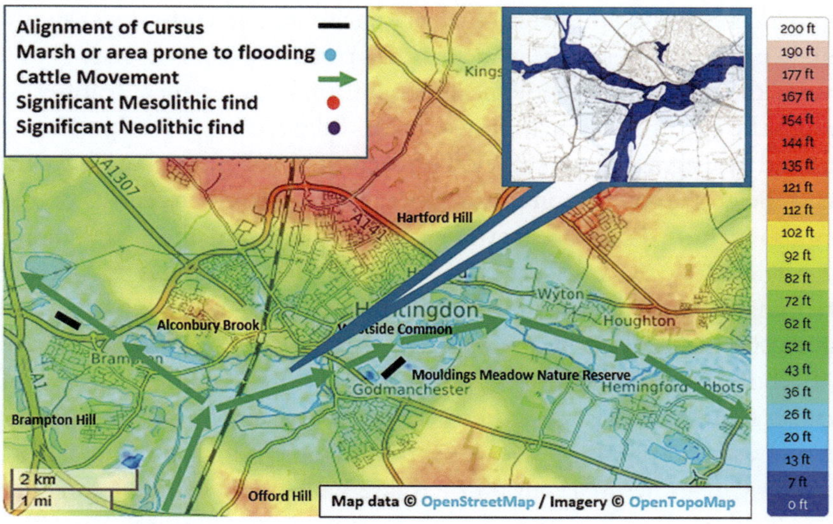

Map 14: Cattle movement across the Godmanchester & Brampton Cursuses (Main map data created using © OpenStreetMap / Imagery © OpenTopoMap. Inset reproduced with the permission of the British Geological Survey ©UKRI. All rights reserved)

The Godmanchester Cursus lies in the Ouse Valley, to the south-east of Westside Common, where extensive alteration to the surrounding landscape has occurred through quarrying in the 1980s, leading to the creation of the Mouldings Meadow nature reserve and the use of the landscape as landfill, with the result that it is now impossible to accurately identify potential routes for cattle movement. However, wider landscape investigation implies that pastoralists travelling with their domestic cattle across the region would have moved in a northerly direction along the Ouse Valley before appearing to split at Offord Hill (OS Grid Ref TL 232688), where one route moved in an easterly direction towards the Godmanchester Cursus. This route would have been subject to significant further restriction between Hartford Hill (OS Grid Ref TL 270740) and the hill where the Wood Green Animal Centre is now situated (OS Grid Ref TL 262684). It would, therefore, appear that the direction of travel along the valley floor would have been in accordance with George et al.'s (2007, p. 5) recommendations for the movement of cattle, again potentially suggesting that the Godmanchester Cursus commenced life as some form of droveway. Malim's (1999, p. 83) suggestion that 'the monument terminated at a tributary of the River Ouse' could suggest this was again perhaps to lead cattle to water.

Malim's (1999, p. 84) indication that causeways existed on the north-western ditch of the Godmanchester Cursus would appear to support Harding's (1999, p. 31) theory that the monument was acting as some form of 'barrier', implying that some form of control was being asserted to sideways movement across the landscape, thereby asserting some form of control over access to the spring meadows. This seems to be further supported by the fact that the monument adopted the western boundary of a massive horned (trapezoidal) enclosure as its terminal, while the northern ditch appears to have been associated with a rectangular enclosure, suggesting the Godmanchester Cursus acted in a similar fashion to the Eynesbury Cursus.

Brampton Cursus

To the north of Brampton, in Cambridgeshire, lies the Brampton Cursus (OS Grid Ref TL 203716), the south-eastern terminal having been identified by cropmarks on aerial photographs in 1962 (St Joseph 1965). The cursus monument is orientated north-west to south-east, the cropmarks being visible for approximately 300 m. However, as the north-western terminal was destroyed by quarrying before the existence of the monument was known, it is no longer possible to establish the full length of the monument.

The parallel ditches are approximately 25 m apart and disappear to the north-west of another elongated cropmark enclosure, a probable mortuary enclosure that appears to have shared the same alignment, while a palaeochannel has been identified as running in a north-easterly to south-westerly direction between the two. The chronological relationship between these features is currently uncertain. Although the cursus monument has not been fully excavated, the presence of the ditches has been confirmed by magnetometer survey. Two section trenches which were cut across the outer ditch that ran parallel to the cursus monument appear to have identified gently sloping ditches about 1 m wide. They also identified a squared terminal at the eastern end and a bank along the southern side. Although no dating evidence was recovered from the cursus monument, a separate monument at its eastern terminal, interpreted as a mortuary enclosure, was excavated in 1991 (Malim 1993) and dated to around 2580–2149 cal BC by a *terminus ante quem* from two radiocarbon dates from oak charcoal found as burnt deposits within the mortuary enclosure. Presuming the mortuary enclosure predated the cursus monument, a phasing seen between the cursus and enclosure at Godmanchester, this might indicate a similar phasing between the two features at Brampton.

The Brampton Cursus lies in a spur off the Ouse Valley, on a gentle sloping gravel terrace to the west of the Alconbury Brook. Potential cattle movement in a northerly direction up the Ouse Valley appears to split at Offord Hill (OS Grid Ref TL 232688) with one route moving north-west in the direction of the Brampton Cursus. This would have required any herd movement to follow the alignment of the natural topography which runs parallel to the cursus monument. Potential cattle movement would have been further restricted in the immediate area of the cursus monument, as at this point floodplains lay to the west of the Alconbury Brook, while further west it appears the herd would have been hemmed in by the rising slopes of Brampton Hill (OS Grid Ref TL 180700). It would therefore appear that the direction of travel along the valley floor would have been in accordance with George et al.'s (2007, p. 5) recommendations for the movement of cattle.

Significant narrowing of the original landscape through both extreme event flooding and through first influx fluvial flooding could also support Harding's (1999, p. 31) theory that the monument was acting as some form of 'barrier', implying some form of control being asserted to sideways movement across the landscape. Field observation suggests this would have enabled Neolithic pastoralists to assert some form of control over access to the spring meadows in the landscape between the monument and the Alconbury Brook, enabling them to move their cattle on to the floodplain in the early spring, which would have enabled herd sizes to increase.

Springfield Cursus

In the Chelmer Valley in Essex, the Springfield Cursus lies on a north-sloping gravel terrace above the floodplain of the Chelmer, 1.2 km to the east of a river confluence where the River Cam joins the River Chelmer (OS Grid Ref TL725067 to TL 735084). It was orientated in a north-easterly direction, on an alignment with the Springfield Lyons causewayed enclosure (OS Grid Ref TL 735081). However, the cursus monument has been destroyed during the recent development of Chelmer Village, a suburb of Chelmsford.

Excavated in 1979 (Hedges and Buckley 1981), the monument was approximately 670 m in length and 40 m wide, with several gaps along the course of both side ditches. Both terminals were closed by transverse ditches which met the side ditches at right angles. Within the eastern terminal was a circular arrangement of postholes, potentially representing a timber circle 26 m in diameter. Some small pits within the interior also contained charcoal and burnt flints, and one featured a small quantity of cremated animal bone. Hedges and Buckley's (1981) excavations focused on the terminal areas but several additional places along the length of the monument were also examined. Later excavation work in 1984 discovered various linear and pit-like features within the confines of the monument, and potentially the remains of a barrow beyond its eastern end.

Buckley et al. (2001, pp. 101–62) put forward that the alignment of the Springfield Cursus potentially coincided with the direction that groups of people took as they passed through the area to disperse up on to the boulder clay plateau, perhaps in the spring and summer to return in the autumn and winter. This reasoning could also be applied to either the seasonal migration routes of aurochs passing through the area or the movement of Neolithic pastoralists with their herds of cattle.

Evidence for earlier human activities that would appear to be directly associated with hunting include the worked flint implements recovered on the northern bank of the River Cam to the east of where it joins the Roxwell Brook. A further collection of Mesolithic flints were recovered in Moulsham Street, while fifty-two unretouched blades and flakes were discovered on the southern floodplain of the River Chelmer and a further 400 Mesolithic flints were discovered at the confluence of the River Cam and the River Chelmer. It is interesting that these find spots appear to align with potential wild cattle movement, appearing to form a funnel towards the landscape upon which the later cursus monument was constructed. This suggests that this landscape had for millennia been an ideal environment for supporting cattle, whether that was for hunting or herding, and this is supported by Buckley et al.'s (2001, p. 147) investigations which suggest that 'bone fragments identifiable to domestic cattle have included four sesamoids, part of an ulna, two fragments of first phalanx, three fragments of vertebra and several fragments of unidentified long bone'.

The section of the Chelmer Valley where the Springfield Cursus was constructed appears to have been regularly affected by flooding on both the southern bank to the east of the river confluence and directly south of the cursus monument. However, canalling of the river in the late 1790s appears to have been successful, as the valley is now only affected by intermittent flooding on its northern bank.

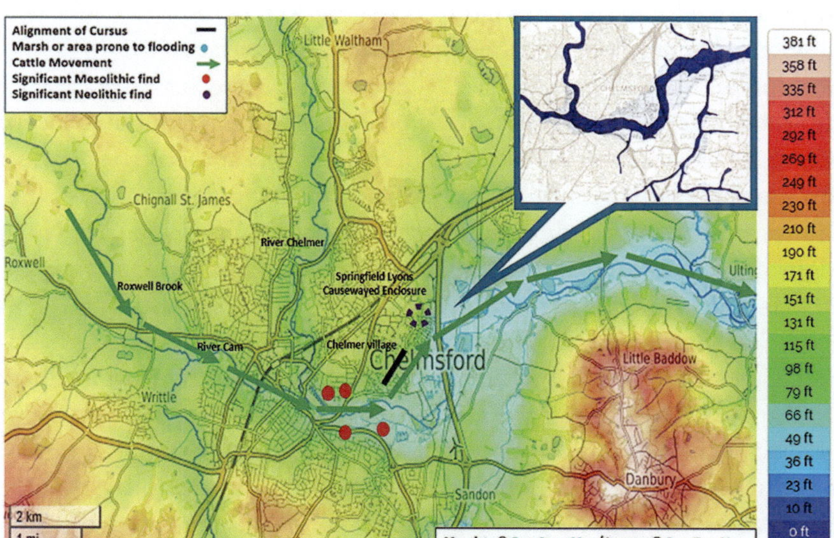

Map 15: Cattle movement across the Springfield Cursus (Main map data created using © OpenStreetMap / Imagery © OpenTopoMap. Inset reproduced with the permission of the British Geological Survey ©UKRI. All rights reserved)

It appears that pastoralists travelling with their domestic cattle towards Chelmsford would have been moving in a south-easterly direction along the course of the Roxwell Brook, which follows the route of the A1060 to where it joins the River Cam (OS Grid Ref TL 675074). From here, cattle movement would have changed direction, moving along the northern bank of the River Cam until its confluence with the River Chelmer (OS Grid Ref TL 712066). Any potential cattle movement would have been significantly restricted after this point, due to regular flooding of the River Chelmer during the Neolithic period. This would potentially have caused the herd to divert in a north-easterly direction, moving on to the gravel terrace along a direct alignment with the Springfield Cursus and in direct accordance with George et al.'s (2007, p. 5) recommendations for cattle movement, once again suggesting that the Springfield Cursus Complex commenced life as some form of droveway.

Due to the development of Chelmer Village, which resulted in the destruction of the cursus monument, it is no longer possible to accurately determine potential routes for cattle movement. However, significant narrowing of the landscape which occurs through both extreme event and first influx fluvial flooding and the fact that Pastscape National Monument Number 879395 indicates several gaps, possible causeways, along the course of both side ditches of the monument would appear to support Harding's (1999, p. 31) theory that 'the monument was acting as some form of barrier', implying that some form of control was being asserted to sideways movement across the landscape. Field observation suggests that this would have enabled Neolithic pastoralists to assert some form of control over access to the spring meadows between the monument and the River Chelmer, enabling them to move their cattle on to the floodplain in the early spring, which would have enabled herd sizes to increase.

Stanwell Cursus Complex

Initially identified through aerial photographs (Poulton 1978 and RCHME Heathrow Mapping Project 1995), the Stanwell Cursus complex appears to have consisted of four, potentially five (the latter being detected as a cropmark outside the area of excavation) cursus monuments and several small circular enclosures. These were extensively excavated prior to the construction of Heathrow Airport Terminal 5 between 2002 and 2007. The site was in the Middle Thames Valley on a flat landscape to the eastern edge of the River Colne (OS Grid Ref TQ 055756), adjacent to an ancient palaeochannel system that ran alongside the Hounslow Heath Terrace.

The monuments at Stanwell seem to have fallen into two categories that appear to have represented different scales of human endeavour and involvement. The cursus monument complex of long linear enclosures together with their banks or central mounds appear to have been constructed within a relatively short time period between 3600 and 3300 BC, while the construction of the small circular enclosures and various styles of ring ditch appear to have been constructed over a longer period between the middle of the fourth millennium and the early second millennium BC.

Evidence from pollen and molluscan data (Lewis et al. 2010) suggests that the western half of the excavated area was primarily open, although some oak, hazel and lime woodland existed upon the drier ground. Later excavations from the eastern section of the area identified that any clearance that had occurred appeared to be much greater than just a corridor along the cursus monument path. The further discovery of burnt humic topsoils potentially suggested that the management of grazing land by fire had occurred during the Mesolithic/Neolithic transition period as relict organic matter, possibly dung, was observed within thin soil sections from the western cursus ditch.

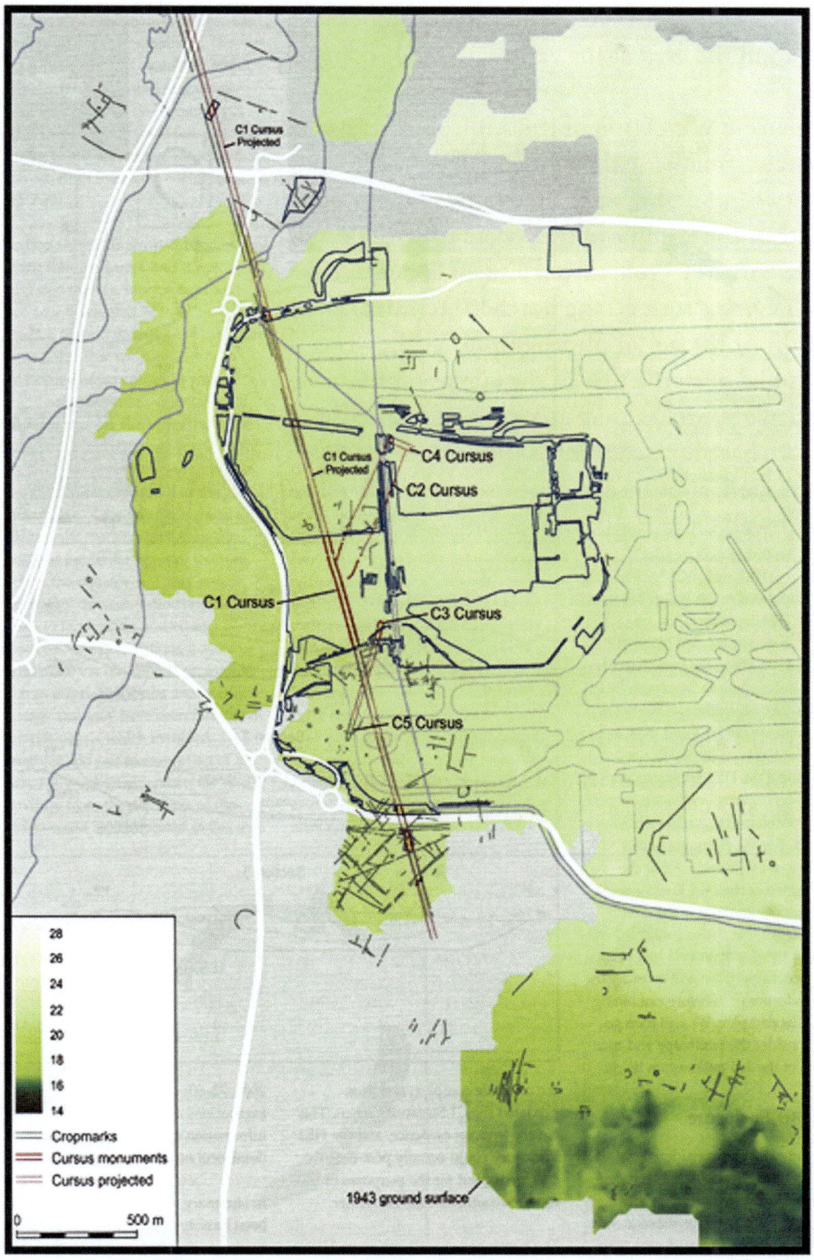

Map 16: The Stanwell Cursus Complex (After Lewis et al. 2010. Reproduced courtesy of Framework Archaeology © Oxford Archaeology and Wessex Archaeology)

The C1 Stanwell Cursus was first identified through aerial photography (Cotton 1990, pp. 29–32). It was initially interpreted as a Roman road, as its two parallel ditches were only about 20 m apart. It continued for between 3.6 and 3.8 km, at least five times longer than the other monuments, and was aligned on a north-north-west by south-south-east orientation. The cursus monument ran through earlier eighth to sixth millennium pit complexes, terminating at its northern end in the Colne Valley. It was unusual compared with the other cursus monuments, in that it had been constructed with only a single central mound.

The monument followed the course of the 22 m ordnance datum contour which separated the Colne Valley floodplain from the Taplow terrace. This led Lewis et al. (2010, p. 75) to suggest that 'the C1 Stanwell Cursus had been constructed along an older pre-existing pathway which physically linked numerous important places such as the remnants of the late Mesolithic midden and posts'.

This is supported by the fact that, as the cursus ditches began to silt up, additional posts appear to have been erected in the area of the earlier posthole complex, potentially suggesting a reaffirmation of this location.

At the same time as erection of the additional posts occurred, construction of the C2 Stanwell Cursus commenced, following the more usual arrangement of internal banks adjacent to the ditches. These varied between 80 and 90 m apart, more than four times as wide as any of the other cursus monuments within the complex. They were orientated north-north-east by south-south-west and continued for 480 m. The C1 Stanwell Cursus appears to have served as the southern terminal while the C4 Stanwell Cursus formed the northern terminal.

Investigation of the aerial photographs led Lewis et al. (2010, p. 69) to suggest that 'the C3 Stanwell Cursus had been the first of the cursus monuments within the complex to be constructed'. Its 230 m length, with ditches 19 m apart, was also orientated north-north-east by south-south-west, where it appears to have formed a north-eastern extension to the C5 Stanwell Cursus which, although currently unexcavated, appears to have existed prior to construction of the C1 Stanwell Cursus.

The C4 Stanwell Cursus survived as only a short 82 m length of twin ditches. These appear to have been approximately 21 m apart and to have had only a single terminal which appeared to be a later addition. However, the majority of this monument had been destroyed by the construction of a large sludge lagoon in the 1980s, making it difficult to classify.

The Stanwell Cursus complex lay adjacent to an ancient palaeochannel that ran alongside the Hounslow Heath Terrace. It was flanked to the north by the start of the Chiltern Hills and to the south by the North Downs. Both of these factors would have allowed only minimal entry and exit points for cattle movement, as herds potentially moved on to the floodplain from the south through the Wey and Mole River Valleys. Lewis et al. (2010, p. 103) believed that the alignment of the initial C5/C3 Stanwell Cursus suggest that this movement, which originates upon the Colne floodplain, was 'pointing the way onto the Heathrow Plateau'. Although they have no firm evidence, the orientation of these monuments strongly suggests that a route out of the Colne floodplain was being formalized. This appears to have been pioneered from the beginning of the fourth millennium within the less densely wooded Colne and Thames floodplains before spreading on to the increasingly cleared Heathrow Terrace. It is noticeable that each of the three different alignments for the C1 Stanwell Cursus, for the C2, C3 & C5 Stanwell Cursus monuments and for the C4 Stanwell Cursus appear to have pointed to a routeway that would have taken herds of cattle along preferred routes through the valley to the distant hills that ran to the north and north-west of the cursus complex.

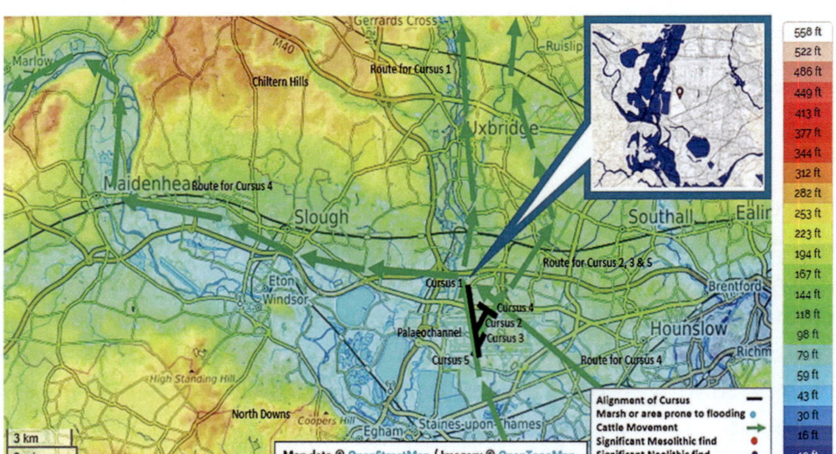

Map 17: Cattle movement across the Stanwell Cursus Complex (Main map data created using © OpenStreetMap / Imagery © OpenTopoMap. Inset reproduced with the permission of the British Geological Survey ©UKRI. All rights reserved)

Initial investigation of the larger hills and valleys surrounding the area seems to indicate that a herd would have travelled in accordance with George et al.'s (2007, p. 5) recommendations for the movement of cattle, which would potentially suggest that the Stanwell Cursus complex potentially commenced life as numerous droveways. However, due to the destruction of the previous ancient landscape, initially by the creation of a sludge lagoon in the 1980s and more recently by the development of Heathrow Terminal 5, it is now only possible to surmise the topography of the local landscape through archaeological records.

Extreme levels of both extreme event and first influx fluvial flooding highlight the complicated nature of this landscape. Framework Archaeology had four years (2006–10) to attempt to explain the landscape prior to the construction of Terminal 5 and had to accept that excavation of the cursus monument had given them little in the way of answers. However, the nature of the floodplains, although further complicated by the ancient palaeochannels, together with Lewis et al.'s (2010, p. 31) identifications of at least four causeways in the Stanwell 1 Cursus, at least one in the Stanwell 2 Cursus and a further four in the Stanwell 3 Cursus, appear to support Harding's (1999, p. 31) theory that the monument was acting as some form of 'barrier', implying that some form of control was being asserted to sideways movement across the landscape. This could have enabled Neolithic pastoralists to assert some form of control over access to the spring meadows between the monument and the river, enabling them to move their cattle on to the floodplain in the early spring, which would have enabled herd sizes to increase. It is also possible that Stanwell Cursus monuments 1, 2, 3 and 5, together with the palaeochannel, potentially created some form of cattle holding area or corral.

Biggleswade Cursus

In the Ivel Valley to the north of Biggleswade in Bedfordshire, the Biggleswade Cursus (OS Grid Ref TL 197466 to TL 189466) lies on a gentle eastern-sloping gravel terrace above the floodplain of the River Ivel, 500 m south of Bunkers Hill (OS Grid Ref TL 189479). The cursus monument, initially identified by cropmarks on aerial photography in 1946 (NMR TL 1946/1/321–2), was excavated by Albion Archaeology in 2004, who investigated a 33-m section. The cursus monument appears to have been orientated east-south-east by west-north-west and had a squared terminal at the eastern end.

The cursus monument appears to have been part of a Neolithic complex which includes a line of three ring ditches, one possibly hengiform, that lie adjacent to the south-eastern corner of the cursus monument. The full extent of the complex has not been identified, as the western end disappears under a sewage works to the west of Furzenhall Farm. This suggests that the cursus monument could potentially have had an overall length in excess of 750 m. The ditches do not appear to have taken the form of a continuous circuit but appear to have been cut as a series of ditch lengths between 20 m and 180 m long which formed the rectangular shape. The surviving cursus monument is approximately 750 m in length and around 70 m wide. However, construction of the Great Northern Railway in 1850, and later the sewage works, has obscured the west-north-western terminal and resulted in extensive alteration to the landscape surrounding the Biggleswade Cursus, although the ditch at the southern end of the cursus monument appears to have been recut after partial silting of the earlier ditch.

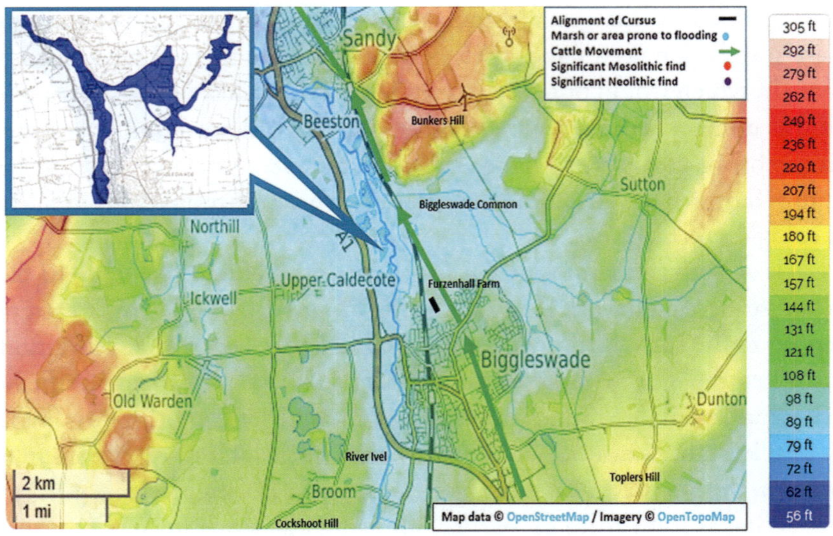

Map 18: Cattle movement across the Biggleswade Cursus (Main map data created using © OpenStreetMap / Imagery © OpenTopoMap. Inset reproduced with the permission of the British Geological Survey ©UKRI. All rights reserved)

It appears that pastoralists travelling with their domestic cattle along the Ivel Valley in a northerly direction would have been restricted by two hills in the region of Langford, between which the River Ivel flows. These are Toplers Hill (OS Grid Ref TL 215403) to the east of the River Ivel and Cockshoot Hill (OS Grid Ref TL 145404) to the west of the river. After crossing the area where construction of the later cursus monument occurred, cattle movement would have been significantly restricted between the eastern bank of the River Ivel and Bunkers Hill (OS Grid Ref TL 192481). It should be noted that this thin stretch of flat landscape is the area where the now dismantled Great Northern Railway was constructed and is also the current route of the East Coast Railway. This suggests this section of landscape has been suitable as a thoroughfare for many years, again seeming to indicate that herds would have travelled along the valley floor in accordance with George et al.'s (2007, p. 5) recommendations for the movement of cattle. This also suggests that the Biggleswade Cursus potentially commenced life as a droveway, the alignment of the monument possibly suggesting this was to lead herds of cattle to water at the River Ivel. This again perhaps identifies an initial practical function of the landscape prior to any possible later ritual importance.

Although the development of both railways and later the sewage works has resulted in extensive alteration to the landscape surrounding the Biggleswade Cursus, the extreme levels of extreme event and first influx fluvial flooding highlight the extensive floodplain area, supporting Harding's (1999, p. 31) theory that the monument was acting as some form of 'barrier', implying that some form of control was being asserted by the monument over access to the surrounding landscape.

Cardington Cursus

To the east of Bedford in the Ouse Valley, the Cardington Cursus lies on the south-eastern sloping gravel terrace above the floodplain of the River Great Ouse. The cursus monument was initially identified by cropmarks on aerial photographs (Oblique aerial photograph reference number CUCAP ADO70 06-JUL-1961) and later by excavation during work to complete the Bedford southern bypass. The cursus monument is orientated south-west by north-east and measures about 75 m by 15 m. It seems to have been a straight-sided enclosure with rounded ends. It had been constructed with a wide entrance gap along the southern side and may have included an external bank where the southern ditch terminal of the gap curved inwards. The Cardington Cursus appears to have aligned with a causewayed enclosure located approximately 1 km north-east of Cardington and 1.5 km from the River Great Ouse, and with a further three isolated, irregular enclosures.

The Cardington Cursus lies within the Ouse Valley between the River Great Ouse and the steep ridgeline of Hammer Hill (OS Grid Ref TL 095428). Pastoralists travelling with their domestic cattle in a north-easterly direction along the Great Ouse Valley would have been required to pass through this gap before changing direction due to the ridge around Shortstown (OS Grid Ref TL 074468) and the slight ridge upon which Cople was later constructed (OS Grid Ref TL 105485). Floodplains to the south-west of Cople would have further restricted cattle movement in alignment with the Cardington Cursus.

This indicates that herds would have travelled in accordance with George et al.'s (2007, p. 5) recommendations for the movement of cattle, again suggesting that the Cardington Cursus potentially commenced life as a droveway. The alignment of the monument with the river suggests this was possibly to lead herds of cattle to water at the River Ivel, again perhaps identifying an initial practical function of the landscape prior to any later possible ritual importance.

Although the creation of a nature park, including a lagoon, has resulted in extensive alteration to the immediate landscape which surrounds the Cardington Cursus, levels for extreme event and first influx fluvial flooding highlight the extensive floodplain area. This, together with Loveday's (2006, p. 30) identification of two possible causeways in the side ditches of the Cardington Cursus, appear to support Harding's (1999, p. 31) theory that 'the monument was acting as some form of barrier', implying that some form of control was being asserted by the monument over access to the surrounding landscape.

Cople Cursus

To the east of Bedford in the Ouse Valley, the Cople Cursus (OS Grid Ref TL 093500) lies on the eastern-sloping gravel terrace above the floodplain of the River Great Ouse. The cursus monument appears to have been initially identified by cropmarks on aerial photographs (Google Earth.com 1 January 2006, assessed 31 August 2016), where it measures 130 m by 20 m. It appears to have been orientated south-west by north-east, however the south-western terminal is not visible as a cropmark as this area has been obscured by a trackway and the former course of the Bedford to Cambridge railway. The site appears to be part of a larger monument complex located north-west of Octagon Farm, where a probable ring ditch and a Bronze Age ring ditch are located nearby. The visible sections of the parallel ditches are between 15 and 17 m apart and between 1 and 2.5 m wide.

The Cople Cursus lies within the Ouse Valley between the River Great Ouse and the steep ridgeline of Hammer Hill (OS Grid Ref TL 095428). Pastoralists travelling with their domestic cattle in a north-easterly direction along the Ouse Valley would have been required to pass through this gap before changing direction due to the ridge around Shortstown (OS Grid Ref TL 074468) and the slight ridge upon which the village of Cople was constructed (OS Grid Ref TL 105485). After passing through the landscape upon which the Cardington Cursus was later constructed, the herd would once again have needed to change direction to pass between Mox Hill (OS Grid Ref TL 127468) and the floodplains of the River Great Ouse to the west. This would have further restricted cattle movement, causing it to move in alignment with the Cople Cursus. This seems to indicate the herd would have travelled in accordance with George et al.'s (2007, p. 5) recommendations for the movement of cattle. This could suggest that the Cople Cursus potentially commenced life as a droveway, the alignment of the monument possibly suggesting this was to lead herds of cattle to water at the River Great Ouse.

Although the development of a trackway and the Bedford to Cambridge railway has resulted in extensive alteration to the immediate landscape surrounding the Cople Cursus, extreme event and first influx fluvial flooding highlight the extensive floodplain area, supporting Harding's (1999, p. 31) theory that the monument was acting as some form of 'barrier', implying that some form of control was being asserted by the monument over access to the surrounding landscape.

Ivinghoe Beacon Cursus

Gover (2000) discovered a small cursus monument while undertaking geophysical investigation of the Ivinghoe Beacon (OS Grid Ref SP 961168). It appears to lie on a flat eastward slope adjacent to the scarp of the Chiltern Hills. Situated 13 km east of Aylesbury, Ivinghoe Beacon is the crossroads for a number of ancient routes. It lies at the start of the Ridgeway, which has led Taylor (1979, p. 2) to suggest that it 'could have begun life as a migratory animal route around 8000 BC'. This location, having been separated from the main chalk ridge by the Tring Gap to the south-west and by the subsidiary Gade Gap to the north-east, is situated 140 m above the Vale of Aylesbury, rising to a maximum height of 230 m, giving excellent views of the surrounding countryside.

In the centre of a flat surface, Gover (2000) discovered an elongated structure with curved ends that measured 140 m on its long axis by 30 m on its short axis. He believes this to have been either a short cursus monument or, less likely, a mortuary long barrow enclosure. Within the interior of the cursus monument lies at least one 20 m-diameter ring that appears to honour the cursus monument boundaries. This appears to have been either contemporary or constructed while the cursus monument was still visible.

Holgate (1995, p. 3) took molluscan evidence from the region to suggest that 'a mixed deciduous wooded environment occurred in the area between the sixth and the fourth millennia BC'. He believed that radiocarbon dates of associated charcoal identified 'woodland clearance did not start within the valleys of the Chiltern Hills until the early second millennium BC'. This was further supported by an English Heritage earthwork survey report carried out by Brown and Field in November 2001, which suggests that only the tops of the Chiltern Hills were open during the Mesolithic/Neolithic transition period.

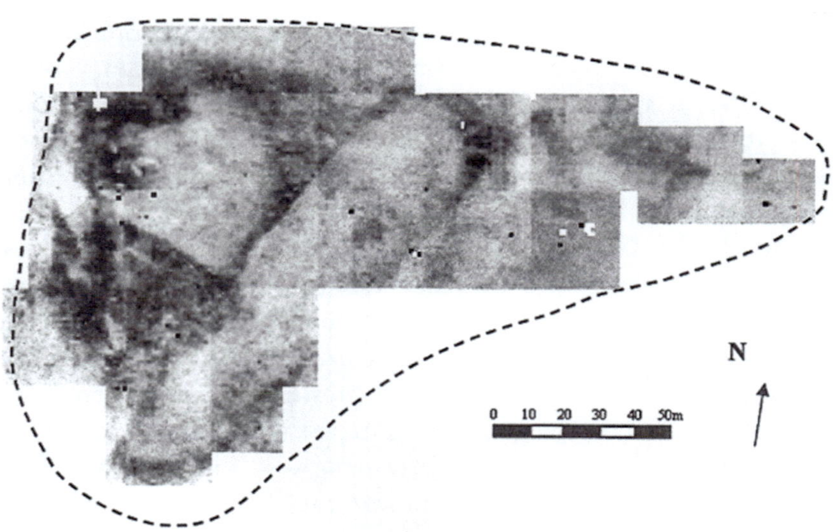

Figure 4: Ivinghoe Beacon in-site raw resistivity (Reproduced courtesy of John Gover 2000)

The Chiltern Ridge stretches from the Goring Gap in Oxfordshire to Royston in Hertfordshire. It is bounded to the north-west by the steep Chiltern escarpment, while to the south-east the dip slope inclines gently, extending as far as the mid Thames Valley. The area is dissected by a series of river valleys that run in a south-easterly direction to drain into the Thames or one of its tributaries. The paths of two ancient routeways, the Ridgeway and the Icknield Way, converge upon the cursus monument at Ivinghoe Beacon. If, as Holgate (1995, p. 3) suggests, 'the hilltops consisted of open landscape while the valleys were still heavily wooded', this could potentially have been the route taken by herds of cattle in the direction of the springline and the River Ouzel at Edlesborough, which seems to indicate the herd travelled along the ridgeline pathways in accordance with George et al.'s (2007, p. 5) recommendations for the movement of cattle. This suggests that even the upland Ivinghoe Cursus potentially commenced life as a droveway, again perhaps identifying an initial practical function of the landscape prior to any possible later ritual importance.

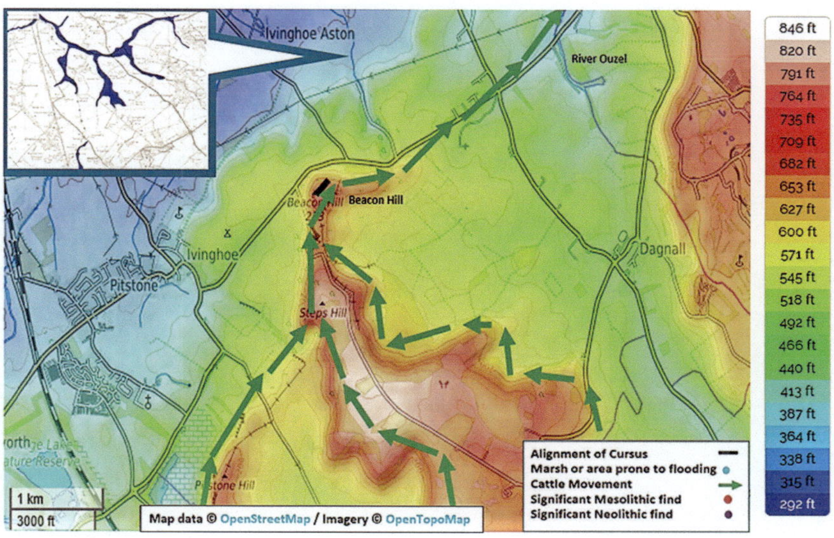

Map 19: Cattle movement across the Ivinghoe Beacon Cursus (Main map data created using © OpenStreetMap / Imagery © OpenTopoMap. Inset reproduced with the permission of the British Geological Survey ©UKRI. All rights reserved)

Wolverton Cursus Complex

Between 2007 and 2011 the Cambridge Archaeological Unit discovered a significant Neolithic monument complex on the southern gravel terraces of the Great Ouse floodplain during a series of excavations at the Hanson Aggregates Manor Farm Quarry, Old Wolverton, Milton Keynes (OS Grid Ref SP 808422). They were nestled between low-lying limestone hills at the confluence of the River Great Ouse and the River Tove. The floodplain appears to have been transformed into a ceremonial complex through the construction of a Neolithic mortuary enclosure and up to four cursus monuments.

Hogan (2013b, p. 5) believes that 'the changing course of the Great Ouse has been integral to our understanding of the prehistoric archaeology and its environmental setting', as current excavations identified broad distinctions between the earlier prehistoric landscape and the later landscape. In the earlier landscape, two major braided palaeochannel systems appear to have bracketed the Neolithic monuments to the north and to the south. Following excavation, these palaeochannels were left open, creating the Ouse Valley Park Floodplain Forest Nature Reserve, becoming a series of pits and pools planted with a combination of self-set willow scrub and weedy species such as willow herb and stinging nettle. To prevent the landform from becoming totally overgrown, thereby impoverishing it for nesting birds and other wildlife, twenty head of grazing cattle were introduced. Martin Kincaid, Senior Biodiversity Officer, The Parks Trust (2018, personal communication, September) outlined that, although the original proposal was for the nature reserve to solely use Konik horses for the grazing management of the site, the reserve needed to supplement them with cattle during the spring and summer months.

The four cursus monuments were categorized by the presence of internal banks and staggered causeways. A fifth monument, yet to be fully excavated, may represent the remains of a mortuary enclosure or bank barrow. This appears to be the earliest monument in the landscape. This earlier feature is significantly narrower than the other cursus monuments and is further distinguished by its rounded western terminal which appear to have been breached by a single causeway. However, the remnant internal banks associated with its ditches appeared similar to those associated with the other cursus monuments.

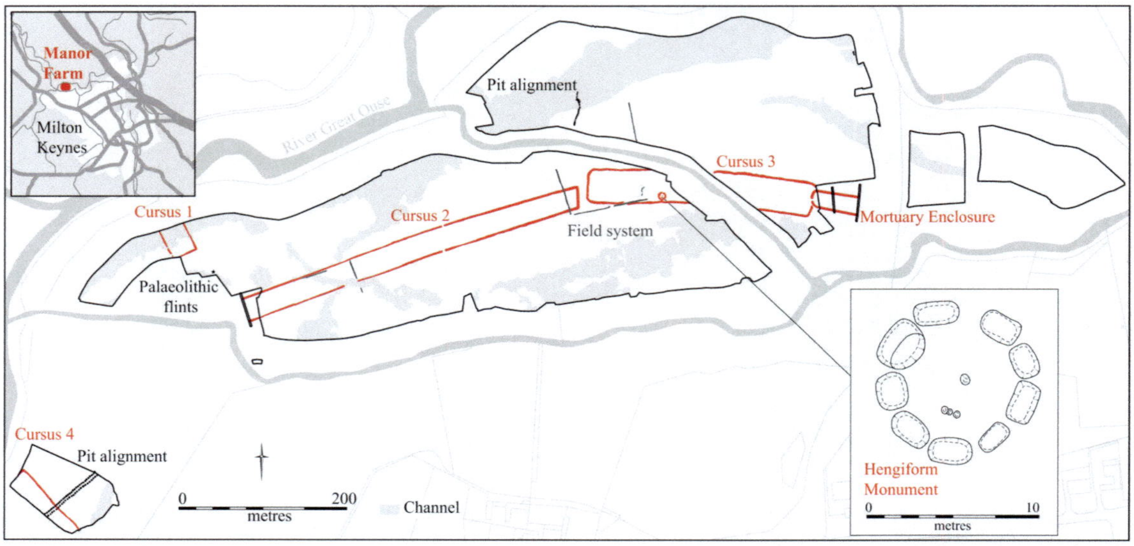

Figure 5: Cambridge Archaeological Unit's drawing of the Wolverton Cursus complex (Reproduced courtesy of Cambridge Archaeological Unit)

The square-ended southern terminal of the Wolverton 1 Cursus was exposed south of the confluence with the River Tove. It appears to align with the River Tove and runs perpendicular to the River Great Ouse. Investigations revealed the northern palaeochannel system to have been active from the Mesolithic period, which could indicate a direct relationship between the river and the monument. Hogan (2013b) identified a possible two causeways in the Wolverton 1 Cursus.

The Wolverton 2 Cursus was discovered slightly to the south of the Wolverton 1 Cursus, extending across a higher ridge of sand and gravel. A total of 423 m were uncovered, indicating a north-east by south-west alignment roughly parallel to the River Great Ouse. Hogan (2013b) identified a possible four causeways in the Wolverton 2 Cursus. To the east of the Wolverton 2 Cursus, the western terminal of a third rectangular cursus monument (Wolverton 3 Cursus) was also discovered. However, unlike the Wolverton 2 Cursus, which remains straight along its entire course, the third enclosure has a gentle curve to its form that appears to mirror the bend in the northern palaeochannel system.

A total of forty-three sherds (277 g) of prehistoric pottery were recovered from the Wolverton 2 and Wolverton 3 cursus monuments, the majority appearing to comprise of a single dump of twenty-seven sherds of Peterborough Ware pottery in the upper silts of the Wolverton 2 Cursus. This stack of material potentially suggests that an in-situ dump occurred rather than the material was inadvertently incorporated. Knight (2013, p. 34) therefore suggests that the cursus monument should be assigned to the Middle Neolithic by association. However, a fifth monument, yet to be fully excavated, may represent the remains of a mortuary enclosure or bank barrow, which would suggest it was the earliest monument in the landscape.

In terms of cursus monument development, at least three phases appear to have taken place. The fifth monument appears to be the earliest, the second phase being represented by the Wolverton 3 Cursus, while the third phase appears to be represented by both the Wolverton 1 and 2 Cursus monuments. However, without definite dating evidence across each of the monuments, this phasing can only be inferred by their alignment, positioning and morphology (Loveday 2006). The physical relationship between the five enclosures implies that a succession of monument construction occurred from east to west, running roughly parallel to the general present course of the Great Ouse and the route of the northern palaeochannel system. Faunal remains from Manor Farm indicate that the as-

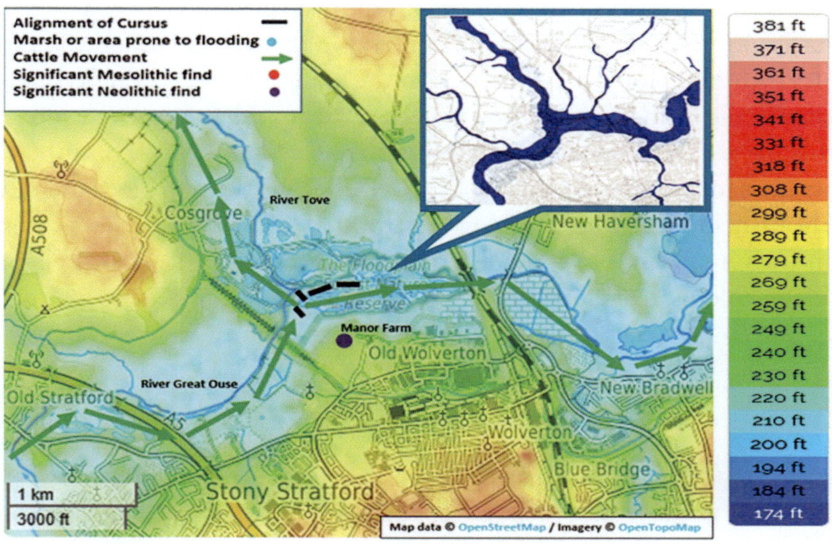

Map 20: Cattle movement across the Wolverton Cursus Complex (Main map data created using © OpenStreetMap / Imagery © OpenTopoMap. Inset reproduced with the permission of the British Geological Survey ©UKRI. All rights reserved)

semblage was dominated by both wild and domestic cattle, suggesting that the landscape had been used by cattle for millennia (Hogan 2013b, p. 39).

The Wolverton Cursus complex lies at the confluence of the River Tove and the River Great Ouse. At this point, the wide river valley to the south narrowed to pass through higher river terraces between Old Wolverton and Cosgrove. The southern gravel terraces of the floodplain upon which the cursus monument complex was situated are currently a floodplain forest nature reserve, designed to flood during the winter months. A prominent springline (OS Grid Ref SP 805419) is situated midway between the protruding limestone scarp upon which Manor Farm is located and the southern gravel terraces of the floodplain. The two braided palaeochannel systems revealed repeated slow-moving channel phases, interspersed with fast-flowing dynamic phases. This suggests that that the area has often been prone to flooding, which would have further concentrated any herd movement occurring through this section.

Rajkovaca's investigations of the animal bone recovered from the fieldwork at Manor Farm (in: Hogan, S. 2013b, p. 38) resulted in the recovery of 231 assessable fragments of animal bone. From these, forty-eight were identifiable to species, of which twenty-seven pieces (56 per cent) were from domestic cattle. A further forty-seven appear to have been cattle sized but further identification of the specimen was not possible.

The Wolverton 1 and 4 Cursus monuments appear to have aligned directly with the direction that pastoralists travelling with their domestic cattle would have been required to take in order to travel north-north-west into the Tove river valley. The Wolverton 2, 3 and 5 Cursus monuments appear to have aligned with the northern and southern palaeochannels. This would result in any herd movement being parallel to the current direction of the River Great Ouse. This seems to indicate that herds of domestic cattle would have travelled in accordance with George et al.'s (2007, p. 5) recommendations for the movement of cattle, suggesting that the Wolverton Cursus complex potentially commenced life as a droveway. The alignment of the Wolverton 1 and 4 Cursus monuments with the River Tove also possibly suggests this was to lead cattle across the River Great Ouse, enabling them to follow the Tove Valley. This again perhaps identifies an initial practical function of the landscape prior to any possible later ritual importance.

Extreme levels of both extreme event and first influx fluvial flooding together with Hogan's (2013) identifications of a possible two causeways in the Wolverton 1 Cursus, a possible four causeways in the Wolverton 2 Cursus and the single causeway in the earliest monument, the Wolverton 5 Cursus, supports Harding's (1999, p. 31) theory that the monument was acting as some form of 'barrier', implying that some form of control was being asserted to sideways movement across the landscape. This would have enabled Neolithic pastoralists to assert some form of control over access to the spring meadows between the monument and the river, allowing them to move their cattle on to the floodplain in the early spring, which would have permitted herd sizes to increase. It is also possible that Wolverton Cursus monuments 1, 2 and 4 together with the palaeochannel potentially created some form of cattle holding area or corral.

Benson Cursus

In the Thames Valley to the south-east of Benson in Oxfordshire, the Benson Cursus runs obliquely uphill from the second gravel terrace of the River Thames, 1 km away, where it has a possible connection with the Winterbrook stream at Wallingford. The cursus monument (OS Grid Ref SU 624910 to SU 629919) was initially identified by cropmarks on aerial photographs in 1933 by Major G.W. Allen. The remains of this site now lie beneath RAF Benson, situated between Benson and Ewelme, a couple of miles to the north-east of Wallingford, on the eastern side of the River Thames. It was one of the earliest cursus monuments to be discovered as a direct result of the many aerial surveys undertaken in southern England by Major Allen and subsequently described by Leeds (1934) in an *Antiquaries Journal* article.

The monument appears to have been orientated north-north-east by south-south-west, the southern terminal running 1 km from the River Thames while the northern terminal lies 300 m from the Winterbrook stream. The unexcavated cursus monument is 1,090 m in length by 65 m in width and is of the type Loveday (1985) classified as Bi. It has been constructed with wide causeways approximately 300 m from the northern terminal, while the surrounding area includes an oval barrow to the west of the northern terminal and two ring ditches to the west and south-west of the southern terminal.

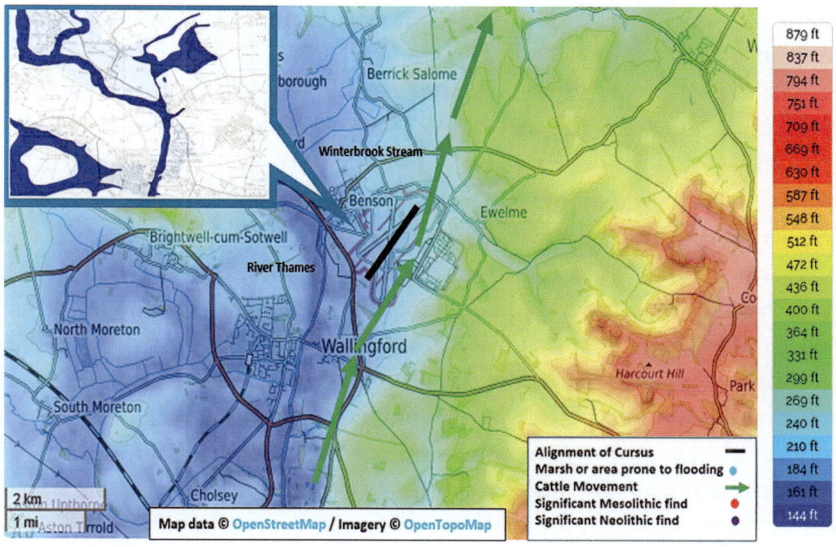

Map 21: Cattle movement across the Benson Cursus (Main map data created using © OpenStreetMap / Imagery © OpenTopoMap. Inset reproduced with the permission of the British Geological Survey ©UKRI. All rights reserved)

The Benson Cursus lies across the northern downward slope of the Chiltern Hills between Harcourt Hill (OS Grid Ref SU 665890) and the River Thames (OS Grid Ref SU 613913). It would appear that potential cattle movement along the Thames Valley initially would have been in a north-north-easterly direction. This would have required pastoralists travelling with their domestic cattle to follow the later ancient route of the Ridgeway, which runs parallel to the river. The herd would have been forced to change direction to the north-east, due to marshland situated close to the River Thames around the hamlets of Crowmarsh Gifford (OS Grid Ref SU 612889) and Preston Crowmarsh (OS Grid Ref SU 615908). This would have required the herd to move across the second gravel terrace of the River Thames in a direction parallel to the cursus monument. This seems to indicate that herds of domestic cattle would have needed to travel across the downward slope in accordance with George et al.'s (2007, p. 5) recommendations for the movement of cattle. This could suggest that the Benson Cursus potentially commenced life as a droveway, again perhaps identifying an initial practical function of the landscape prior to any possible later ritual importance, the alignment with the Winterbrook stream perhaps suggesting this was to lead cattle to water.

Extreme levels of both extreme event and first influx fluvial flooding, together with Barclay and Hey's (1999, p. 72) identification of at least two causeways in the western ditch and at least two, possibly three, causeways in the eastern ditch of the Benson Cursus, appear to support Harding's (1999, p. 31) theory that the monument was acting as some form of 'barrier', implying that some form of control was being asserted over sideways movement across the landscape, perhaps controlling access to the spring meadows. The alignment of the Benson Cursus with the Winterbrook stream, the River Thames and the floodplain possibly suggests some form of cattle corral was being applied to this section of the landscape.

Buscot Wick Cursus

In the Thames Valley to the south-east of Lechlade-on-Thames, the Buscot Wick Cursus (OS Grid Ref SU 217989 to SU 222985) lies on a floodplain gravel terrace at the confluence of the Rivers Thames, Leach and Coln. This unexcavated cursus monument was initially identified by cropmarks on aerial photographs in 1969 and appears to have measured 750 m in length by 50 m in width. It is orientated north-west by south-east and is of the type classified by Loveday (1985) as Bi. The site appears to have been part of a larger monument complex as it has a double ring ditch 100 m from the cursus monument's northern terminal and eight ring ditches and a linear long enclosure close to the southern terminal.

It appears that pastoralists with their domestic cattle would have been required to travel in a north-easterly direction along the Thames Valley. This route appears to have been initially restricted between Horcott Hill (OS Grid Ref SP 155002) and a steep ridge to the east of the River Coln (OS Grid Ref SP149023), forcing the herd to change direction on to a more easterly route. From here the route continues parallel to the River Coln, through an area that is currently the Cotswold Water Park. This was also the previous route of a dismantled railway line. The herd would again be forced to change direction, this time to the south-east, to pass between the high ground of Little Lem Hill (OS Grid Ref SP 208012) and Bury Hill (OS Grid Ref SU 233969). This would the herd to move towards the restricted floodplain of the river confluence in a direction parallel to the cursus monument. This seems to indicate that herds of domestic cattle would have travelled along the valley floor in accordance with George et al.'s (2007, p. 5) recommendations for the movement of cattle. Extreme levels of both extreme event and first influx fluvial flooding together with Barclay and Hey's (1999, p. 72) identification of a causeway in both the southern and northern ditches of the Buscot Cursus appear to support Harding's (1999, p. 31) theory that the monument was acting as some form of 'barrier', implying that some form of control was being asserted to sideways movement across the landscape, perhaps controlling access to the spring meadows.

Dorchester-on-Thames Cursus

In the Thames Valley to the north of Dorchester-on-Thames, the Dorchester-on-Thames Cursus (OS Grid Ref SU 569958 to SU 581948) lies upon the second floodplain gravel terrace about 350 m from a significant bend in the River Thames. The cursus monument's southern terminal lies within 250 m of where the River Thames converges with the River Thame, while the northern terminal lies less than 100 m to the east of where a stream also converges with the River Thames. Initially identified by cropmarks on aerial photography by Crawford in 1927, the cursus monument has been excavated between 1944 and 1952 by Atkinson (Atkinson et al. 1951), in 1981 by Bradley and Chambers (1988) and in 2008 by Gill Hey and David Griffiths as part of the University of Oxford, School of Archaeology (Hey 2016, personal communication, February), who discovered the position of the northern terminal. The cursus monument measures 1.6 km in length by 60 m in width. It is orientated north-west by south-east and appears to have rounded terminals. The site appears to have been part of a larger monument complex, as a D-shaped enclosure forms part of the southern terminal, while three hengiform ring ditches lie within 100 m of the southern cursus ditch and a further fourteen ring ditches lie within 1 km of the monument. The monument has been largely destroyed by gravel extraction, with only the northern terminal remaining, situated under an allotment. The south-eastern segment of the monument, which is interrupted by a central entrance, appears to have been constructed on a slightly different alignment from the rest of the monument, potentially following the course of the floodplain.

It appears that pastoralists with their domestic cattle would have initially been required to travel in an easterly direction as they moved along the Thames Valley floodplain. However, restriction of this route in the area to the south-east of Round Hill (OS Grid Ref SU 545977) would have forced the herd to change direction to the south-east on to a route that ran parallel to the alignment of the Dorchester-on-Thames Cursus. This seems to indicate that herds of domestic cattle would have travelled along the valley floor in accordance with George et al.'s (2007, p. 5) recommendations for the movement of cattle. This would once again suggest that the Dorchester-on-Thames Cursus potentially commenced life as a droveway. Extreme levels of both extreme event and first influx fluvial flooding, together with Loveday's (1999, p. 51) identification of a possible six causeways in the south-western ditch and a possible three causeways in the north-eastern ditch of the Dorchester-on-Thames Cursus, appear to support Harding's (1999, p. 31) theory that the monument was acting as some form of 'barrier', implying that some form of control was being asserted over sideways movement across the landscape, perhaps controlling access to the spring meadows. The placement of the monument between the River Thame and the stream that converges with the River Thames potentially suggests that the droveway led cattle to water.

Drayton St Leonard Cursus

In the Thame Valley to the north-east of Drayton St Leonard, the Drayton St Leonard Cursus (OS Grid Ref SU 601971 to SU 602972) lies on the second floodplain gravel terrace to the north of a significant bend in the River Thame. The unexcavated cursus monument was initially identified by cropmarks on aerial photography (RCHME/EH/HE Aerial photographer's comment Small, 21 October 1996/RCHME: Drayton St Leonard Enclosure Project), which also showed an ancient stream passed within 50 m of the monument. The monument measures 390 m in length by 40 m in width. It is orientated north-north-east by south-south-west and is of the type classified by Loveday (1985) as Bii. The site appears to be part of a larger monument complex, as a hengiform double ring ditch appears to have been centrally placed within the monument approximately 200 m from the northern terminal, while a further two ring ditches and a D-shaped enclosure lie to the west, as does the cropmark of a potential long barrow.

Pastoralists with their domestic cattle would appear to have been required to travel in a northerly direction as they moved up the Thame Valley between Primrose Hill (OS Grid Ref SU 611960) and Newbury Hill (OS Grid Ref SU 612977) on the eastern valley side and Windmill Hill (OS Grid Ref SU 553985) on the western valley side, a route that runs parallel to the alignment of the Drayton St Leonard Cursus. This seems to indicate that herds of domestic cattle would have travelled in accordance with George et al.'s (2007, p. 5) recommendations for the movement of cattle. This could suggest that the Drayton St Leonard Cursus potentially commenced life as a droveway, again perhaps identifying an initial practical function of the landscape. Extreme levels of both extreme event and first influx fluvial flooding, together with the fact that the monument appears to cut directly across a significant bend in the River Thame, appear to support Harding's (1999, p. 31) theory that the monument was acting as some form of 'barrier', implying that some form of control was being asserted to sideways movement across the landscape, perhaps controlling access to the spring meadows. However, to date no causeways have been identified within the ditches of the Drayton St Leonard Cursus.

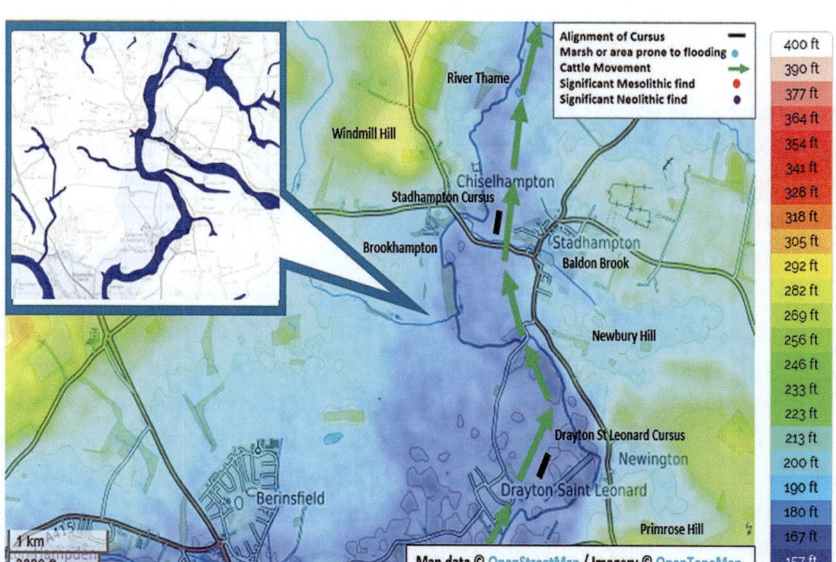

Map 22: Cattle movement across the Drayton St Leonard and Stadhampton Cursuses (Main map data created using © OpenStreetMap / Imagery © OpenTopoMap. Inset reproduced with the permission of the British Geological Survey ©UKRI. All rights reserved)

Stadhampton Cursus

At Stadhampton in the Thame Valley, the Stadhampton Cursus (OS Grid Ref SU 598986 to SU 599991) lies on a north-north-east by south-south-west orientation upon the second gravel terrace of the floodplain. The cursus monument is one of the latest to be discovered, being identified by cropmarks on aerial photography in 1986 (RCHME Air Photography Unit). It is situated upon the floodplain to the north of a confluence of the Baldon Brook and the River Thame, the southern terminal lying approximately 50 m from the brook and 150 m from the river. The cursus monument is 400 m in length by 45 m in width and has terminals that are of the type Loveday (1985) classified as Bi. The eastern ditch appears to have been cut by a 10 m-wide causeway 270 m from the southern terminal. To the east of the cursus monument (OS Grid Ref SU 598986) is an incomplete rectangular enclosure, defined by a single ditch. It measures 45 m by 15 m, suggesting it may have been a mortuary enclosure. An oval barrow ditch that lies mostly outside the cursus monument intersects the cursus ditch on both sides of the causeway. There is currently no evidence for earlier human activity in the area.

It appears that pastoralists with their domestic cattle would have travelled from the south-south-west into the Thame Valley. Entering the valley at the River Thame confluence with the River Thames which occurs between Dorchester and Warborough (OS Grid Ref SU 578932), the herd would have moved in a northerly direction passing Primrose Hill (OS Grid Ref SU 611959) on its eastern side, before being forced to change to a more north-westerly direction at Newington (OS Grid Ref SU 609965) to pass between the River Thame floodplain and Newbury Hill (OS Grid Ref SU 613976). At Brookhampton (OS Grid Ref SU 601980) the herd would once again have been required to change direction to circumnavigate the floodplain, moving back on to a north-north-easterly route parallel to the cursus monument. This seems to indicate that herds of domestic cattle would have travelled in accordance with George et al.'s (2007, p. 5) recommendations for the movement of cattle, again perhaps identifying an initial practical function of the landscape.

Significant levels of both extreme event and first influx fluvial flooding could appear to support Harding's (1999, p. 31) theory that the monument was acting as some form of 'barrier', implying that some form of control was being asserted over sideways movement across the landscape, perhaps controlling access to the spring meadows, while the placement of the monument between the River Thame and the two small streams that converge with the river potentially suggests that the droveway led cattle to water.

Drayton North Cursus

In the Thames Valley to the east of Drayton in Oxfordshire, the Drayton North Cursus (OS Grid Ref SU 490941 to SU 492950) lies on the first floodplain gravel terrace at a significant bend in the River Thames. Initially identified by cropmarks on aerial photography (RCHME/EH/HE Aerial Photographer's comment, Brown/11-MAR-1993/RCHME: Thames Valley NMP), the cursus monument was excavated in 1977 by members of the Abingdon Area Archaeological Society and again in 1985 in advance of gravel extraction.

The monument measures 650 m in length by 75 m in width. It is orientated north-north-east by south-south-west and is of the type classified by Loveday (1985) as Bi. The site appears to be part of a larger monument complex, as a ring ditch intersects the western cursus ditch, while another ring ditch lies 30 m from the eastern end of the east cursus ditch and the Drayton South Cursus lies 500 m further to the south. The cursus monument's northern terminal has been destroyed during gravel extraction, but earlier aerial photographs appear to show it had a rounded north-eastern corner. Excavations of the western cursus ditch showed it appeared to have been dug in 5 m sections, as there were wiggles, potentially suggesting realignment was required at frequent intervals.

Evidence for early human activity within the area included a high proportion of Mesolithic flintwork found on the ground surface during excavation of the cursus monument, which has led Holgate (1986, p. 12) to suggest that 'a seasonal occupation occurred at this location', perhaps for the hunting of wild cattle, identifying that the landscape had been suitable for the grazing of cattle for millennia. A significant factor with regard to the Drayton Cursus complex could be the recovery of twenty-two Early Neolithic arrowheads. Although these were actually found within the Drayton South Cursus, this would support Trantalidou and Masseti's (2015, p. 72) suggestion that this could potentially indicate some form of trophy hunting occurred post monument construction at the site.

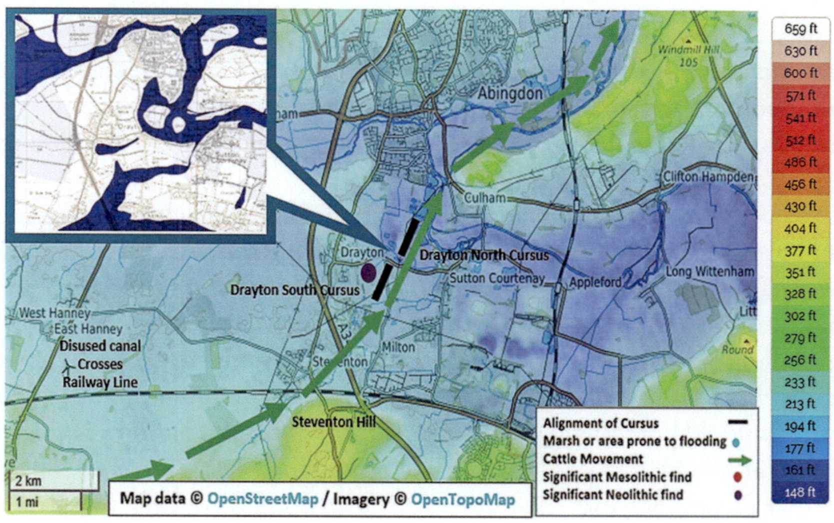

Map 23: Cattle movement across the Drayton North and Drayton South Cursuses (Main map data created using © OpenStreetMap / Imagery © OpenTopoMap. Inset reproduced with the permission of the British Geological Survey ©UKRI. All rights reserved)

Investigation on the bone fragments recovered from the Drayton North Cursus by Ayres and Powell (in: Barclay et al. 2003, p. 159) identified a total of 248 bone fragments which were recovered from the east cursus ditch. Around half of these were identifiable, the majority (100) being domestic cattle sized. Although in smaller numbers, the west cursus ditch also provided similar results, with a predominance of cattle bones (20) in the identifiable material.

Pastoralists with their domestic cattle would appear to have been required to travel in an easterly direction as the herd moved along the lee of the Chiltern Hills. This route runs parallel to both the later Wilts & Berks Canal and the current railway line. The herd would potentially have passed to the north of Chain Hill (OS Grid Ref SU 405871), continuing in that direction until it was forced to change direction to the north-east due to restrictions around Steventon Hill (OS Grid Ref SU 474910). From this point, the herd could move on to the floodplain, following the alignment of both the Drayton South Cursus and the Drayton North Cursus. This seems to indicate that herds of domestic cattle would have travelled along the valley floor in accordance with George et al.'s (2007, p. 5) recommendations for the movement of cattle.

Extreme levels of both extreme event and first influx fluvial flooding, together with Barclay et al.'s (2003, p. 9) identification of at least four causeways in the western ditch and at least three causeways in the eastern ditch of the Drayton North Cursus, appear to support Harding's (1999, p. 31) theory that the monument was acting as some form of 'barrier', implying that some form of control was being asserted to sideways movement across the landscape, perhaps controlling access to the spring meadows.

Drayton South Cursus

In the Thames Valley to the south-east of Drayton in Oxfordshire, the Drayton South Cursus (OS Grid Ref SU 485935 to SU 489941) lies on the second gravel terrace of the River Thames. Initially identified by Major Allen in 1933 from cropmarks on aerial photography, the cursus monument was extensively excavated in 1934 by Leeds, who had been excavating a Saxon village in the area between 1921 and 1937 prior to gravel extraction. The cursus monument was excavated again in 1994 prior to the construction of a highways depot. The cursus monument, which lies between the Mill Brook and the Thames Valley floodplain, measures 750 m in length by 70 m in width. It is orientated north-north-east by south-south-west and is of the type classified by Loveday (1985) as Bi. The site appears to be part of a larger monument complex, as a large ring ditch lies 400 m from the cursus monument's southern terminal, while another ring ditch lies 10 m from the south-western corner of the monument's southern terminal and a small ring ditch lies close to the eastern cursus ditch. The cursus monument's northern terminal had been destroyed during gravel extraction prior to Leeds' investigations.

Evidence for early human activity within the area included a high proportion of Mesolithic flintwork found on the ground surface during excavation of the cursus monument, which has led Holgate (1986, p. 12) to suggest that 'a seasonal occupation occurred at this location', perhaps for the hunting of wild cattle, identifying that the landscape had been suitable for the grazing of cattle for millennia. A significant factor with regard to the Drayton Cursus complex could be the recovery of twenty-two Early Neolithic arrowheads. This would support Trantalidou and Masseti's (2015, p. 72) suggestion that this could potentially indicate some form of trophy hunting occurred post monument construction at the site.

It appears that pastoralists with their domestic cattle would have been required to travel in an easterly direction as the herd moved along the lee of the Chiltern Hills. This route is parallel to both the later Wilts & Berks Canal and the current railway line. The herd would potentially have passed to the north of Chain Hill (OS Grid Ref SU 405871), continuing in that direction until it was forced to change direction to the north-east due to restrictions around Steventon Hill (OS Grid Ref SU 474910). From this point, the herd could move on to the floodplain, following the alignment of both the Drayton South Cursus and the Drayton North Cursus. This seems to indicate that herds of domestic cattle would have travelled in accordance with George et al.'s (2007, p. 5) recommendations for the movement of cattle. This would again suggest that the Drayton South Cursus potentially commenced life as a droveway, again perhaps identifying an initial practical function of the landscape prior to any possible later ritual importance.

Extreme levels of both extreme event and first influx fluvial flooding, together with Barclay et al.'s (2003, p. 9) identification of at least one causeway in the eastern ditch of the Drayton South Cursus, appear to support Harding's (1999, p. 31) theory that the monument was acting as some form of 'barrier', implying that some form of control was being asserted over sideways movement across the landscape, perhaps controlling access to the spring meadows.

Lechlade Cursus

At Lechlade, in the Thames Valley, the Lechlade Cursus (OS Grid Ref SP 212002 to 212005) lies upon the second gravel terrace of the floodplain along a north-north-west by south-south-east orientation. The southern terminal of the cursus monument appears to have been situated at an oblique angle about 850 m from the River Thames, while the northern terminal ended approximately 1.5 km from the River Leach. The cursus monument was initially identified from cropmarks on aerial photographs in 1944 (N Riley). It was later excavated by Vatcher in 1965, who dug three trenches across the south-western ditch, finding a number of postholes in the inner ditch edge, and by Moore, who excavated approximately 600 sq m in 1985. The cursus monument appears to have been 300 m in length by 45 m in width and is of the type classified by Loveday (1985) as Bi. The site appears to be part of a larger monument complex. Seven ring ditches, including one that appeared as a triple ditch hengiform, are situated 280 m to the south-west. There is currently no evidence for earlier human activity within the area. Investigations by Ayres and Powell (in: Barclay et al. 2003, p. 207) of the animal bone recovered from the Lechlade Cursus identified a total of 148 fragments, the majority unidentifiable. Most of the bone was recovered from the upper fills of the east cursus ditch and, of the identifiable fragments, the majority (28) appear to have been cattle sized.

It appears that pastoralists with their domestic cattle would have been required to travel in a north-easterly direction along the Thames Valley. This route appears to have been initially restricted between Horcott Hill (OS Grid Ref SP 155002) and a steep ridge to the east of the River Coln (OS Grid Ref SP149023), which would have forced the herd to change direction on to a more easterly route. From here, the route continues parallel to the River Coln, through an area that is currently the Cotswold Water Park. This was also the previous route of a dismantled railway line. The herd would once again have been forced to change direction, this time to the south-east to pass between the high ground of Little Lem Hill (OS Grid Ref SP 208012) and Bury Hill (OS Grid Ref SU 233969). This would have required the herd to move towards the restricted floodplain of the river confluence in a direction that was parallel to the cursus monument.

This seems to indicate that herds of domestic cattle would have been required to travel along the valley floor in accordance with George et al.'s (2007, p. 5) recommendations for the movement of cattle, perhaps suggesting that the Lechlade Cursus commenced life as a droveway. Extreme levels of both extreme event and first influx fluvial flooding could support Harding's (1999, p. 31) theory that the monument was acting as some form of 'barrier', implying that some form of control was being asserted to sideways movement across the landscape, perhaps controlling access to the spring meadows.

North Stoke Cursus

At North Stoke, in the Thames Valley, the North Stoke Cursus (OS Grid Ref SU 611856 to SU 611858) lies along a north-south orientation. The cursus monument was initially identified from cropmarks on aerial photography by Major Allen in 1933. The southern terminal lies about 500 m from the River Thames, to the south of a springline that flows from the lower slopes of Wicks Hill (OS Grid Ref SU 614863). The cursus monument, which had a central gravel mound, is 240 m in length by 20 m in width and has squared, type B terminals. The cursus monument appears to have later become part of a larger complex, as numerous ring ditches surround the monument, mainly to the south and west between the monument and the River Thames.

The North Stoke Cursus lies on the second gravel terrace of the River Thames floodplain between the river and the steep escarpments of Wicks Hill (OS Grid Ref SU 614863), White Hill (OS Grid Ref SU 609845) and Watch Folley (OS Grid Ref SU 620846). It is noticeable that at this point the Ridgeway diverts away from the banks of the River Thames to move along higher ground, potentially to circumnavigate a springline. It appears that pastoralists with their domestic cattle would have entered this section of the Thames Valley from a southerly direction, where it seems to have been initially restricted between the steep-sided valleys on both sides of the River Thames at the Goring Gap (OS Grid Ref SU 595805). From this point, the herd would have followed the ancient route of the Ridgeway along the valley bottom until it diverted towards higher ground away from the River Thames in the area around the South Stoke Cursus. After the South Stoke Cursus, the route once again runs adjacent to the banks of the River Thames until being restricted for a second time just to the south of the North Stoke Cursus. At this point, the route diverts to run parallel to the cursus monument between the river and White Hill. This seems to indicate that herds of domestic cattle would have travelled in accordance with George et al.'s (2007, p. 5) recommendations for the movement of cattle. This could suggest that the North Stoke Cursus commenced life as a droveway, again perhaps identifying an initial practical function of the landscape prior to any possible later ritual importance.

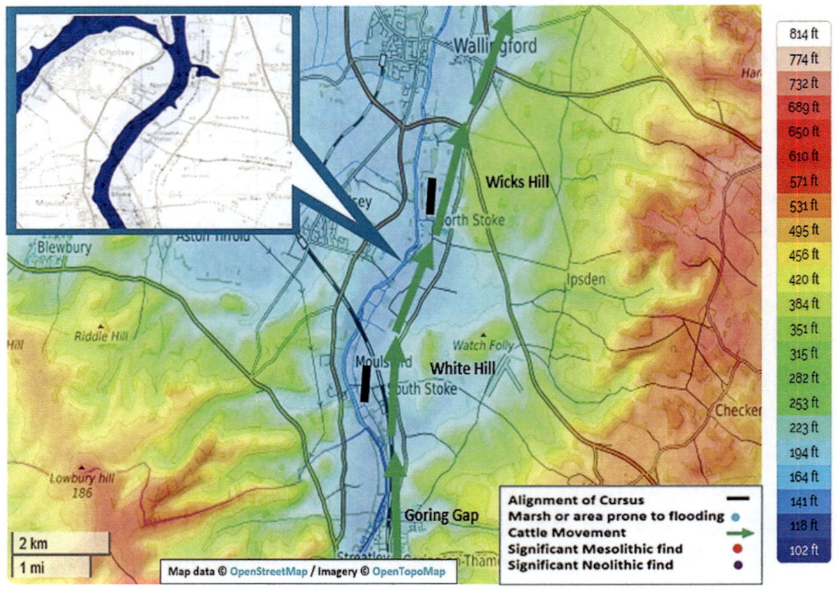

Map 24: Cattle movement across the North & South Stoke Cursuses (Main map data created using © OpenStreetMap / Imagery © OpenTopoMap. Inset reproduced with the permission of the British Geological Survey ©UKRI. All rights reserved)

Significant levels of both extreme event and first influx fluvial flooding, together with Loveday's (2006, p. 94) identification of at least two possible causeways in the side ditch of the southern terminal of the North Stoke Cursus, appear to support Harding's (1999, p. 31) theory that the monument was acting as some form of 'barrier', implying that some form of control was being asserted to sideways movement across the landscape, perhaps controlling access to the spring meadows. The significant easterly expansion of the first influx fluvial flooding at the northern terminal of the North Stoke Cursus potentially suggests that this, in combination with the monument, completed enclosure of the meadows, perhaps acting as some form of cattle corral.

South Stoke Cursus

At South Stoke, in the Thames Valley, the South Stoke Cursus (OS Grid Ref SU 595831) lies on a north-south orientation upon the second gravel terrace of the floodplain. The cursus monument was initially identified from cropmarks on aerial photographs by Major Allen. The northern terminal lies about 50 m from the River Thames adjacent to a spur within the steep side of the valley. The cursus monument appears to have been 240 m in length by 30 m in width and has rounded, type A terminals.

The South Stoke Cursus lies at a point where the Ridgeway diverts away from the banks of the River Thames to move along higher ground, potentially to circumnavigate the floodplain. It appears that the route for pastoralists with their cattle entering this section of the Thames Valley from a southerly direction would have been initially restricted between the steep-sided valleys on both sides of the River Thames at the Goring Gap (OS Grid Ref SU 595805). From this point, it appears the herd would have followed the route of the Ridgeway along the valley bottom until it diverted away from the River Thames adjacent to the later location of the South Stoke Cursus which lay between the river and White Hill (OS Grid Ref SU 609845). This seems to indicate that herds of domestic cattle would have travelled along the valley floor in accordance with George et al.'s (2007, p. 5) recommendations for the movement of cattle.

Extreme levels of both extreme event and first influx fluvial flooding appear to support Harding's (1999, p. 31) theory that that the monument was acting as some form of 'barrier', implying that some form of control was being asserted to sideways movement across the landscape, perhaps controlling access to the spring meadows, while a slight easterly expansion of the first influx fluvial flooding at the southern terminal of the South Stoke Cursus could suggest that this completed the enclosure of the meadow, perhaps acting as some form of cattle corral.

Sonning Cursus

At Reading, in the Thames Valley, the Sonning Cursus (OS Grid Ref SU 767760) lies on an east-west orientation upon the second gravel terrace of the floodplain. The cursus monument was initially identified by cropmarks on aerial photographs by St Joseph in 1959. The western terminal lies about 250 m from the River Thames, approximately 500 m south of a confluence where the St Patrick's Stream joins the River Thames to form an island that dissects the corner of the river. The cursus monument is 250 m in length by 45 m in width and has terminals classified by Loveday (1985) as type Bi.

The Sonning Cursus lies on the second gravel terrace of the Thames floodplain approximately 1 km north-east of a confluence between the River Kennet and the River Thames. It appears that pastoralists with their domestic cattle would have been required to travel in an easterly direction as they moved along the Thames Valley, being restricted between the river and the high ridgeline that ran adjacent to the northern bank of the River Thames. In the area around Sonning Eye (OS Grid Ref SU 751761) the herd would have been further restricted by the springline of Pool Spring (OS Grid Ref SU 746767) which would have forced it to change to a north-easterly direction before moving directly east to pass to the south of Borough Marsh, adjacent to the later cursus monument (OS Grid Ref SU 771779). This would have resulted in the herd moving on to the floodplain following the alignment of the Sonning Cursus. This seems to indicate that herds of domestic cattle would have travelled in accordance with George et al.'s (2007, p. 5) recommendations for the movement of cattle, perhaps suggesting that the Sonning Cursus again commenced life as a droveway.

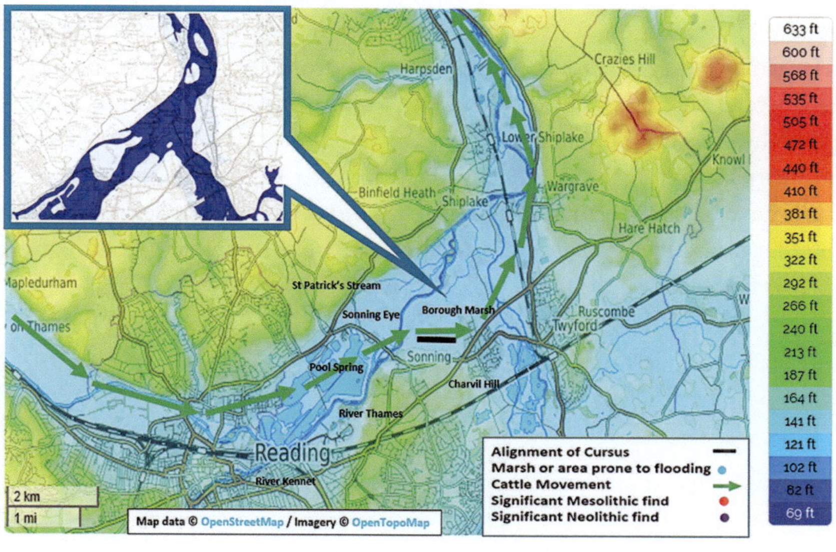

Map 25: Cattle movement across the Sonning Cursus (Main map data created using © OpenStreetMap / Imagery © OpenTopoMap. Inset reproduced with the permission of the British Geological Survey ©UKRI. All rights reserved)

Significant levels of both extreme event and first influx fluvial flooding, together with Ford's (1987) identification of the site being markedly rectangular with an entrance gap, possibly a causeway, at its far eastern end, would appear to support Harding's (1999, p. 31) theory that the monument was acting as some form of 'barrier', implying that some form of control was being asserted to sideways movement across the landscape, perhaps controlling access to the spring meadows, while the placement of the monument between the River Thames, the St Patrick's Stream and Borough Marsh potentially suggests that the droveway led cattle to water.

Stonehenge Greater Cursus

Orientated upon an east-west alignment, the Stonehenge Greater Cursus (OS Grid Ref SU109429 to SU 137431) lies approximately 1 km north of the earlier Mesolithic postholes (Vatcher and Vatcher 1973) and the later stone circle and 500 m south-east of the potentially earlier Stonehenge Lesser Cursus (Richards 1990). Initially described in 1723 by William Stukeley, who believed it to have been an old Roman racecourse for charioteers, the Stonehenge Greater Cursus was the first cursus monument to be discovered. Although Stukeley traced the full 2.7 km of its course and initially appeared to appreciate the square-ended enclosure form, by the time he published his findings (1740) he had falsified his drawings to coincide with his Roman chariot racetrack theory.

The Stonehenge Greater Cursus, together with its banks and ditches, between 100 and 130 m wide, traverses the Stonehenge Plain between the Fargo Plantation on Winterbourne Stoke Down in the west to the Amesbury 42 long barrow on the King Barrow Ridge in the east. Along its route, it crosses the intervening north-south valley of Stonehenge Bottom. Although this valley is currently dry and has been depicted as such since Ordnance Survey mapping from the 1880s, transposing the varying water table height identified during excavations within the Wilsford Shaft (Ashbee et al. 1989) on to the ordnance datum height along the entire length of the Stonehenge Bottom valley floor identifies that the valley potentially acted as a winterbourne river joining the River Avon at Lake during the Neolithic period (Saunders 2015 unpublished), especially during the period around 3650–3600 BC when, as Crane (2016, p. 96) suggests, 'a drop in solar intensity produced colder and stormier winters'.

It should be noted that the upper arc of the Stonehenge Bottom river valley, to the north of the Stonehenge Greater Cursus, is the precise point used for the disposal of rubbish in the 1920s for clearance of the local military airfield at Stonehenge. This means that the present shape and profile of the landscape no longer represent its original natural form. However, investigation using an 1850s Ordnance Survey map indicates that the original valley profile was identical to that within the adjacent section of Stonehenge Bottom. It would also appear that this section of Stonehenge Bottom which transverses the Greater Stonehenge Cursus, although not flat, contains the gentlest slope gradient along the entire section of the Stonehenge Bottom valley. Use of the same 1850s map highlights the fact that wells have been constructed upon high ground next to this section of the valley, potentially indicating that the valley itself would probably have been subject to some degree of flooding or marshland, a factor that field observation still identifies today.

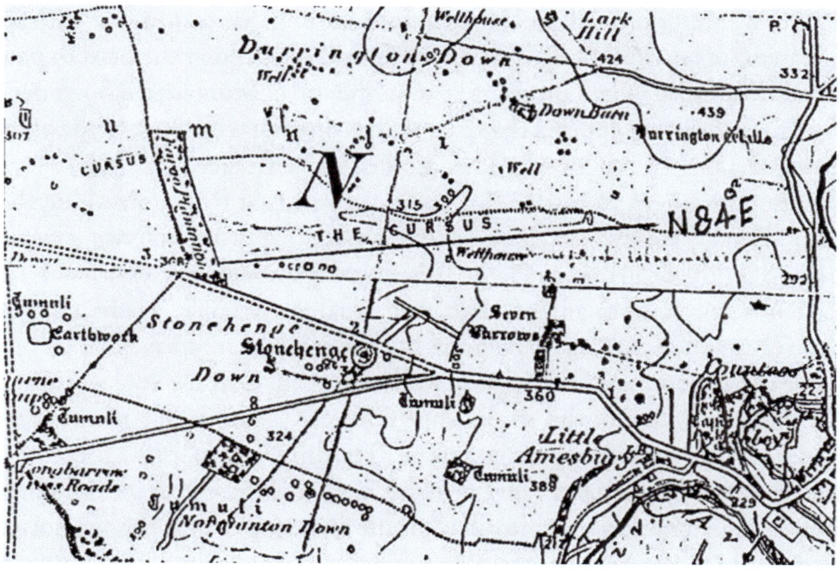

Map 26: Stonehenge Bottom Ordnance Survey map of 1850 (Reproduced with kind permission of Ordnance Survey)

Much has been said regarding the potential alignment of the Stonehenge Greater Cursus, such as by Stone (1947, p. 18) who, while undertaking excavations at the cursus monument, noted that the axis, if projected 1.5 km to the east, would strike Woodhenge, passing the Cuckoo Stone en route. This idea was endorsed by archaeologists Hedges and Buckley (1981), while Loveday (2006, p. 136) suggested that 'the alignment potentially goes much further, aligning on the lower northernmost prominence of Beacon Hill, 8 km away'.

Lidar and Ordnance Survey maps of the wider Stonehenge landscape identify that the east-west path across the Stonehenge area appears to have been cut by a series of steep valleys which were formed during previous glacial periods. At the end of the last Ice Age, although the actual glaciers did not reach Stonehenge, meltwater from the retreating ice sheets caused torrential fast-flowing post-glacial rivers that carved out the river valleys seen today. Research undertaken by numerous eco-archaeologists such as French, Allen, Scaife and Green (in French et al. 2012) has identified that, although the central area of the Stonehenge environment appears to have mainly consisted of open landscape, the slopes away from this section were probably still covered by woodland during the Early/Middle Neolithic period. This would potentially have resulted in pastoral movement of cattle being restricted by the natural topography of the landscape. Nick Branch (2019, personal communication, January) has indicated that the final part of closed woodland to be removed from the Stonehenge Plain appears to have lain between the Lesser and Greater Cursus monuments, perhaps being the reason behind the construction of two monuments in such close vicinity to each other.

Further research undertaken by Bowden et al. (2015, p. 65) using a GIS software package to highlight slope gradients in excess of 40 degrees strengthens the theory put forward by Saunders (2019, pp. 173–200) that only three west-east routes were available for cattle movement across the plain to the River Avon. The first route passed through a narrow gap just to the north of Durrington Walls. The second route moved south along Stonehenge Bottom before following the later avenue to the River Avon at West Amesbury. And the third route followed the landscape's natural downward curve from the eastern end of the Stonehenge Greater Cursus to a small gap where Countess Roundabout would later be placed. Investigations across the wider Stonehenge landscape has enabled Saunders (2019, p. 185) to identify that, apart from the three small gaps identified above, a 16-mile north-south section of the landscape running from Salisbury to Upavon was potentially impassable for cattle movement.

The primary route for pastoralists with their domestic cattle to travel into the Stonehenge area appears to have been from the south-west, entering the plain at the Fore Down enclosure (Grid Ref SU 082423). Their journey would have continued in a north-easterly direction between the valleys of Fore Down and Winterbourne Stoke Down before turning due east in line with where the later Stonehenge Greater Cursus would be constructed. This would require the herd to pass to the south of the post-glacial river valley which splits into two at the top of Stonehenge Bottom and the area of Durrington Down. It therefore appears that pastoralists moving with their herds of domestic cattle would have travelled along the valley profile in accordance with George et al.'s (2007, p. 5) recommendations for the movement of cattle. This could suggest that this route along the Stonehenge Greater Cursus potentially commenced life as a droveway, perhaps to circumnavigate the final section of closed woodland on the Stonehenge Plain. This would again perhaps identify an initial practical function of the landscape prior to any possible later ritual importance, where it appears that herds of cattle were traversing the Stonehenge plain on their way to water at the River Avon in the region of where the later construction of Durrington Walls occurred. Gaffney et al.'s (2012) identification of causeways in the side ditches of the Stonehenge Greater Cursus during the Stonehenge Hidden Landscapes Project could initially appear to support Harding's (1999, p. 31) theory that the monument was acting as some form of 'barrier' to the marshlands to the north of the monument. However, this would require the assumption that the valley profile at the top of Stonehenge Bottom held water.

A secondary route for pastoralists travelling with their domestic cattle along the Stonehenge Greater Cursus appears to occur at the mid-point of the monument. At this point, the herd appears to have had the opportunity to move in a southerly direction along Stonehenge Bottom, eventually joining the Stonehenge Avenue to continue to the River Avon at West Amesbury. This would again suggest that pastoralists would have been moving their herds of domestic cattle in accordance with George et al.'s (2007, p. 5) recommendations, perhaps on their way to water at the River Avon in the region of West Amesbury.

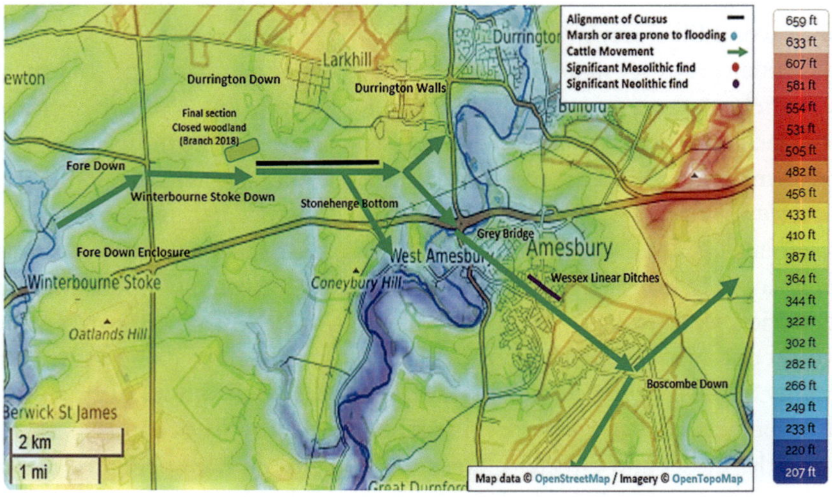

Map 27: Cattle movement across the Stonehenge Greater Cursus (Main map data created using © OpenStreetMap / Imagery © OpenTopoMap. Inset reproduced with the permission of the British Geological Survey ©UKRI. All rights reserved)

The third route would have been used by pastoralists travelling with their domestic cattle across the Stonehenge Plain into the wider Boscombe Down area. After entering the area in the same way as the other two routes, the third route follows the landscape's natural downward curve from the eastern end of the Stonehenge Greater Cursus to a small gap at what would later become Countess Roundabout, before crossing the River Avon at the Grey Bridge. The herd would have continued upon open grassland (Wessex Archaeology 2004) until finally splitting at Boscombe Down, the site of a Mesolithic posthole, perhaps indicating that the routeway had been used for cattle movement (wild and domestic) for millennia. It is also of note that this route would take the herd past an area that incorporated a pair of Wessex linear ditches.

Raymond et al. (1994) suggest that 'understanding for linear ditches appears to have advanced little in recent years'. This resulted in them setting up a team to simultaneously investigate both the environmental and the artefactual subjects as an integrated aspect of material culture associated with Wessex linear ditch systems. Through this they were able to identify that the function of the linear ditches system seems to have changed with regard to demography and social interaction as the character and perception of the landscape changed over time. This in turn appears to have affected both the subsistence productivity and the scale of domestic production of the immediate landscape. Darvill (2013, p. 13), in a research framework for the wider Stonehenge landscape, identifies a watching brief that was carried out during the construction of a new accommodation block and the laying of a new fire hydrant system at Boscombe Down Airfield in 2008, the latter revealing a section of this Wessex linear ditch. The Wessex linear ditch system traverses the airfield landscape, appearing on several historical aerial photographs, as recently as 1917 as an earthwork and later, at least until 1940, as a cropmark. It would therefore appear that the third route includes a landscape feature that could have assisted the movement of cattle.

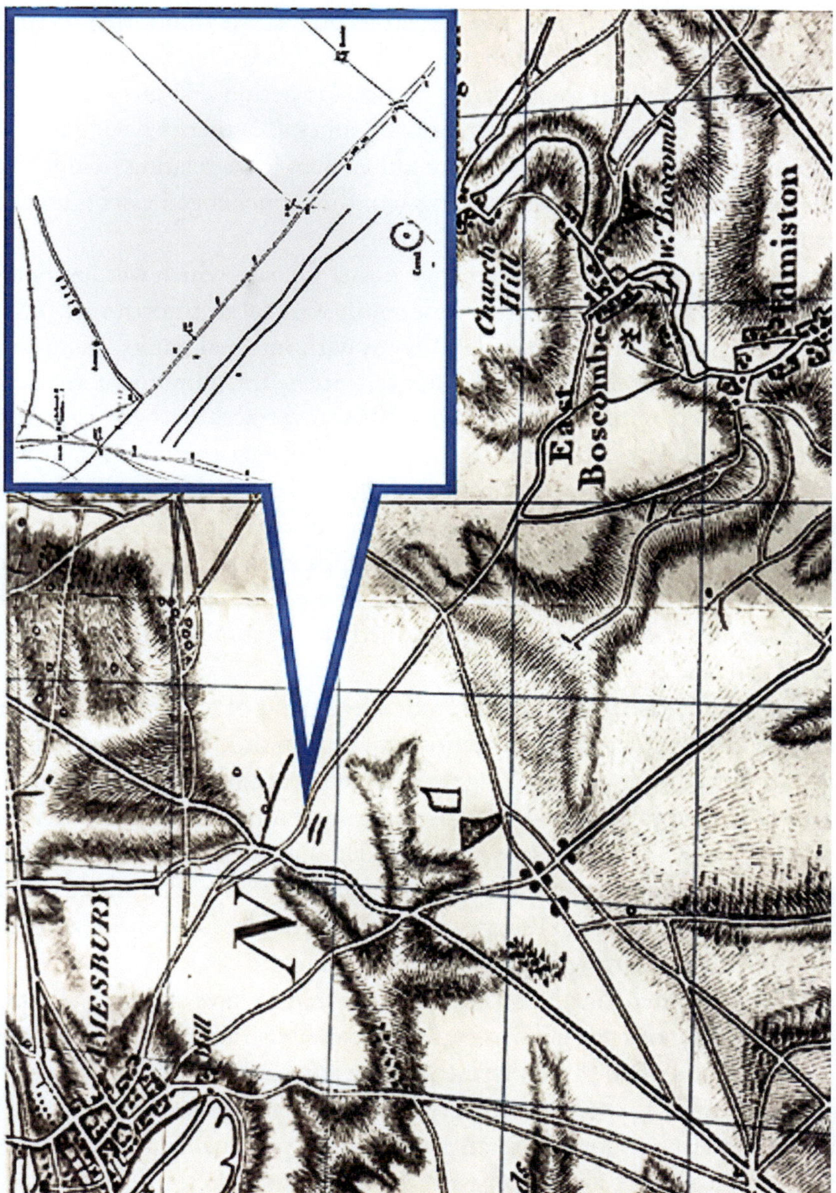

Map 28: 1811–1817 Historical map of Salisbury and the Plain with inset of Digimap Historic Roam map from 1880 identifying potential Wessex linear ditches (Produced with kind permission of Ordnance Survey)

All three routes would therefore seem to indicate that herds of domestic cattle would have travelled in accordance with George et al.'s (2007, p. 5) recommendations for the movement of cattle. This could suggest that the Stonehenge Greater Cursus potentially commenced life as a droveway, perhaps again identifying an initial practical function of the landscape prior to any possible later ritual importance. Two of the routes appear to have led the herd directly to water, while the third route enabled the herd to circumnavigate the wider Stonehenge Plain, perhaps passing yet another parallel ditch structure en route.

Stonehenge Lesser Cursus

Richard Colt Hoare (1812, pp. 157–58) was the first to recognize that the Stonehenge Lesser Cursus lies along the summit of a flat ridge top, approximately 600 m to the north-west of the western terminal of the Stonehenge Greater Cursus. As with Stukeley's (1740) earlier classification of the Stonehenge Greater Cursus, Colt Hoare initially gave the cursus monument a Roman origin. However, drawing evidence from the numerous prehistoric tombs that scattered the area of Salisbury Plain allowed Fergusson (1872) to place both the Stonehenge Greater and the Stonehenge Lesser Cursus monuments within their correct prehistoric period. He put forward that they were possibly dug to mark out battle lines from prehistoric times, this theory having been influenced by the notion of prehistoric barbarism and warfare rather than ideas relating to any cult of the dead. This resulted in him extending the supposed length of the Stonehenge Lesser Cursus to match that of the Stonehenge Greater Cursus.

The 400-m-long and 60-m-wide Stonehenge Lesser Cursus, which was levelled by ploughing between 1934 and 1954, runs approximately west-south-west by east-north-east (OS Grid Ref SU 103434 to SU 107435). It consists of parallel ditches with internal banks, the western end comprising a closed terminal while the eastern end appears open. The monument appears to have been effectively cut in half by a cross ditch which potentially shows traces of a bank to its eastern, exterior side. Excavation undertaken by Richards in 1983 identified that the monument appears to have been constructed in two phases, the intersection of the Phase 1 ditch being totally removed by the subsequent monument extension. Richards (1990, p. 72) notes that Phase 1 comprised of a slightly trapezoidal enclosure 200 m by 60 m, where the ditch appears to have been recut on more than one occasion and to have been deliberately back filled. In Phase 2 the earlier enclosure was remodelled by elongating the whole structure eastwards by a further 200 m, the extension consisting of only two parallel-sided ditches.

Richards (1990, p. 76) discovered heavily eroded fragments of red deer antler from the primary chalk rubble of the Phase 1 ditch, which appear to have been in situ. These gave a radiocarbon date of 3496–3042 BC, while further substantial fragments of red deer antler, including picks, found on the floor of the Phase 2 ditch appear to have been used during its original construction (Richards 1990, p. 77). Apart from these red deer antlers, which included both shed and unshed examples, very few bones were discovered. However, from a total of 178 fragments, 38 were classified as unidentified large mammal, which Richards (1990, p. 83) suggested were potentially from either aurochs or domestic cattle while a further 15 fragments were definitely from domestic cattle.

Until recently, the Stonehenge Lesser Cursus appeared to have stood in isolation within the Stonehenge landscape, previous research having failed to establish any direct link between the monument and the Robin Hood's Ball causewayed enclosure lying approximately 2 km due north of the cursus monument. However, the recent discovery of a causewayed enclosure at Larkhill (OS Grid Ref SU 141443) by Wessex Archaeology (Alistair Barclay 2016, personal communication, November), which appears to be on a direct alignment with the Stonehenge Lesser Cursus, could potentially transform our understanding of this important monumental landscape. This, together with Nick Branch's (2019, personal communication, January) suggestion that the last section of closed woodland within the Stonehenge landscape lay between the Lesser and Greater Cursus monuments, only finally being cleared by the Late Neolithic period, perhaps sets the foundations for an understanding of the different alignments of the two monuments.

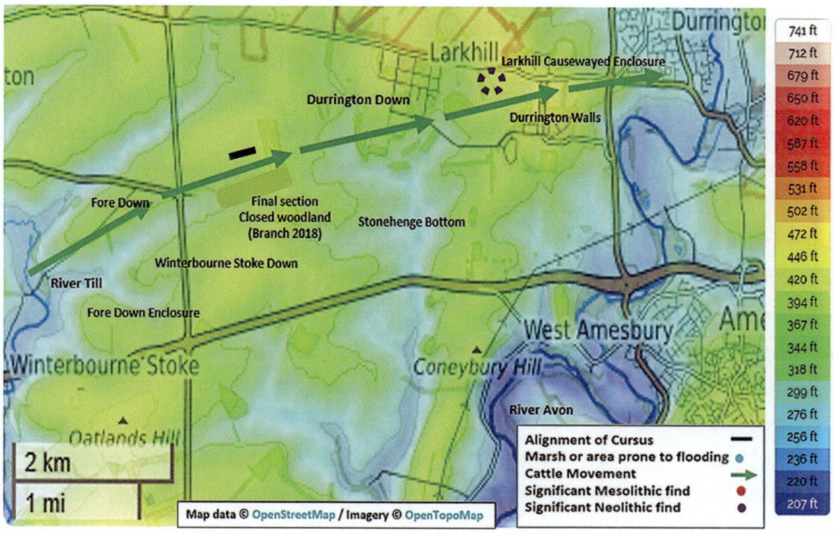

Map 29: Cattle movement across the Stonehenge Lesser Cursus (Main map data created using © OpenStreetMap / Imagery © OpenTopoMap. Inset reproduced with the permission of the British Geological Survey ©UKRI. All rights reserved)

The Stonehenge Lesser Cursus lies along the summit of a flat ridge top, which would have required herds of cattle approaching from the west to cross the River Till at the Fore Down enclosure (Grid Ref SU 082423). Their journey would have continued in a north-easterly direction to the area of the cursus monument on a slightly rising slope between the valleys of Fore Down and Winterbourne Stoke Down. As the herd continued through the landscape to the point where the Stonehenge Lesser Cursus was later constructed, it would have been forced to pass between Durrington Down and the post-glacial river valley that splits into two just north of Stonehenge Bottom. Although this area currently appears to consist of a gentle slope, this point of the valley was used for the disposal of rubbish from the clearance of local military installations at the end of the First World War. Investigation of an 1850s Ordnance Survey map highlights that the original profile of this section of the valley had steep sides which included numerous springs. This, together with Saunders' (2015, unpublished) investigations into the Wilsford Shaft and the theoretical resultant heights of the chalk aquifer, suggest that Stonehenge Bottom potentially became a marshland environment at certain times of the year.

It seems, therefore, that pastoralists moving with their herds of domestic cattle would have travelled along the valley profile in accordance with George et al.'s (2007, p. 5) recommendations for the movement of cattle. This could potentially suggest that the Stonehenge Lesser Cursus commenced life as a droveway, perhaps to circumnavigate the final section of closed woodland upon the Stonehenge Plain. It appears that herds of cattle would have traversed the Stonehenge plain past the recently discovered Larkhill causewayed enclosure on their way to water at the River Avon in the region of Durrington Walls. Richard's (1990) identification of causeways during excavations undertaken as part of the Stonehenge Environs Project appears to support Harding's (1999, p. 31) theory that the monument was acting as some form of 'barrier', implying some form of control was being asserted to sideways movement across the landscape.

Yatesbury Cursus (Avebury Cursus)

The Marlborough Downs form the western end of the Berkshire Downs, a large block of high chalk landscape approximately 50 km wide that runs from Reading in the east to the Bristol Valley in the west. They are characterized by a series of high, steep escarpments and gentler dip slopes, which consist of up to 150 m of Late Cretaceous chalk deposited between 72 and 65 million years ago. The characteristic steepness of the slopes was produced as a result of uplift caused by the ripple effect during the formation of the Alps about 50 million years ago. These Late Cretaceous chalk deposits can be subdivided into three distinct layers classified as lower, middle and upper chalk formations, which can be distinguished not only by the rock characteristics but by the fossil content, where new species appeared, and older ones died out with time (Brown et al. 2005, p. 58). The upper chalk can be characterized by the presence of flints, occurring mainly in layers parallel to the bedrock, while few flints are found in the middle chalk and there appears to be no evidence of flint within the lower chalk layer.

A significant feature of the geology of the Marlborough Downs is the numerous concentrations of naturally occurring sarsen stone found in areas to the east and to the south of Avebury, between the western edge of the Marlborough Down ridge and in the Avebury valley to the east of the River Winterbourne. These weathered stones, which appear to have littered the prehistoric landscape, are remnants of a sandstone layer formed approximately 50 million years ago that once overlaid the chalk of the downland regions. Although there is no evidence glacial ice ever reached the Marlborough Downs, inter-glacial thawing of the ice and the associated eroding of the chalk through meltwaters resulted in the sandstone fracturing into these irregular-shaped stones which were carried downhill into the region. These numerous concentrations of naturally occurring sarsen drifts would have significantly restricted cattle movement along the western edge of the Marlborough Down ridge and in the Avebury valley to the east of the River Winterbourne. By fitting cattle with GPS collars, Lyons and Machen (2001) have identified that they tend to move away from the presence of these areas of loose and imbedded rock.

An early aerial photograph of cropmarks in Yatesbury Field (Allen 1933), about 3 km to the west of Avebury, identified the potential ditches of a cursus monument on a slight ridge of high ground just to the west of an alignment between Knoll Down and Windmill Hill (OS Grid Ref SU 070705) (Pollard and Reynolds 2002, p. 70). The potential Yatesbury (Avebury) Cursus appears to have been situated approximately 500 m to the south-west of Windmill Hill at the confluence of two winterbourne rivers, the River Winterbourne and the River Oslip. Both rivers are still fed by numerous springs, the Oslip springs covering an extensive area of about eight acres. To the south of Avebury, the Galteemore springline feeds into the Beckhampton Stream, which forms braids as it crosses the Silbury meadow, expanding during winter to become a lake, while south-east of Silbury more winterbourne springs, the Swallowhead, the Pan and the Waden, flow into the River Kennet.

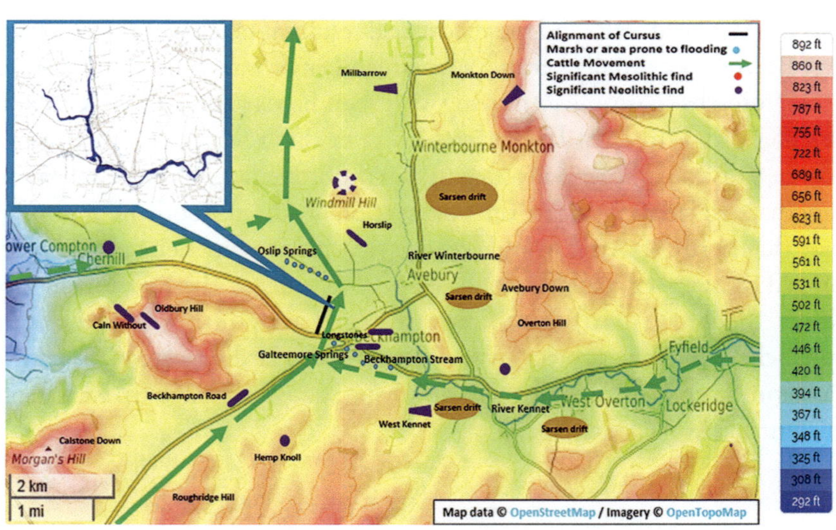

Map 30: Cattle movement across the Yatesbury (Avebury) Cursus (Main map data created using © OpenStreetMap / Imagery © OpenTopoMap. Inset reproduced with the permission of the British Geological Survey ©UKRI. All rights reserved)

The primary route for pastoralists with their domestic cattle to travel into the Avebury area appears to have been from the south-west, where they would have been significantly restricted on their western side by the sharp escarpment of Calstone Down and on their eastern side by Roughridge Hill. After this point, the natural topography of the landscape widens until the valley again becomes restrictive to cattle movement due to the River Kennet, the Beckhampton Stream and the Galteemore Springs. As the herd approached the Galteemore springline, there would have been only a 200-m gap through which it could have proceeded. Moving to the north of the Galteemore springline, the herd would have had to pass between the eastern edge of the Oslip Springs and the western side of Windmill Hill, their movement being further restricted by the springline of the River Winterbourne and the extensive sarsen drifts to the east of Avebury.

A secondary route for pastoralists with their domestic cattle travelling into the Avebury area appears to have been in a westerly direction along the Kennet Valley. Wainwright's (1960, p. 201) comparison of microlith industries from Somerset and Bodmin Moor to microlith industries in Essex and at Thatcham led him to conclude that they all had close affinities, suggesting communities travelled for millennia from the east coast, across southern England to Cornwall, possibly using the Kennet Valley as a routeway, providing access to the River Avon and River Severn in the west. Domestic cattle were perhaps driven along this same route, which runs to the south of the Galteemore springline before turning on to the primary route between the Galteemore and Oslip springlines. Once past this section, which runs parallel to the later potential Yatesbury (Avebury) Cursus, the herd could follow routes to the western side of Windmill Hill or continue into the Severn Valley via Cherhill.

A third route for pastoralists travelling with their domestic cattle into the Avebury area could have been from the Severn Valley past Cherhill on to the Avebury plain. All three routes would seem to indicate that herds of domestic cattle would have travelled in accordance with George et al.'s (2007, p. 5) recommendations for the movement of cattle. This could suggest that the Yatesbury (Avebury) Cursus potentially commenced life as a droveway, again perhaps identifying an initial practical function of the landscape prior to any possible later ritual importance.

Avebury is sited adjacent to a confluence of two winterbourne rivers, the Winterbourne and the Oslip. The spring source of the River Winterbourne lies at Berwick Bassett, about 3 km north of Avebury, where it flows through the villages of Winterbourne Bassett and Winterbourne Monkton before it is joined by the smaller, east-flowing River Oslip to the west of Avebury (Marshall 2016, p. 24). The River Oslip starts its journey approximately 6.5 km north-west of Avebury where it drains large expanses of flat, featureless farmland through Yatesbury until it becomes spring-fed as it reaches the area south of Windmill Hill. The Oslip springs cover a huge area of about 8 acres and may have been the main water supply for prehistoric communities feasting on Windmill Hill (Marshall 2016, p. 25).

Although levels for both extreme event and first influx fluvial flooding do not appear severe, this could perhaps be due to the fact that the confluence of the Winterbourne and Oslip, which appears to be the start of the River Kennet, was canalized in the sixteenth or seventeenth century, perhaps to create water meadows, where regular flooding still causes the river to revert to its natural prehistoric course. To the south of Avebury, the Beckhampton Stream flows eastwards towards Silbury Hill, forming braids as it crosses the meadow at the point where it would have joined the Old Kennet. Fed by the Galteemore springs, just to the east of Beckhampton, this would have resulted in a significant restriction point to any potential cattle movement across the area.

Dorset Cursus

The Dorset Cursus, on Cranborne Chase, is just under 10 km long. It runs roughly south-west by north-east between Thickthorn Down and Martin Down alongside the current A354, following a sinuous course across the chalk downland where it crosses the winterbourne valley of the south-easterly Allen stream in the region of Down Farm. The Dorset Cursus is, in fact, two cursus monuments laid end to end. The earlier cursus is the south-western Gussage portion (OS Grid Ref ST 969125 – SU 018160) which was constructed between 3360 and 3030 BC. It terminates at Bottlebush Down, while the later north-eastern Pentridge Cursus (OS Grid Ref SU 018160 – SU 040193) continues on a slightly different alignment, adjusting its course slightly before terminating on Martin Down. A series of naleds created towards the end of the glacial period due to water percolating deep within the chalk, resulting in the subsequent partial collapse of subterranean cave systems, produced a cratered landscape to the north and north-east of Down Farm. These naleds extend as far as the Pleistocene river cliff (OS Grid Ref SU 006147) (Green 2000, p. 59).

The Dorset Cursus appears to have been first observed by William Cunnington during the early part of the nineteenth century. He recorded it in his manuscript notes, which are preserved in the library of the Society of Antiquaries. After Cunnington's death, Sir Richard Colt Hoare (1812, pp. 157–58) later publicized this recognition of the Dorset Cursus in his second volume of Ancient Wiltshire, even though it lay outside the boundary of the county. However, later excavations by Atkinson (1955), Bradley (1986 & 1991) and the continuing investigations by Martin Green potentially make this one of the most studied areas in southern Britain, second only to Stonehenge.

Evans (1972, pp. 219 & 256) has identified that from the earliest Mesolithic Boreal period this section of Wiltshire chalk consisted primarily of open woodland showing evidence of manipulation through man's use of fire. Later, Michael Allen (2000, pp. 36–49) identified that the cursus monument appears to have been dug through a level of Mesolithic soil which has then eroded back into the ditch. He suggests that this was a landscape consisting of deciduous woodland together with small openings or clearings. Allen (2000, p. 42) also identifies that progressively larger clearances occurred within the woodland around 4350–4000 BC, potentially indicating one of the first recorded woodland clearances in southern England. This would support Legge's (in: companion volume of Barrett et al. 1991, p. 20) earlier investigations of the animal bone, which identifies that fauna from the Neolithic period appear to have been completely dominated by species such as cattle and pig. The Cranborne Chase valley bottom bends in a natural north-easterly curve to meet the River Avon, consisting of a wooded easterly ridge and a westerly escarpment. Allen (2000, pp. 36–49) identifies that this section of the landscape appears to have consisted of large sections of open landscape, ideal for the grazing of large herds of cattle.

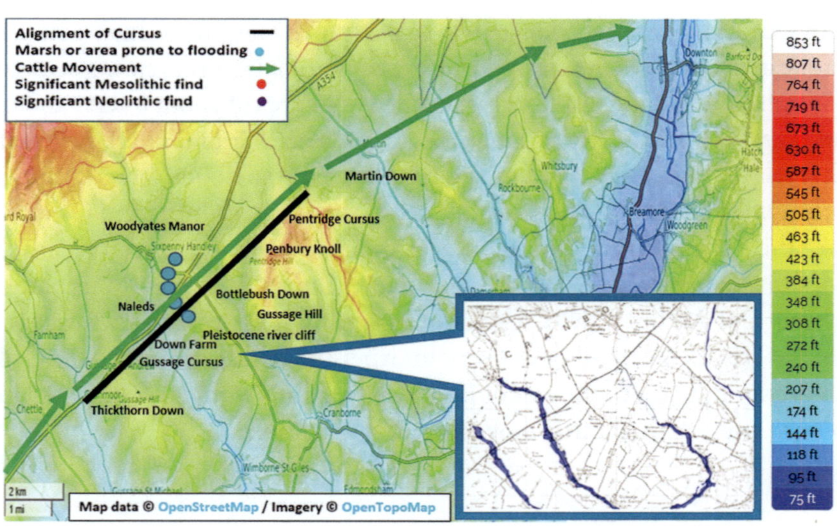

Map 31: Cattle movement across the Dorset Cursus (Main map data created using © OpenStreetMap / Imagery © OpenTopoMap. Inset reproduced with the permission of the British Geological Survey ©UKRI. All rights reserved)

It appears that pastoralists with their domestic cattle would have travelled into the Down Farm area from the south-west, where they would have been significantly restricted on their eastern side initially by Gussage Hill and later by the sharp escarpment of Penbury Knoll. Although significant narrowing of the valley appeared to occur in the area of Down Farm due to the Pleistocene river cliff and the series of naleds, it was not until Martin Green (2016, personal communication, June) explained the effect that winter floodwaters, extending from Woodyates Manor to the most westerly point of the naleds (Endless Pit), would have on north-easterly cattle movement that the landscape started to make sense. This seems to indicate that herds of domestic cattle would have travelled along the edge of the valley in accordance with George et al.'s (2007, p. 5) recommendations for the movement of cattle, potentially suggesting that the Dorset Cursus commenced life as a droveway, perhaps to circumnavigate the floodwaters around Down Farm.

The direction of both extreme event and first influx fluvial flooding, which appear to cut across the monument in three places, would suggest that the droveway led cattle to water, while Gill's (2019, p. 30) recent geophysics survey of the northern ditch during November 2018 has identified two small causeways, which would appear to support Harding's (1999, p. 31) theory that the monument was acting as some form of 'barrier', implying some form of control was being asserted to sideways movement across the landscape, perhaps controlling access to the spring meadows.

CHAPTER 6

Cursus Monuments: A Statistical Evaluation

Current Theories

Although it is not the intention of this book to disprove any current cursus monument theory put forward by others, it does seem evident that few previous interpretations appear to be immediately transferable across all other cursus monument sites. The majority of previous theories appear to have been based on a mere handful of sites, the results then being extrapolated across the rest without sufficient testing of the hypothesis being undertaken to determine just how well the theory actually works across multiple cursus monument sites. Therefore, to establish a datum for my theory within the realms of previous research, I believe it is important not only to place my own results within the context of these previous theories, but also to undertake a full statistical evaluation across each of the fifty individual cursus monuments found on or adjacent to the English chalkland belt that established my data set.

While I appreciate that the selection of this geographical area for the study group omitted several large cursus monument sites such as the Thornborough Cursus and the Scorton Cursus in North Yorkshire, the East Adderbury Cursus near Banbury and the Potlock Cursus in Derbyshire, by investigating each individual site rather than extrapolating data it became one of the largest studies of cursus monuments undertaken to date.

The combination of aerial survey, digital mapping and identification of slope gradients useful to the movement of cattle, together with research into floodplain and springline data and good old-fashioned boots-on-the-ground investigation enabled the study to determine precisely how each cursus monument interacts with its surrounding landscape. This enables us to establish that cursus monuments appear to have been sited within the sort of landscape that would permit the type of cattle movement identified by George et al.'s (2007, p. 5) investigations. The study ascertains that a combination of natural topography together with winter flooding of springs and rivers could have led to these points becoming prime cattle feeding grounds during the early spring, which would have enabled Neolithic pastoralists to retain larger herds over the winter period. Therefore, the placement and alignment of cursus monuments by later generations at these precise locations could possibly identify a correlation between the movement of Neolithic domestic cattle and the construction of Neolithic cursus monuments.

The study undertook a basic statistical analysis for each of the various components that make up the layout of each individual cursus monument and the functions that have been put forward for their potential uses. It used univariate analysis, examining each variable on an individual basis, highlighting the distribution factors as a percentage of the total study group, outlining the potential importance and ranking of each factor. To add weighting to each assumption identified through this methodology, calculations were undertaken using Pearson's correlation coefficient, thereby establishing the potential correlation between the various data sets and identifying the potential strength of each relationship. To coincide with the 5 per cent convention of statistical significance within archaeological hypothesis testing, the study used the significance level (alpha) of 0.05 establishing the confidence level at the conventional setting of 95 per cent. The study then undertook identical calculations from the data within my own fieldwork, thereby establishing precise measurements for each individual relationship, undertaking a comparison with previous research and hopefully starting to reduce the current 'enigma' status of cursus monument research.

The succession of cursus monument discoveries, almost all identified through cropmarks, together with the excavation of some of these sites, has enabled a gradual refinement of the cursus monument question. But while this appears to have developed a clearer awareness of their distribution, Brophy (2016, p. 21) questions whether 'this increase in data and the dataset did not necessarily increase understanding of cursus monuments in terms of their use, their chronology or their role in Neolithic society'. The extra data appears only to have maintained the status quo with what Loveday (2006, p. 11) described as the cursus 'enigma'. Prof Tim Darvill, of Bournemouth University, is one of the leading experts on Neolithic Britain and the editor of *The Concise Oxford Dictionary of Archaeology*, which states (Darvill 2008, p. 120):

> Cursus: a kind of Neolithic ceremonial monument that comprises of a rectangular enclosure which is defined by a bank with external ditches. Typically, cursus monuments are between 500 metres and 3,000 metres in length and up to 80 metres in width. Dating mainly to the third millennium BC, their purpose is unknown, although they are widely believed to have been ceremonial pathways that, when used for processions, structured the participants' vision of the surrounding landscape and the monuments within it.

While the dictionary could not be classified as a prime source of reference, I believe it is an excellent tool to highlight many of the problems associated with previous efforts to overcome what Loveday (2006, p. 11) describes as the cursus 'enigma'.

Design of Cursus Monument

One of the many problems encountered by the first archaeologists investigating cursus monuments was that of scale. Several of the largest cursus monuments extend for many kilometres, such as the Dorset Cursus on Cranbourne Chase, the Rudston D Cursus in Yorkshire and the Stonehenge Greater Cursus in Wiltshire, while others appear tiny in comparison, such as the Stanwell 4 Cursus which is only 82 m in length, although excavations have found that the majority of the Stanwell 4 Cursus may have been destroyed by the earlier construction of a large sludge lagoon in 1980, making it difficult to establish whether it is in fact rectangular in plan (Lewis et al. 2010). Although this raises the question, were monuments of such length really cursus monuments or some other kind of long mortuary enclosure, Loveday (2006, pp. 35–62) has suggested that appropriate divisions can be made within each of the various groups of cursus monuments based upon their mathematical length.

He has therefore categorized cursus monuments into four broad groupings:

- *Long mortuary enclosures which extend up to 150 m in length and 25–30 m in width.*
- *Minor cursuses which extend 180–800 m in length.*
- *Major cursuses which extend 1,000–2,000 m in length and achieve 40–100 m in width.*
- *Mega cursuses which range from 2,700 to 5,640 m in length and demand distinction from the former group.*

However, it should be noted that, while Loveday's (2006, pp. 35–62) initial categories covered the mathematical lengths of all cursus monuments at the time of his classification, the discovery of new cursus monuments has resulted in gaps within the definition. For example, the discovery of two cursus monuments at Clifton Reynes in Buckinghamshire, one of which measures 160 m, raises the question, is it a very long 'long mortuary enclosure' or a very short 'minor cursus'?

Further investigation into Loveday's (ibid., p. 26) broad groupings for cursus monument classification identifies one of the first problems encountered with the cursus enigma, this being that the groupings do not appear to fit within the typical length of cursus monuments as defined by Darvill (2008, p. 120), which states that 'typically cursus monuments are between 500 and 3,000 m in length and up to 80 m in width'. If these lengths are typical, it would perhaps be expected that the majority of cursus monuments would have been constructed between these lengths. However, while the majority appear to fall between the distances classified as major (26 per cent) and minor (46 per cent) cursus monuments within Loveday's (ibid., p. 26) classification, indicating that that 72 per cent of cursus monuments initially appear to fall within Loveday's broad groupings (2006, p. 26), when the monuments that are between 500 and 3,000 m in length are analysed, the figure falls to 34 per cent, which should probably no longer be defined as typical.

Brophy (2016, p. 6) believes there are no typical cursus monuments, suggesting that 'cursus monuments remain for many archaeologists, enigmatic, empty spaces, upon which many theories, not always convincing, have been thrust'. Our investigation highlights that the standard cursus monument length is potentially between 180 and 2,000 m. It also appears, when looking at the distribution of each major and minor cursus monument site, that the standard cursus monument is predominantly found outside the southern downlands of Wiltshire. Within Wiltshire, only the Stonehenge Lesser Cursus and the potential Yatesbury (Avebury) Cursus appear to fit the criteria described by Darvill (2008, p. 120) as 'typical'. The distribution pattern for the major or minor cursus monument sites which would meet the definition described as 'typical' by Darvill (2008, p. 120) appears to be evenly spread throughout the various regions included within the study – although obviously the southern downlands of Wiltshire would also include the mega cursus monuments of both the Stonehenge Greater Cursus and the Dorset Cursus which would not be described as 'typical' within Darvill's (2008, p. 120) dictionary definition. This would suggest that they are atypical, or perhaps, as Brophy (2016, p. 3) states, 'the weirdest type of field monument in the country'.

The distribution of mega cursus monument sites does, however, appear to follow a more, 'typical' pattern, where 83 per cent appear to have been situated upon the chalklands of either southern Wiltshire or the Yorkshire Wolds. The only exception to this appears to be the Stanwell C1 Cursus, which is situated upon the silty and sandy clay of the Eocene Thames Group (Harwich and London Clay fms). It is also noticeable that the Stanwell 1 Cursus is the only cursus monument that has a central mound rather than the normal ditches with internal banks (Lewis et al. 2010, p. 32). The distribution pattern for cursus monuments classified by Loveday (2006, p. 26) as long enclosures only appears to fall within a few regions of my study group – the pit alignments found within the Lincolnshire Wolds, the potential small upland cursus monument discovered by Gover (2000) by geophysical survey at the summit of Ivinghoe Beacon and the ditch enclosures found between Bedford and Milton Keynes.

Using Darvill's (2008, p. 120) dictionary definition to describe a typical cursus monument highlights that only 34 per cent of the study group could be defined as 'typical'. However, by using Loveday's classifications (2006, p. 26) of major and minor cursus monuments, this figure rises to 72 per cent. This perhaps identifies the importance of, firstly, asking the right question and, secondly, asking it in the correct way, as it appears the data can alter significantly, according to how the question is asked.

Terminals

Loveday (2006, p. 26) found that, even when he used his concept of categorising cursus monuments into groups based upon their mathematical length, a significant variation still appeared to occur across the various cursus monument groups. This resulted in Loveday (2006, pp. 28–31) proposing numerous subdivisions, based on the terminal form of the monuments, which appear to have differing degrees of curvature. Loveday (2006, p. 28) therefore defined the cursus monument terminals as rounded or squared (type A and B) and included further categories for distinctive subgroups which appear to have been laid out with geometric precision, which he termed Bi and Bii sites.

One difficulty with Loveday's terminal categorization is that it appears only to take into account the plan view of the ditches gained either through the identification of cropmarks or through the excavation of ditch segments, a view no Neolithic community would ever have seen. It does not appear to take into consideration the various styles of terminal bank that could have been constructed from this ditch fill. For example, were the terminal banks constructed in a high, thin manner to gain a maximum visual effect towards the ends of the monument, or was the ditch fill spread wider to ensure the monument bank height maintained the same level as the side banks?

Loveday himself (2006, pp. 23–24) states that the style of terminal 'carries no implication of date, cultural affiliation or function, simply being a record of observable difference, sufficiently marked to suggest a positive choice occurred on the part of the builders', while Brophy (2016, p. 28) believes that perhaps more problematic is that 'such typological schemes are based on an assumption that these monuments should be defined by the full extent of the traces that we find today'. Investigation of the study group indicates that 62 per cent of cursus monument sites appear to have terminals which could be categorized as squared (32 per cent) or Bi (geometric precision squared terminals) (30 per cent). Inclusion of the pit alignments would increase this number to 68 per cent, while only 20 per cent fall within the category that Loveday (2006, p. 28) terms as rounded. However, as I have not been able to determine the terminal type across 12 per cent of the monuments investigated due to destruction of the landscape upon which they were situated, therefore this figure could increase further.

But this does not appear to clarify the situation totally. For although it seems that the majority of cursus monument complexes such as the Rudston Cursus complex, the Wolverton Cursus complex and the mega cursus sites of Wiltshire consisted of squared terminals, at the Stanwell complex only the Stanwell Cursus 3 monument has squared terminals, the remainder of the Stanwell complex consisting of rounded terminals. It is also notable that the terminals of the Rudston C Cursus have not been discovered, having potentially been destroyed during the construction of the village.

Although it initially appears that the Bi-terminal cursus monument sites were mainly distributed throughout the Thames Valley, the Greater Ouse Valley and up into Cambridgeshire, a number of cursus monument sites within both the Thames Valley and the Great Ouse Valley also contain monuments that were constructed using rounded terminals, some of which are part of the same monument complex as those sites that consisted of squared or Bi terminals. It therefore appears that the terminal type is not dependent upon the location of the monument. However, it should be noted that a number of cursus monument sites evenly distributed throughout the study were situated upon landscapes that have been recently destroyed, preventing identification of the style of terminal.

Straightness of Ditches

Investigation into the style of their ditches does not appear to provide any clarification of the enigmatic nature of cursus monuments. While it is certainly true that the majority of cursus monuments had two ditches, each with an internal bank, the Scorton Cursus, the Cleaven Dyke and the Stanwell 1 Cursus only appear to have had a single central mound. Although analysis of both cropmarks and excavation plans within this investigation indicates that 74 per cent of cursus monument sites appear to have straight ditches, closer inspection shows approximately 24 per cent of the ditches to be decidedly irregular. Anomalies are found at the Dorset Cursus in Wiltshire, the Rudston A Cursus in Yorkshire, the Drayton Cursus in Oxfordshire, the Springfield Cursus in Essex, the Fornham All Saints Cursus in Suffolk, the cursus monument complex at Stanwell in the London Borough of Hillingdon and the cursus monument complex at Wolverton in Buckinghamshire, all of which appear to have ditches which included distinctive sinuous or angular sections.

And it appears to have been not only the straightness of the ditch profile that varied across the different cursus monument sites. Atkinson (1955, p. 9) noted that 'the ditches on different sides of individual cursus monuments also altered'. At the Dorset Cursus, for example, one side of the ditch appears inferior to the other, with one side being prominent and straight while the opposite side appears to follow an irregular course. Loveday (2006, p. 120) suggests that 'interestingly, when an earlier monument is incorporated, it often appears that the straighter ditch is on the other side of the cursus'. McOmish (2003, p. 12) has suggested that 'the ditch imbalance was intentional, where one side was deliberately built on a more monumental scale'. This could suggest that the side ditches of a cursus monument may have acted in a similar way to the first and second-line trench systems developed during the First World War – the second-line trench normally being constructed in a more prominent manner, on the basis that any initial attack would be on the first-line trench, which would be expected to be overrun, while the second-line barrier would hold firm. Could a similar thought process have been used millennia earlier to prevent cattle movement across the terrain where monuments were constructed? It also appears that not only did the ditches on opposite sides of the monument vary but, as Julian Richards (1990, p. 80) discovered when he undertook excavations of several trenches at both the Greater and Lesser Stonehenge Cursus monuments, 'the cursus ditch profile, width and depth all varied as he moved along the monument'.

Pryor's (1982) excavation of the Maxey Cursus near Peterborough and Richard's (1990) excavation of the Stonehenge Cursus showed that they both appeared to have been built in stages and backfilled virtually straight after construction. This has led Pryor (1998, p. 364) to suggest that the Maxey Cursus appeared more like a project in progress rather than a structure built to a predetermined plan. He (ibid., p. 364) believes that people may have visited the site on a seasonal basis, adding new bits to the structure year after year, indicating that the actual construction activity appeared to be more important than the actual cursus architecture.

The distribution of cursus monuments that contained sinuous cursus ditches appears to be evenly spread throughout the area of the study, with examples occurring in Yorkshire, the Great Ouse Valley, the Fenlands, the Thames Valley and the Wiltshire chalklands, although there does appear to have been a cluster of cursus monuments with sinuous ditches around the Eynesbury, Godmanchester and Brampton area. Only three cursus monuments within the study appear to have been constructed using what Loveday (2006, pp. 117–18) describes as angular cursus ditches. These appear to have occurred at random locations across the study group, where two of the monuments, the Rudston D Cursus and the Wolverton 3 Cursus, are part of cursus monument complexes. Therefore, from a statistical point of view there does not appear to be the evidence to enable me to make any supposition with regards to the location of cursus monuments which were constructed with angular ditches.

A Typical Cursus Monument

Statistical analysis of the fifty cursus monuments constructed on or adjacent to the English chalkland belt appears to indicate that the typical cursus monument would have been constructed as a minor cursus having an overall length between 180 and 800 m. It would have had squared terminals and straight ditches. However, only three cursus monument sites within this study group, which are not part of cursus monument complexes that include other categories, actually meet these criteria. They are the Hanworth Cursus in Norfolk, the Stratford St Mary Cursus in Suffolk and the North Stoke Cursus in Oxfordshire. Pearson's correlation coefficient identifies that the composition of the typical cursus monument design contains only a negligible positive relationship. It therefore appears that there is probably no such thing as a typical cursus monument. Although cursus monuments do appear to follow a basic design, the actual layout of the individual monument was potentially of secondary importance and therefore perhaps the prime element in cursus monument design was more to do with its function.

The Cursus Monument as a Processional Pathway

Stone (1947, p. 18) appears to have been the first person to put forward the theory that cursus monuments could have acted as processional pathways, when he applied Sir Norman Lockyer's (1906, p. 311) suggestion that 'as with Bronze Age stone rows on Dartmoor, the Stonehenge Greater Cursus could have been used as a processional way'. Atkinson (1955, p. 9) later appeared to combine Stone's theory with that of the original Roman theory put forward by Stukeley (1740, p. 4), when he outlined the Roman festival of Lupercalia to suggest that 'the Dorset Cursus may have had a religious or ceremonial rather than domestic function'.

Although others, such as Tilley (1994, p. 198) and Bradley (1991, pp. 209–19), have attempted to attribute processional meaning to these enclosures, using phenomenology to interpret the Dorset Cursus as a path along which rites of passage could be made tangible, leading Tilley to suggest that 'it was bodily movement along the monument which made it meaningful to the participants' (ibid., p. 198), it does appear that Atkinson's (1955, p. 9) original theory was supported primarily through the lack of archaeological evidence.

Thomas (1999, p. 48) suggests that 'the development of Neolithic monumental architecture can be interpreted as demonstrating increasing efforts to regulate the ways in which particular spaces were moved through, and thus experienced'. Thomas (ibid., p. 42) believes that the initial area upon which these monuments were constructed consisted of ancient pathways through which originally wild cattle and then later people and their domestic cattle moved. He (ibid., p. 52) further suggests that 'the construction of monuments in earlier Neolithic Britain introduced discontinuity into the landscape, by establishing boundaries around secluded and differentiated places' which constrained movement towards a specified location, where activities took place associated with the deposition of the dead.

McOmish (2003, p. 13) further suggests that 'an association with artefacts such as the arrowheads discovered within the Amesbury 42 long barrow at the eastern terminal of the Stonehenge Greater Cursus, might imply that the procession consisted of some form of trophy hunting where archery was part of the test, as such sites could have served as proving grounds for young men'. It is interesting to note that Neolithic arrowheads have been found within the actual archaeological context of 50 per cent of cursus monument sites within my study group, a figure which would increase to 75 per cent if those arrowheads found in the surrounding area were to be included and would further increase to 93 per cent with the inclusion of the mixed flint industry of Neolithic date found throughout the Wolverton Cursus complex.

Table 5: Neolithic arrowheads found at cursus monuments sites suggesting hunting continues (See Map 3: distribution of cursus monuments within my study group)

Id No	Cursus monument Site	Neolithic arrowhead type
1	Rudston A Cursus	Arrowhead at South Side Mound
4	Rudston D Cursus	16 leaf-shaped arrowheads from North Burton
6	Fimber Cursus	Arrowheads from Towthorpe
7	Kirby Underdale Cursus	4 leaf arrowheads in Kirby Underdale area
9	Harlaxton	Leaf and barbed and tanged arrowheads were found in a spring area on high ground at Harlaxton
15	Eynesbury Cursus	Barbed and tanged arrowhead
18	Maxey Cursus	Transverse and barbed and tanged arrowheads from enclosure ditch tertiary fills
21	Stanwell Cursus 1	Barbed and tanged arrowheads in Cursus 1 ditch
22	Stanwell Cursus 2	Arrowhead junction with C1 Cursus
23	Stanwell Cursus 3	Arrowhead in western ditch
30–34	Wolverton Cursus 1–5	Oblique arrowhead found in hengiform and leaf arrowhead found in another context
35	Benson Cursus	Barbed and tanged arrowheads found at Benson Hill
36	Dorchester Cursus	Wessex-type arrowhead & lozenge-type arrowhead from upper ditch fill
37	Drayton St Leonard	Lozenge-type arrowhead
38	Drayton North Cursus	22 arrowheads from Cursus ditch fill
39	Drayton South Cursus	Chisel arrowhead within Cursus ditch
41	Lechlade Cursus	Leaf-shaped arrowhead upper ditch fill
42	North Stoke Cursus	Flint arrowhead from ditch fill
46	Stonehenge Greater Cursus	Numerous arrowheads found in Amesbury 42
47	Stonehenge Lesser Cursus	Numerous worked flint from the plough soil or upper colluvial ditch fills
49	Gussage Dorset Cursus	Numerous arrowheads at terminal ends
50	Pentridge Dorset Cursus	Numerous arrowheads at terminal ends

Others, such as Johnston (1999, pp. 39–48), however, hold a different view. Johnson believes that the conditions required to meet the concept of a cursus monument being generally ceremonial in nature have not been met and that the construction of the monument actually appears to infringe upon any processional movement. He suggests (ibid., p. 46) that 'the construction of the monument created a boundary that passed the landscape into the more sacred realms of the ancestors at a time when human processions no longer took place'. This is supported by Parker Pearson and Ramilisonina (1998) who indicate that, if the route was meant only for ancestors, the actual cursus construction may have stopped people accessing the area. Brophy (2016, p. 30) also questions the theory that cursus monuments were used as processional pathways, when he suggests that 'it may be that cursus monuments were not physically suited for the control and definition required for focused ceremonial procession', and this is supported by Loveday's (2006, p. 126) earlier work suggesting that 'Major cursuses seem too wide for processions and minor ones too short'. McOmish (2003) notes that the almost obsessively straight nature of cursus monuments does give encouragement to the idea that any movement along the cursus monument occurred from a start point to a finishing point while the apparent visual focus upon one end of the monument potentially suggests that movement occurred in one direction only.

Although the use of numerical statistics within my methodology does not appear to support the theory that cursus monuments were constructed as ceremonial processional pathways, perhaps this element of the cursus debate cannot be proved using this type of methodology. Bradley (1991, pp. 209–19) and Tilley (1994, pp. 143–201) have endeavoured to overcome this type of problem through the use of phenomenology, attempting to understand how ritual participation may have occurred within the interior of the Dorset Cursus. Rather than use a statistical yes/no scenario, they have attempted to identify potential relationships to the cursus monument's distinctive architecture, associated burial monuments and immediate topography. However, despite Bradley and Tilley both separately walking the entire length of the Dorset Cursus recording spatial relationships that potentially offered evidence of a structured procession through the enclosure, their interpretations remain problematic, as it is evident that these interpretations are not immediately applicable to other sites, although this obviously does not invalidate their methodology.

Alignment with Earlier Monuments

Pennick and Devereux (1989, p. 51), in their book *Lines on the Landscape*, were some of the first to investigate cursus monument alignment in association with other aspects of the surrounding landscape. They were unable to investigate all known cursus monuments but believed their sample size was large enough to represent a preliminary study, which included site visits to seventeen monuments. Although much of their research was to identify cursus monument alignment with later features, such as churches and abbeys, they were able to identify cursus monument alignment with earlier features such as long barrows at the Stonehenge Greater Cursus (ibid., p. 52) and the Dorset Cursus (ibid., p. 56–57) together with an alignment with a causewayed enclosure at the Maxey Cursus (ibid., pp. 70–71) as well as an alignment with the distant Beacon Hill at the Stonehenge Greater Cursus (ibid., p. 52). I intend to use the data from my own fieldwork to examine their findings.

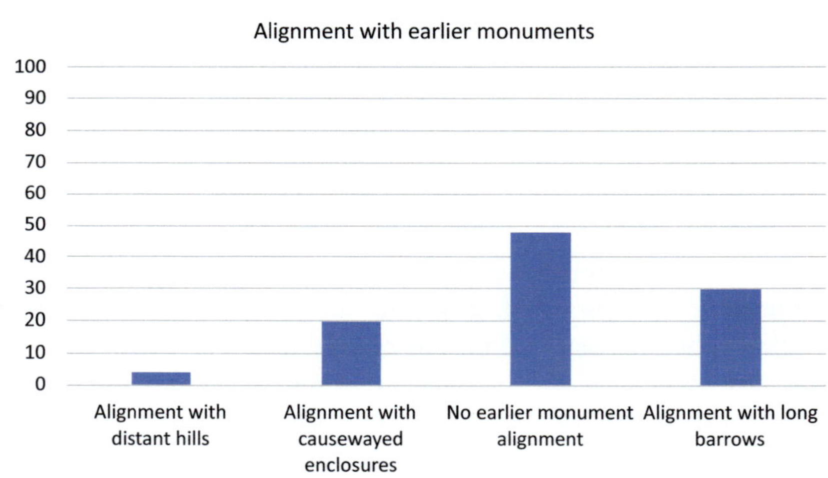

Figure 6: Cursus monument alignment with other aspects within the landscape

Alignment with Long Barrows

Pearson and Field (2011, p. 36) suggest that 'cursus monuments potentially linked earlier monuments and routes together', where they indicated a plausible link with the Greater Stonehenge Cursus and the earlier Amesbury 42 long barrow. They suggest that, 'the fact that the cursus monument terminated just short of the long barrow could potentially have indicated a deep measure of respect for the earlier monument' (ibid., p. 36). This would be supported by Parker-Pearson's (2012, p. 145) suggestion that 'the two Stonehenge Cursus monuments might have been constructed as a pair, with the Greater Cursus leading east to the Amesbury 42 tomb and the Lesser Cursus leading west to its own tomb'. The Amesbury 42 long barrow survives as a virtually ploughed-out earthwork, 20 m east of the Stonehenge Greater Cursus monument's eastern terminal. The 20 m-wide bank and 11 m-wide ditches were excavated by Thurnam in 1866, who discovered the skeletons of two infants and a crouch adult inhumation which appeared to be secondary internments and animal bones including what he identified as an ox skull, probably domestic.

Tilley's (1994, p. 175) earlier use of phenomenology across the Dorset Cursus appears to encourage this emphasis on the placement of cursus monuments in alignment with long barrows, where he believes that 'the symbolic importance of the Dorset Cursus terminal is marked out in the landscape through its relationship with four barrows and its intervisibility with a number of others that lie to the east' (ibid., p. 178).

However, Loveday (2006, p. 118) counteracts the alignment with long barrows argument by pointing out that 'although there is evidence for the use of earlier long barrows for siting purpose, such as at the southern extremity of the Gussage arm of the Dorset Cursus where the monument ran straight to the long barrow on Gussage Hill, the dog-leg correction to the cursus alignment on the northern side of the hill does not appear to have reflected drift caused by the constructers having lost sight of the barrow in the lee of the hill, as the cursus monument starts to drift away from the long barrow before the barrow disappears from sight'.

Although Loveday (ibid., p. 118) identifies that other cursus monuments appear to have been aligned directly upon earlier monuments, such as the Dorchester-on-Thames Cursus with the Site VIII long enclosure and the Stadhampton Cursus upon a trapeziform enclosure, he suggests that 'very few cursus monuments actually appear to link existing monuments' (ibid., p. 126). Brophy (2016, p. 139) also suggests that 'although similar arguments can be made for the sites in Scotland, there are not as many spatial relationships between cursus monuments and other rectangular structures as one might expect'. Brophy (2016, p. 139) suggested that 'in part this may be because shorter structures may have been converted into cursus monuments'. The data from this study appears to highlight that only approximately 30 per cent of cursus monuments had a potential alignment with earlier long barrows. Statistical analysis using Pearson's correlation coefficient identifies that, while there is a moderate positive relationship between the alignment of cursus monuments and the earlier placement of long barrows, the confidence interval of this relationship does not appear to be significant. Therefore, it appears there is no evidence to support a correlation between long barrow construction and later cursus monument alignment, although it is interesting to note that this potential alignment of cursus monuments with earlier long barrows does appear to be evenly distributed throughout the study area.

Alignment with Causewayed Enclosures

Whittle et al.'s (2011) chronological sequence identified that the four causewayed enclosures (Hambledon Hill, Maiden Castle, Whitesheet Hill and Robin Hood's Ball) did not exist prior to c. 3700 *cal* BC, thereby establishing an earlier construction than for the cursus monuments within this study group. Although there have been many explanations for the function of causewayed enclosures, ranging from them being used as enclosed settlements (Oswald et al. 2001, pp. 120–32) or cattle kraals (Piggott 1954, pp. 29–30) to exchange centres (Whittle et al. 1999, p. 354) and necropolis or cult centres (Pryor 1998, pp. 363–71), all have tended to agree with Thomas (1999, p. 38) who suggests that 'they existed at the edge of inhabited areas'. Whittle (et al. 2011, p. 11) believes this suggests 'established routeways already existed prior to the causewayed enclosure construction', a proposal which supports the earlier work by Bradley (1978, p. 103), Gardiner (1984, p. 21) and Evans et al. (1988, pp. 73–84).

Examination of Neolithic soils beneath the long barrows at Beckhampton Road, West Kennet, and Wayland's Smithy and of Neolithic soils beneath the Windmill Hill causewayed enclosure has led Evans (1967 & 1972) to suggest that these provide representative examples for the wider landscape. But this has been challenged by Allen and Gardiner (2009, pp. 49–66) who suggest that 'there is significant variation regarding the presence of Neolithic woodland at a number of these locations'. They believe that 'the main landscape environment of the chalkland regions of Wiltshire and Dorset consisted of open post-glacial woodland, which only developed into denser woodland in a few places'. This had been earlier indicated by French et al.'s (1992, p. 166) investigations of soil profiles beneath the Neolithic causewayed enclosure and cursus monuments at Etton and Maxey in the lower Welland Valley, which enabled him to identify that 'much of the monument landscape had been cleared of woodland prior to the construction of the Etton causewayed enclosure'.

This has led Tilley (2010, pp. 42–44) to suggest that 'Neolithic monuments such as causewayed enclosures and cursus monuments became part of a changed perspective within the visual landscape'. However, this could only happen if the overall visual perception of the landscape had been extended and widened through some form of forest clearance. Did this result in cursus monument alignment with earlier causewayed enclosures? Probably not. Thomas (1999, p. 74) indicates that causewayed enclosures stretched throughout north-western Europe, from Poland to Ireland. Yet with the exception of a few examples in the Rhine and Mosel Valleys in Germany, cursus monuments appear to have been exclusive to the British Isles and Ireland. If there was any correlation between these two features, we might reasonably expect cursus monument construction to have extended over a far greater geographical area than it appears to have done. However, although the relationship appears to be too rare to potentially have any great significance, the question arises as to why the same suite of monuments should exist everywhere.

This appears to be supported by the statistical data from my study group. It appears to be evident that only about 20 per cent of cursus monuments had any potential alignment with earlier causewayed enclosures. While this figure obviously represents only a fifth of the cursus monuments investigated, it is interesting to note that there appears to be an even distribution of cursus monument alignment with earlier causewayed enclosures throughout the study area.

Alignment with Distant Hills

Pennick and Devereux (1989, p. 55), in their book *Lines on the Landscape*, identify that 'extension of the eastern alignment of the north bank of the Stonehenge Greater Cursus embraces both the Cuckoo Stone and the central area of Woodhenge'. Although these features were later than the cursus monument, a further extension of this alignment to a few miles east of Woodhenge brought into view Beacon Hill and its ridgeline of long barrows. They claim the ridgeline was highly visible from Woodhenge, although it disappears from view as you move westwards along the cursus monument, only reappearing as the west end of the monument is approached.

Although the Greater Stonehenge Cursus does appear to have been aligned with the lower northern prominence of Beacon Hill and the Dorset Cursus appears to have roughly aligned with Penbury Knoll, Loveday (2006, p. 136) believes that 'it is rare indeed for cursuses to point to significant landscape features'. He suggests that no other natural target appears to exist across other chalkland sites. Brophy (2016, p. 166) supports this argument, suggesting that although 'the Cleaven Dyke does seem to generally point towards a low, rounded hill, Benachally, to the north-west and Northballo Hill on the more distant south-east horizon' and the 'Broich aligns on the Knock of Crieff to the north', it was very clear, when looking at the alignment of Scottish cursus monuments that there is no overall trend identifying any recurring association. Although Barclay and Maxwell (1998, p. 117) suggest that 'the Cleaven Dyke appears to align with the Herald Hill long barrow', other examples are few and far between.

The statistical data taken from the fifty cursus monuments that made up the study group appears to support both Loveday (2006, p. 136) and Brophy (2016, p. 166) in that only 4 per cent of cursus monuments seem to have any potential alignment with regard to distant hills. Therefore, from a statistical point of view, there does not appear to be the evidence to support the theory that cursus monuments were constructed in an alignment with distant hills.

Alignment with Celestial Events

Loveday (2006, p. 132) questions whether those constructing cursus monuments were aligning the monuments with a greater cosmological concept, suggesting that, without the knowledge as to which was the 'business end' of the monument it is impossible to determine which of the opposing alignments would have been considered important. The fact that cursus monuments were overwhelmingly constructed upon flat gravel terrace landscapes would tend to lead to unimpressive skyline features. This could mean that the use of celestial alignments as a basis for explanation could be deceptive.

Lockyer (1906, p. 311) appears to have been the first person to suggest a celestial alignment with regard to cursus monuments when he indicated that 'the Stonehenge Greater Cursus aligned towards the rising Pleiades', while Barrett et al. (1991, p. 50) later argued that if one was to stand at the correct location within the Dorset Cursus and look to the west, 'the midwinter sun sets over a skylined long barrow situated within the cursus'. Brophy, however, (2016, p. 33) points out that 'the fact cursus monuments are linear will inevitably mean that they point towards something, such as astronomical events'.

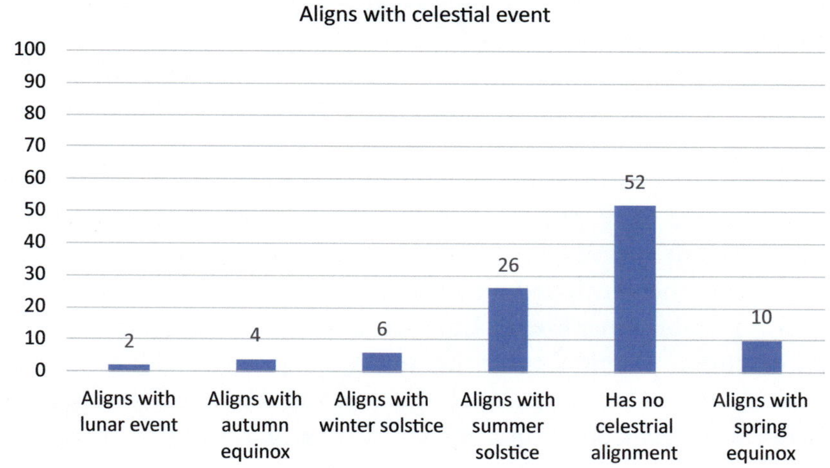

Figure 7: Cursus monument alignment with celestial events

Nonetheless, as the nature of this research is primarily to ascertain whether a combination of natural topography and winter flooding created spring meadows resulting in the areas becoming prime feeding grounds for cattle during the early spring, I believe it is important not only to concentrate on the summer and winter solstices but also to investigate both the spring and autumn equinoxes, to establish what percentage of cursus monuments within the study align with these celestial events.

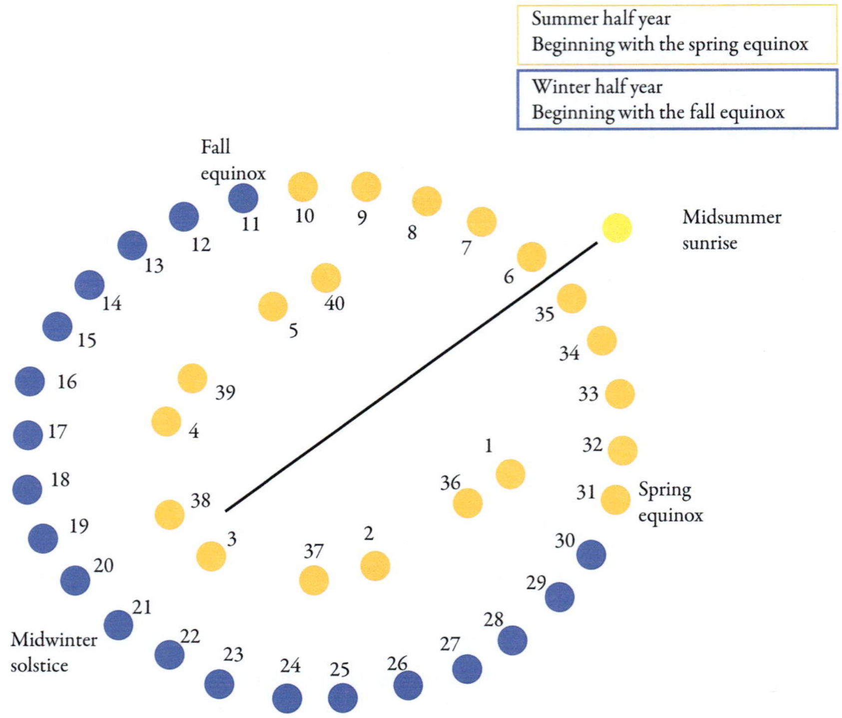

Figure 8: Direction of seasonal equinox at Stonehenge

Although the initial impression is that 40 per cent of cursus monuments seem to have some form of celestial alignment, it is apparent that this can take many forms. Although various cursus monuments appear to align with the summer (26 per cent) and winter solstice (6 per cent) and others appear to align with the spring (10 per cent) and autumn equinox (4 per cent), the question must be, are any of these alignments significant? To enable this study to evaluate the potential significance of cursus monument celestial alignment, it has undertaken statistical analysis across each aspect using Pearson's correlation coefficient. This initially appears to identify only a moderate positive relationship between the alignment of cursus monuments and any individual celestial event, a factor supported by the confidence intervals of these relationships, which do not appear to be significant.

Before dismissing a theory of cursus monument alignment with celestial events, where previous investigation appears to have been solely the result of extrapolation of data from only a few monument sites, this study examined each individual cursus monument for possible alignments with any celestial event. These included the summer and winter solstices, the spring and autumn equinoxes as well as possible lunar events. Loveday (2006, p. 137) suggests that 'caution is therefore vital' when looking at potential cursus monument alignments with celestial events. Even if you constrain the rising and setting points of the sun and moon, there appear to be countless stars within the northern hemisphere upon which cursus monument alignment could potentially have been based. Yet when Penny and Wood (1973) set out the case for the Dorset Cursus having been a Neolithic astronomical observatory, only two out of their six proposed alignments were actually contained within the monument itself, those being from the Bottlebush terminal to the skylined long barrow on Gussage Hill which appeared to align with the midwinter sunset and from the long barrow to the south-western terminal on Thickthorn Down which aligned with the minimum midwinter moonset. To achieve any kind of accuracy with this latter alignment required the authors to suggest that a bonfire had been placed at the terminal because it is not actually on the skyline.

Alignment with Summer Solstice

Claims for solar alignment have also been made for the Greater Stonehenge Cursus, however Loveday (2006, p. 137) again suggests 'real caution is needed since the monument is clearly not laid out on a true east-west course'. Loveday (ibid., p. 138) believes that 'the Dorchester-on-Thames Cursus appears more accurately related to a solar event' due to its alignment with the midsummer sunset from one direction and the midwinter sunrise from the other. However, Loveday (ibid., p. 138) advises that the fact that solar alignment does not appear to have occurred along the entire length of either the Dorset Cursus or the Dorchester-on-Thames Cursus could suggest that 'solar alignment was secondary'.

The data from this study would initially appear to support both Loveday (2006, pp. 137–38) and Brophy (2016, p. 33) in that approximately half (52 per cent) of cursus monuments do not appear to have had any alignment regarding celestial events. Although research into the correlation between cursus monument alignment and the summer solstice obtained from the use of Pearson's correlation coefficient identifies a moderate positive relationship, the confidence interval of this relationship does not appear to be significant.

Alignment with Winter Solstice

While Penny and Wood (1973, p. 47) set out the case for the alignment of the Gussage section of the Dorset Cursus with the midwinter sunset between the Bottlebush terminal and the skylined long barrow on Gussage Hill, the only other cursus monument within the study that would appear to have a similar alignment would be the Hanworth Cursus in Norfolk, although the Dorchester-on-Thames Cursus in Oxfordshire would appear to have an alignment with the midwinter sunrise. Therefore, from a statistical point of view there does not appear to be enough evidence to support the theory that cursus monuments were constructed in an alignment with the winter solstice.

Alignment with Spring Equinox

There does not appear to have been any previous research undertaken regarding cursus monument alignment with the spring equinox. However, the fact that my research potentially identified that the combination of natural topography and winter flooding which created spring meadows could have led to a concentration of cattle at these points during the early spring made it important to investigate potential spring equinox alignments. However, only 10 per cent of the cursus monuments within the study group could be identified as having any possible alignment with the spring equinox. It should be noted, however, that this did include a single monument from within each of the Rudston, Wolverton and Stanwell Cursus complexes.

Pearson's correlation coefficient identifies only a weak positive relationship between the possible correlation of cursus monuments and their potential alignment with the spring equinox, the time period cattle would be leaving their winter forest cover to feed upon the open grassland (Rogers et al. 2018, p. 142), a relationship where the confidence interval does not appear to be significant. It therefore appears that there is little evidence to support a correlation between cursus monument alignment and the spring equinox.

Alignment with Autumn Equinox

Within the study group, only the Etton and Biggleswade cursus monuments could potentially be identified as having any alignment with the autumn equinox. Therefore, from a statistical point of view there does not appear to be the evidence to support the theory that cursus monuments were constructed upon an alignment with the autumn equinox.

Alignment with Lunar Events

Brophy (2016, p. 33) has suggested that cursus monument alignment to lunar events are 'still less common' than with other celestial alignments. Although Ruggles (1999, p. 128) suggested that 'the Godmanchester Cursus appears to have had an alignment with some minor lunar events', Brophy (ibid., p. 33) believes that these are potentially 'coincidences, that were recognized only after monument construction'.

Although the Godmanchester Cursus is the only cursus monument within the study group that would appear to have had any form of possible alignment with a lunar event, Loveday (2006, p.140) does put forward an argument that groups of north-south oriented cursus monuments may have had a possible association with orbiting constellations. However, from a statistical point of view there does not appear to be the evidence to support the theory that cursus monuments were constructed in an alignment with any celestial event.

The fact that just over half (52 per cent) of cursus monuments have no celestial alignment with any seasonal equinox and the fact that only 10 per cent of these monuments have any potential alignment with the spring equinox, when Rogers et al. (2018, p. 142) suggest that cattle left the forest cover to feed on the early spring grasses, indicate there is not the evidence to support a theory that cursus monuments were constructed in alignment with the seasonal solstices.

Alignment with Rivers and Streams

The potential connection between landscape features and cursus monuments appears to have initially been postulated in *A Matter of Time* (RCHME 1960), when it was noted that cursus monuments shared a close proximity to rivers. Therefore, it was suggested that processions along these monuments led to the river, which would seem to indicate a close association between cursus monuments, rivers and streams. Brophy's (1999) original thoughts during his earlier PhD investigation work were that 'as rivers could potentially be seen as both life-giving and dangerous places, the cursus monument could be seen as a symbolic river built in response to this paradox of nature'. However, in his later book (Brophy 2016, p. 159) this changes to one where the spatial connection seems to be with the overall landscape upon which the rivers and streams flow, suggesting that many cursus monuments had actually been planned with topography in mind. Brophy (2016, p. 159) also believes the fact the majority of cursus monuments appear to have occupied low-lying positions running alongside or across river valleys and streams, potentially results in an apparent linking of these watercourses.

Loveday (2006, p. 133) also suggested that 'the juxtaposition of cursuses and rivers is nonetheless striking'. He believes that 'most cursus monuments either appear to have run along gravel terraces parallel to rivers, to terminate near streams, or to cluster close to confluences'. However, he (ibid., p. 133) believes that 'this did not reflect some idealistic intent, as cursus monument construction away from the chalklands would rarely have given Neolithic builders any option but to construct them parallel to the river'. Brophy, however, (2018, personal communication) believes 'this may reflect a presumption that there are no upland cursus monuments, a fact now being challenged in Scotland'.

While the mega chalkland cursus monuments, such as the Dorset Cursus and the Stonehenge Greater Cursus in Wiltshire and the Rudston Cursus complex in the Yorkshire Wolds, appear to follow the natural contours of the surrounding landscape, and the Kirby Underdale Cursus appears to rise steeply uphill following a natural gap that gives access to the highland region of the Yorkshire Wolds, Loveday (2006, p. 133) believes that, when investigating cursus monuments away from the chalkland belt, the apparent prerequisite to construct them across flat land would leave no 'option but to lay them out parallel to the river' in order to avoid the terrace step edge. It should, however, be noted that the rivers at several cursus monuments, such as the Stanwell Cursus complex, the Biggleswade Cursus, the Drayton St Leonard Cursus, the Drayton South Cursus, the Stadhampton Cursus and the potential Yatesbury Cursus all appear to run at or across the end of the monument. He (ibid., p. 133) did, however, warn that 'small streams dissect terraces and would create natural boundaries, while the meandering of larger water courses could furnish a false impression of deliberate alignment'. However, as far as the reasoning for the placement of upland cursus monuments is concerned, additional variations would be required in order to fall in line with Loveday's (ibid., p. 133) suggestions. While the statistical data taken from my study group appears initially to support Loveday's (2006, pp. 133–36) argument, in that almost all (94 per cent) of the cursus monuments appear to have an immediate association with rivers and streams, the variability of the direction of alignment between cursus monuments, rivers and streams raises further questions. Brophy (2016, pp. 159–60) also indicates that this connection between cursus monuments and rivers appears evident in Scotland. He (ibid., p. 159) believes that in some cases the association appears explicit, with the river being more than just near to the cursus monument as 'there are several instances of cursus monuments terminating adjacent to or overlooking the rivers'.

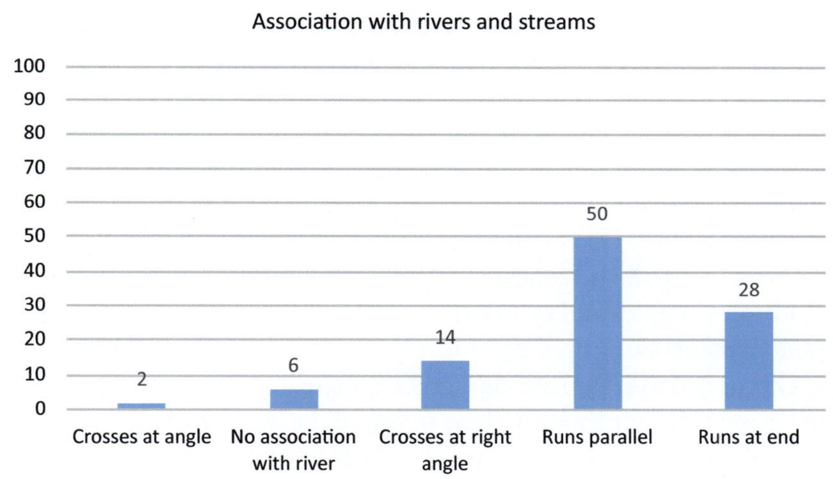

Figure 9: Cursus monument association with rivers and streams

While the flow of water definitely appears to be of significance, the data from this study appears to challenge Loveday (2006, pp. 133–36) due to the fact that there appear to be various angles of alignment between the various cursus monument sites and their associated rivers or streams. While half of the cursus monument sites explored (50 per cent) did have rivers that ran in parallel to the monument, half were associated with a river that ran at a totally different angle to the monument or, in 16 per cent of cases, the river actually crossed the monument. There are also a number of cursus monuments (6 per cent) such as the Fimber Cursus, the Barnack Cursus and the Ivinghoe Cursus which do not appear to have any river association, although it is possible that the Barnack Cursus was associated with the same ancient palaeochannel network that Pryor and French (2005, p. 7) discovered when excavating the Maxey Cursus.

If the angle of the river alignment with the cursus monument was the important factor, one would expect a more constant theme to have developed. Yet at nearly a third of the cursus monument sites explored (28 per cent) the river runs at one end of the monument, potentially determining the monument's length rather than its alignment, while at others (16 per cent) the river actually crosses the monument. While most of these (14 per cent) appear to cross the cursus monument at a right angle, this does not appear to happen in all cases, as the Rudston C Cursus crosses the Gypsey Race at an angle of approximately 45 degrees. The cursus monument's association with rivers and streams would appear to have more to do with the choice of the actual location for the monument rather than how it aligned with the river.

Applying Pearson's correlation coefficient to possible correlations between cursus monuments and their association with rivers and streams running parallel to cursus monument sites, that being the highest water probability factor, identifies a moderate positive relationship where the confidence interval of this relationship does not appear to be significant. It would, therefore, appear that there is little evidence to support any direct correlation between cursus monuments and how they are associated with the actual flow of rivers and streams within the landscape.

This led Loveday (2006, p. 134) to suggest that 'a deeper significance' may have been the siting of cursus monuments at river confluences, where he notes that 'they generally do not appear to have occurred directly at the confluence but some two kilometres or more from it'. This factor appears to be supported by Brophy (2016, p. 160), who suggests that rather than it being a direct visual connection, the importance of cursus monument alignment potentially lay in the general association of the transitory area, where he highlights that these areas appear to have been commonly used as routeways or meeting places.

While Loveday's (2006, p. 134) suggestion that, especially in the Thames Valley, 'a deeper significance' may have been the siting of cursus monuments at river confluences appears initially to be confirmed by the data within this study group, where 80 per cent of the cursus monuments appear to have been sited within 2 km of a river confluence, it is notable that two significant monuments, the Drayton St Leonard Cursus and the South Stoke Cursus, are not. It does appear that 2 km could be a significant distance, due to George et al.'s (2007, p. 5) indication that American range cattle experience a 50 per cent reduction in grazing capacity when more than this distance from water, and that this factor also has a direct effect upon the number of animals that could graze within the herd. While rivers and areas of marshland continue to appear extremely significant in some areas, in others such as at the Rudston Complex and at Stonehenge the confluence theory fails to prove viable.

The cursus monuments along the Great Ouse Valley, which include the Wolverton Cursus complex, the Cople Cursus and the Biggleswade Cursus, are situated close to river confluences, however the Cardington Cursus does not appear to have followed the same criteria even though it is situated within a couple of kilometres of its neighbour at Cople. Several of the Cambridgeshire cursus monuments, such as the Eynesbury Cursus, the Godmanchester Cursus and the Brampton Cursus, appear to have been situated close to river confluences, however those along the Welland Valley do not initially appear to have followed the same criteria, although it should be noted that Pryor and French (2005, p. 7) identified palaeochannels which appear to contain confluences.

The data within the study group initially reveals that just over half of the cursus monuments investigated (56 per cent) appear to have an association with river confluences, although this figure could potentially be significantly skewed (by 16 per cent) as it incorporates data from all cursus monuments within both the Stanwell and Wolverton Cursus complexes. However, investigation into possible correlations between cursus monuments and their association with river confluences does not appear to support the initial suggestion. Pearson's correlation coefficient identifies only a moderate positive relationship between these events, where the confidence interval of this relationship does not appear to be significant. It therefore appears that there is little evidence to support any direct correlation between the association of cursus monuments and river confluences.

Summary of Previous Cursus Monument Function Theories

When investigating the previous theories that have been put forward to explain the function of cursus monuments, the data from this study appears to highlight that many suggestions are only applicable to the individual cursus monument being investigated rather than being applied across all cursus monuments, which could explain why cursus monuments have been seen as such an enigma.

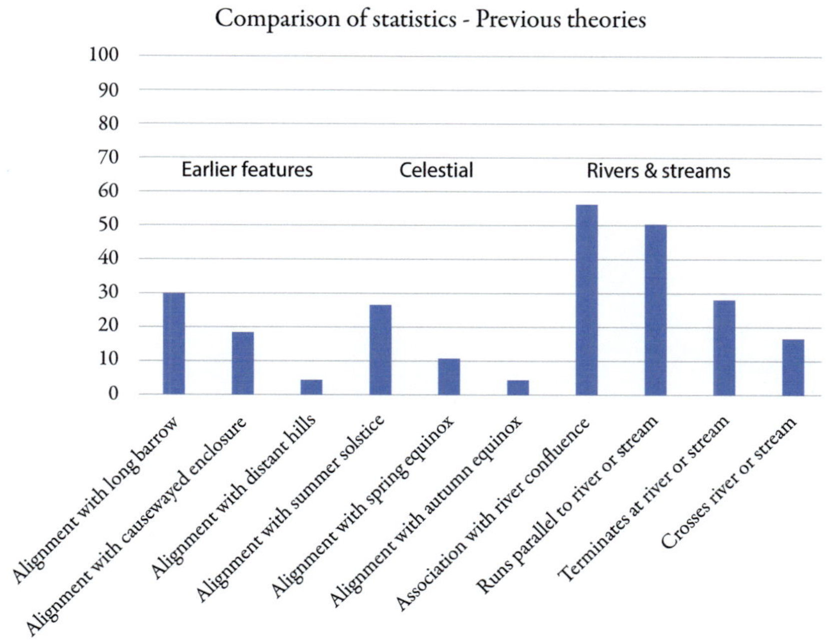

Figure 10: Comparison of statistics of previous cursus monument function theories

From a statistical point of view, when investigating cursus monument links to earlier monuments, the data from the study group appears to highlight that only about 30 per cent of cursus monuments have a potential alignment with earlier long barrows, only about 20 per cent of cursus monuments appear to have a potential alignment with regard to earlier causewayed enclosures and only 4 per cent of cursus monuments appear to have a potential alignment with distant hills. Fifty-two per cent of cursus monuments have no celestial alignment with any seasonal equinox, although just over a quarter of the monuments within my study group appear to have an approximate alignment with the summer solstice. However, only 10 per cent of cursus monuments have any potential alignment with the spring equinox, the period Rogers et al. (2018, pp. 142) suggests that cattle left their winter forest cover to feed on the early spring grasses associated with cursus monument locations. The data from this study group highlights that rivers appear to have played a significant role in the selection of cursus monument sites, as 94 per cent of cursus monuments appear to have been associated with rivers or streams and 56 per cent of cursus monument sites are in the close vicinity of river confluences. However, the actual alignment between the river and the individual cursus monument site varies significantly.

CHAPTER 7

A Statistical Evaluation of my Fieldwork

I will now turn to the analysis of my own theory about cursus monuments, as outlined in Chapter 5.

Topography of Landscape

George et al.'s (2007) research has enabled this study to incorporate scientific evidence for the movement of both wild and domestic cattle, while Saunders' earlier research across the Stonehenge landscape (Saunders 2019, pp. 173–200) places the study in an ideal position to further test and expand the model for cattle movement across a wider region. Investigation of two pieces of aurochs tooth enamel recovered during excavations at Blick Mead (Jacques et al. 2018) has enabled Rogers et al. (2018, p. 142) to establish that 'aurochs appear to have self-managed their annual feeding regime, initially by wintering in the closed canopy forests feeding on acorns and browse then by making use of the early grass growth within spring meadows. This would have enabled larger herd sizes to survive the winter months prior to their moving onto the rich open grasslands to feed throughout the summer months'. Moving into new areas, early pastoralists would need to have known how to make themselves at home in a landscape. This would potentially have included using some of the same open areas and some of the same lifestyle choices for their domestic cattle that had previously been used by the region's wild cattle. This could also have been enhanced by their deliberate intervention within the landscape, the creation or enhancement of clearings significantly improving each area for the grazing of their domestic cattle.

Prior to any initial period of cursus monument construction, pastoralists, or at least their initial ancestors, had successfully transported domestic cattle across the sea from the continent and across most parts of the British Isles. This tends to suggest that they were fully aware of the best ways of moving and looking after cattle. Although discussing a later period of prehistory, this factor appears to be supported through investigations undertaken by Viner et al. (2010, pp. 2812–18) at Durrington Walls which identify 'the relative ease with which cattle could have been moved over long distances'. Greaney et al.'s (2018, pp. 26–31) later research highlights that 'in many cases the cattle investigated had travelled over 145 km to reach the monument, while six cows appear to have travelled from a minimum of 290 km away, two of which had travelled in excess of 320 km'.

The use of scientific evidence from microwear and oxygen isotopes has enabled Henton (2010, pp. 105–14) to identify evidence about the first and last days of the beast's life, providing a history of the style of cattle management individual animals have received. Although the study was only a pilot model undertaken on a small scale, it does highlight probable Neolithic cattle husbandry practices. Henton's (2010, pp. 105–14) use of microwear and oxygen isotopes extracted from cattle tooth enamel obtained across the Fengate and Flag Fen Basin identifies that to some extent the style of cattle management inaugurated by movement between grassland meadows and forest browse continues across millennia, until at least the later stages of the Bronze Age.

The study shows that a dominance of cattle fauna recovered from within a phosphate-rich, hoof-marked area (70 per cent) probably indicates an entry point onto the Flag Fen basin. The fact that 'either cattle could be moved to pasture or the food would be required to be moved to the cattle' (Henton 2010, p. 105) appears to have resulted in the management of well-drained fields and seasonally flooded meadows within the Flag Fen basin. This appears to have provided both pasture and summer grazing for the herd while also allowing time for the fields to recover from poaching and parasite infestation. Henton also identifies that a significant percentage of cattle (50 per cent) were still being winter fed on leafy browse (ibid., p. 105). As other beasts were fed on hay, this potentially indicates a time of resource stress, as the collection of winter browse would have required more work. Tooth microwear was able to provide a direct relationship with the final fodder eaten and is therefore precisely associated with foddering events prior to the animal's death.

Using high microscopic resolution, Henton (2010, p. 109) was able to identify surface marks on the tooth enamel – striations and pits correlating to dietary preference. Dedicated browsers have less total microwear but proportionally more pits, while dedicated grazers have more microwear and proportionally more orientated striations (Solounias and Moelleken 1992). This is because three significantly different wear types are produced by cattle mouth action, depending on the type of fodder being eaten. Tugging on leaf browse with a crunching action produces large, irregular compression pits, whereas ruminating on grasses requires a sideways movement of the jaw producing highly oriented striations, which are significantly increased if the cattle are grazed on overstocked, poached or muddy winter pasture. This is primarily due to an increase in the amount of grit ingested.

Tooth microwear highlights the results of differing mouth actions. Larger pits with a lower incidence of alignment tend to indicate that the cattle had been fed within a woodland environment or fallen back on browse over the winter months, which Henton (2010, p. 107) has interpreted as the herd being left to wander or having been fed a browse supplement during times of stressed conditions, while a mouth action resulting in more striations with a higher incidence of alignment indicates the herd was fed on winter poached pasture consisting of meadows which were either overstocked or situated upon wetter landscapes, indicating what appears to be an organized system, although it still appears to be under stress.

The pilot study looked at the microwear of fourteen mandibular molars from the Fengate assemblage and eight from the Northey Island, the overall pattern identifying that, by the later stages of the Bronze Age, cattle management consisted of an organized system where most cattle were fed in well-drained meadows. There are, however, signs of less well-managed systems still being in operation, with herds being wintered in waterlogged fields or fed on browse, as 50 per cent of the Fen Gate and Northey Island herds show signs of being managed in systems where the subsistence was either less organized or was organized but under stress. This appears to highlight that, even by the later Bronze Age period, a high percentage of cattle farmers were still using the forest cover to winter cattle.

Henton (2010, p. 108) also studied oxygen isotopes to identify the birth area for fourteen head of cattle from Fengate and eight head of cattle from the Northey Island of the Flag Fen basin. Isotopic analysis identified that both assemblages included spring-born and autumn-born calves and that most cattle had died as young adults. However, results from one of the Flag Fen cattle indicated that a significant change of watering regime had occurred during its first year, where it appears to move to, or through, regions with different oxygen isotope levels. The ($\delta^{18}O$) values indicate the animal was probably born in the spring in an upland region to the north of the Peak District. It was droved downhill during the summer months, ingesting increasingly enriched water on the hoof, prior to wintering in a region to the east of Flag Fen. At a later stage, the animal was finally driven to the Flag Fen basin to be butchered. Another animal lived for at least a year more than 150 km from the Flag Fen basin. It was eventually killed in the region at a young age before it was given any opportunity to live a potentially useful working life, suggesting the beast was probably brought to the Flag Fen area solely for butchering, perhaps as part of a ritual process.

This appears to indicate that across at least 4,000 years of the prehistoric period cattle, whether wild or domesticated, were moving both within the immediate landscape as part of a cattle husbandry system and across large distances, perhaps to introduce new blood into the herd. In all the above instances, cattle movement appears to be in accordance with George et al.'s (2007, p. 5) recommendations. This cattle husbandry system would still be recognizable today, as highlighted by the introduction of twenty head of cattle to undertake environmental management at the Wolverton Floodplain Forest Nature Reserve (Martin Kincaid 2018, personal communication, September).

George et al.'s (2007, p. 2) investigations, which highlight factors and practices that significantly influence the distribution of herds of Mid-west American range cattle, suggest that 'range cattle rarely use the same feeding site for more than two consecutive days'. George et al.'s (2007, p. 5) research also highlights the optimum way in which to move cattle around the landscape. It is therefore extremely probable that prehistoric pastoralists moved their herds of cattle in a very similar way. This potentially identifies that some degree of movement could have occurred along the chalkland belt as prehistoric pastoralists moved their cattle to fresh pasture, or undertook longer journeys for purposes such as introducing new bloodstock into the herd, social feasting, ritual practices or simply as a result of their nomadic lifestyle.

During this movement, it appears that cattle would spend one third of their time grazing, one third ruminating and one third resting. Therefore, a typical day would potentially have started just before sunrise with a grazing bout lasting until around midday, when the cows would usually travel to water to drink, rest and ruminate. With the approach of dusk, the herd would begin a new feeding bout, often in a new location. Data from range cattle in the central Sierra Nevada foothills (George et al. 2007, p. 3) shows that cows graze for about eight hours a day, however the time spent grazing appears to be directly dependent upon the availability and quality of forage as well as upon the ambient temperature of the day. George et al. (2007, p. 5) further notes that 'both the terrain over which cattle roam and their distance from water play extremely significant factors in the herd's grazing capacity'. He (ibid., p. 5) identifies that 'cattle moving across slope gradients as low as 10 degrees start to see a significant reduction to their grazing capacity and therefore to the milk yield for lactating cows'. It therefore appears that, when given the choice of grazing on or moving across either gentle or steep terrain, cattle will typically congregate on the gentler terrain, preferring slopes with less than a 10-degree gradient.

Previous research (Saunders 2019, pp. 173–200) highlights that 'cattle moved across the Stonehenge plain in a parallel direction to the Stonehenge Greater Cursus'. It would also appear that the section of Stonehenge Bottom which transverses the Greater Stonehenge Cursus, although not flat, contains the gentlest slope gradient along the whole of the Stonehenge Bottom valley. This would indicate that domestic cattle were yet again moving in accordance with George et al.'s (2007, p. 5) recommendations for the movement of cattle.

By incorporating George et al.'s (2007, p. 5) investigations of cattle movement and their selection of terrain into this study, I have been able to undertake further research across an additional fifty cursus monuments. This identifies a probable alignment between cursus monuments and the style of cattle movement across these sites similar to that which George et al. (2007, p. 5) recommends. It is apparent that the 5 to 10-degree slope gradient within a GIS slope elevation software package has the ability to identify individual slopes within the rolling hills of the English chalkland belt. The study therefore introduced this figure to highlight potential routes for cattle as they moved across the landscape.

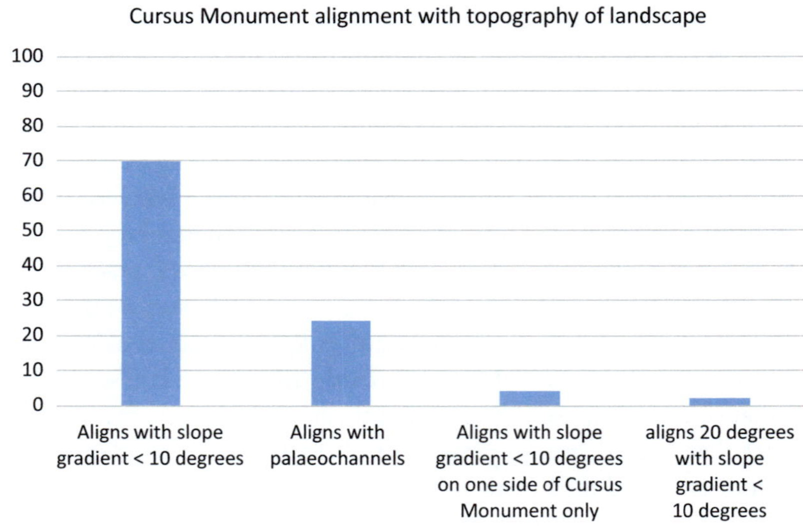

Figure 11: Cursus monument alignment with topography of landscape

Over two-thirds of the cursus monuments investigated in this study (70 per cent) appear to have had a direct association with how cattle would have moved through the landscape in accordance with George et al.'s (2007, p. 5) suggestions. This figure would rise to 94 per cent with the inclusion of additional topographic features such as the palaeochannels excavated at the Maxey Cursus, the Etton Cursus and at the Wolverton and Stanwell Cursus complexes. Although current topographical features are difficult to identify at both the Maxey Cursus and the Etton Cursus, excavations as part of the Etton Landscape Project (Pryor and French 2005, p. 7) clearly show that both cursus monuments appear to align with palaeochannels. Excavations at both the Wolverton and Stanwell Cursus complexes also identified palaeochannels. However, it is also possible to identify current topographical features at both of these complexes that would have assisted cattle movement across the landscapes.

At two other cursus monument sites, the Gussage Cursus and the Pentridge Cursus which together make up the mega Dorset Cursus, it appears that only the eastern side of each monument follows the natural topography of the landscape. Although the Cranborne Chase valley bottom naturally bends in a north-easterly curve that seems to align with the eastern side of the cursus monument, the westerly escarpment appears to be too far away from the monument to affect any alignment. However, a conversation with Martin Green (2016, personal communication, June) helped clarify the situation, when he indicated that winter floodwaters would have extended from the area around Woodyates Manor to the most westerly point of the naleds at Endless Pit, just to the west of the current A354, during the Mesolithic/Neolithic transition period. This would have resulted in a significant narrowing of the Cranborne Chase valley in the area around Down Farm which, together with the natural lake that formed just below the Pleistocene river cliff and the series of naleds, would have significantly concentrated any north-easterly cattle movement. This appears to further increase the number of cursus monuments that have direct associations with George et al.'s (2007, p. 5) suggestions to 98 per cent.

The pit alignment at Harlaxton, in Lincolnshire, appears to have been the only cursus monument within my study group that did not have a direct alignment with the natural topography of the surrounding landscape, as the monument appears to have aligned at an acute angle of 20 degrees to any potential cattle movement along the valley floor. However, the Harlaxton pit alignment's similarity to the multiple pit or post holes that define the Inchbare North and Inchbare South timber cursus monuments in Scotland would question whether this was ever actually an enclosure (Brophy 2016, p. 122 and Thomas 2007, p. 424). Regarding the alignment of the Inchbare North and Inchbare South timber cursus monuments, it is interesting to note that in each case the monument appears to also align at an acute angle to possible animal movement through the landscape, the Inchbare North and South pit alignments running at an angle of 30 degrees.

It would therefore seem that the natural topography of the landscape appears to be a significant factor regarding cursus monument alignment, with 98 per cent of monuments within the study group appearing to align with the potential direction of cattle movement across the landscape. This appears to be supported by Pearson's correlation coefficient which identifies a very strong positive relationship between these events, with the confidence interval of the relationship also appearing to be extremely significant.

This seems to indicate that across at least 4,000 years of the prehistoric period, cattle, whether wild or domesticated, were moving in accordance with George et al.'s (2007, p. 5) recommendations for the movement of cattle, and this type of cattle husbandry is still recognizable, highlighted by the introduction of twenty head of cattle to undertake environmental management at the Wolverton Floodplain Forest Nature Reserve (Martin Kincaid 2018, personal communication, September). This could suggest that cursus monuments potentially commenced life as droveways, perhaps identifying an initial practical function of the landscape prior to its possible later ritual importance as a cursus monument.

However, as construction of the monument could obviously have occurred at any point along the alignment of that cattle movement, the question arises, what other factors influenced the precise location of cursus monument sites? My initial thought was that cursus monuments had perhaps been constructed at places where the 5 to 10-degree slope gradient caused natural points of restriction to potential cattle movement. But data from my study group identifies that this appears to have only been the case with regard to the placement of 38 per cent of cursus monument sites. However, most monuments that meet this criterion appear to have been situated upon the chalk downlands of southern Wessex and the Lincolnshire and Yorkshire Wolds. It therefore appears that, while the natural topography of the landscape and especially the 5 to 10-degree slope gradient are extremely important factors regarding the alignment of each cursus monument, it does not explain the precise location of where the monument was constructed. For this reason, Martin Bell (2016, personal communication, July) and Richard Bradley (2016, personal communication, October) both suggested that I undertake further investigation into areas of potential winter floodwaters.

Cattle Husbandry Determining Cursus Monument Locations

Previous methodologies, such as Loveday's (2006, pp. 133–36) research across England and Brophy's (2016, pp. 159–60) research in Scotland, initially seem to highlight that almost all cursus monuments (96 per cent) have a correlation with the rivers and streams that flow throughout each area. However, while half of the cursus monument sites explored (50 per cent) appear to have rivers that run parallel to the monument, half seem to be associated with a river that runs at a totally different angle, either passing at the end of the monument or actually crossing it. This would suggest that, while water probably plays a part regarding the precise location of cursus monuments, the reason for that specific location is significantly more complicated than can be established by correlating the angle of alignment between the cursus monument and its nearest river. Loveday (2006, p. 134) suggests that river confluences may have held 'a deeper significance'. Yet, this is also challenged by the data within the study group, as only just over half of the cursus monuments investigated (56 per cent) appear to have had any association with river confluences. It therefore seems that, although the course a river takes appears to have been an important factor regarding cursus monument alignment, other factors may have had more influence in determining the precise location for the cursus monument site.

George et al.'s (2007, p. 5) investigation into factors and practices influencing the range of cattle distribution suggests that 'the horizontal distance to water has a strong influence on grazing capacity'. Access to sufficient water to support the herd is potentially the most critical factor, the amount of water required being dictated by factors such as the climatic condition, the activity of the herd and the lactation status of individual animals. A lactating cow can require up to 17 gallons of water per day to produce the 5 gallons of milk required to feed its calf. George et al. (2007, p. 5) suggests that 'if the herd moves more than 1.5 km from water a 50 per cent reduction in grazing capacity occurs and if the herd moves over 3 km from water grazing stops completely'. Close sources of high-quality water are required to create the riparian zones that cattle need, where soil irrigation results in longer growing seasons for the surrounding grasses, resulting in corridors of high-quality forage. These are often shaded, thereby also serving as a refuge from insects.

Figure 12: Riparian zone alongside palaeochannel at Wolverton Cursus complex

George et al. (2007, p. 12) further suggest that the quality of the water is also an important factor. Cattle which have access to cool drinking water, where the temperature is between 4.4°C and 18.3°C, gain between 0.3 to 0.4 lb more per day than those drinking warmer water. However, cows that drink directly from rivers tend to churn up sediments as they move into the water to drink. This forces subsequent animals to move further into the river to get clean water to drink (Jacques and Phillips 2014, p. 22). In deep rivers, this would result in only a couple of the animals receiving clean water, the rest of the herd having to drink water contaminated with sediment, urine and manure. But on shallow floodplains, herds of wild aurochs and the later herds of domesticated Neolithic cattle would all have been able to encroach further into the water, ensuring a cleaner drinking supply for all members of the herd.

Harding's (1999, p. 31) theory that 'cursus monuments act as some form of barrier' appears to have been based on the fact that causeways were identified in each of the individual cursus monuments that made up the Rudston Cursus complex. The fact that 64 per cent of the cursus monuments within the study group had one or more causeways across the side ditches does appear to imply that some form of control was being asserted to sideways movement across either the immediately adjacent landscape or the monument itself. While Pearson's correlation coefficient initially appears to identify only a moderate positive relationship, when cursus monuments that are associated with spring meadows, which would have enabled the herds to feed on early grass growth as indicated by Rogers et al. (2018, p. 142), are combined with cursus monuments where a causeway has been discovered, the Pearson correlation coefficient increases to identify a strong positive relationship between these elements, where the confidence interval of this relationship appears to be significant. The construction of so many parallel ditches, all of which appear to have contained causeways, strongly suggests they were acting as some form of barrier to movement. Controlling points of entry across the monument could have enabled Neolithic pastoralists to manage the herd's movement across landscapes which had been affected by both extreme event and first influx flooding. We have also seen that, regarding a cursus monument's alignment, 98 per cent of monuments within the study group appear to align in accordance with George et al.'s (2007, p. 5) recommendations for the movement of cattle, a factor supported by the present-day introduction of twenty head of cattle to undertake environmental management at the Wolverton Floodplain Forest Nature Reserve (Martin Kincaid 2018, personal communication, September). The evidence potentially suggests that cursus monuments commenced life as droveways, perhaps identifying an initial practical function within the landscape prior to any possible later ritual importance.

But from what or to where was the herd being driven? Of the cursus monuments in this study group, 44 per cent appear to align in a manner that would be associated with moving cattle to water, thereby indicating a moderate relationship using Pearson's correlation coefficient. This would also follow George et al.'s (2007, p. 5) suggestion that 'distance from water has a direct effect upon the number of animals that can actually graze within the herd'.

The siting of cursus monuments across bends in the river or the siting of multiple cursus monuments alongside a single river, as in the case at the Eynesbury Cursus complex where three monuments and the River Great Ouse appear to corral the area, further supports the notion of cursus monuments acting as a barrier within the landscape. Although use of Pearson's correlation coefficient only identifies a weak relationship (24 per cent), little previous investigation into this area appears to have been done.

Table 6: Identified causeways in cursus monuments

Cursus monument site	**Possible Causeways**	**Excavation**
Rudston A Cursus	(Harding 1999, p. 31) Identifies two causeways in the western ditch and at least two causeways in the eastern ditch of the Rudston A Cursus.	1877 Greenwell 1958 C & E Grantham 1988 Manby
Rudston B Cursus	(Harding 1999, p. 31) Identifies one causeway in the southern ditch of the Rudston B Cursus.	
Rudston C Cursus	(Harding 1999, p. 31) Identifies a possible two causeways in the southern ditch and a possible causeway in the northern ditch of the Rudston C Cursus.	
Rudston D Cursus	(Harding 1999, p. 31) Identifies three causeways in the western ditch and at least one causeway in the eastern ditch of the Rudston D Cursus.	
Hanworth Cursus	(HER 18190) Identifies a possible three causeways in the ditches of the Hanworth Cursus.	
Fornham All Saints Cursus	(Loveday, R. 2006, p. 32) Identifies possible causeways in the side ditches at the eastern terminal.	
Eynesbury Cursus	(Malim 1999, p. 79) Identifies at least two causeways in the western ditch and a possible causeway in the eastern ditch of the Eynesbury Middle Cursus.	
Godmanchester Cursus	(Malim 1999, p. 84) Causeways identified along the north-western ditch.	
Maxey Cursus	(Pryor, F. 1998, p. 2) Identifies a possible causeway in the northern ditch of the Maxey Cursus.	1962–63 Simpson 1979–81 Pryor 1982–84 Pryor (et al.) 1985 The Fenland Project, no.1: Archaeology and Environment in the Lower Welland Valley
Etton Cursus	(Pryor 1998, p. 4) Identifies at least three possible causeways in the northern ditch and a further three possible causeways in the southern ditch of the Etton Cursus.	1982–87 Pryor

Table 6: *Continued*

Cursus monument site	Possible Causeways	Excavation
Springfield Cursus	(Pastscape National Monument Number 879395) States there are several gaps along the course of both side ditches – potentially causeways.	1979–84 Hedges and Buckley
Stanwell Cursus Complex	(Lewis et al. 2010, p. 31) Identify at least four causeways in the Stanwell C1 Cursus, at least one in the Stanwell C2 Cursus and a further four in the Stanwell C3 Cursus.	1979–85 O'Connell 2006–10 Framework Archaeology
Biggleswade Cursus	No evidence of causeways.	2004 Albion Archaeology
Cardington Cursus	(Loveday, R. 2006, p. 30) Identifies two possible causeways in the side ditches of the Cardington Cursus.	
Wolverton Cursus Complex	(Hogan 2013b) Identifies a possible two causeways in the Wolverton 1 Cursus and at least two, possibly four, causeways in the Wolverton 2 Cursus.	2008–11 Cambridge Archaeological Unit
Benson Cursus	(Barclay, A. and Hey, G. 1999, p. 72) Identify at least two causeways in the western ditch and at least two, possibly three, in the eastern ditch of the Benson Cursus.	
Dorchester Cursus	(Loveday 1999, p. 51) Identifies a possible six causeways in the south-western ditch and a possible three causeways in the north-eastern ditch of the Dorchester-on-Thames Cursus.	1947–52 Atkinson 1981 Chambers 1988 Bradley and Chambers 2010–17 Hey
Drayton North Cursus	(Barclay et al. 2003, p. 9) identifies four causeways in the western ditch and three causeways in the eastern ditch of the Drayton North Cursus.	1977 Parrington 1979–82 Ainslie and Wallis 1985–86 Oxford Archaeological Unit
Drayton South Cursus	(Barclay et al. 2003, p. 9) Identifies one causeway in the eastern ditch of the Drayton South Cursus.	1921–37 E T Leeds 1994 Oxford Archaeological Unit
Buscot Cursus	(Barclay and Hey 1999, p. 72) Identify a causeway in both the southern and northern ditches of the Buscot Cursus.	
Lechlade Cursus	No evidence of causeways	1965 Vatcher & Vatcher 1985 Oxford Archaeological Unit
North Stoke Cursus	(Loveday 2006, p. 94) Identifies two possible causeways in side ditches at southern terminal.	1982 Case
Sonning Cursus	(Ford 1987) The site was described as having a markedly rectangular end with entrance gap – possibly a causeway – at its far eastern end	
Stonehenge Greater Cursus	(Gaffney et al. 2012) Causeways identified during Stonehenge Hidden Landscapes Project.	1947 Stone 1963 Christie
Stonehenge Lesser Cursus	(Richards 1990) Identified causeways during excavations undertaken as part of Stonehenge Environs Project.	1983 Richards
Gussage Dorset Cursus	(Gill 2019, p. 30) Geophysics survey of northern ditch of Dorset Cursus in November 2018 reveals two small causeways.	1953 Atkinson 1986 Bradley 1991 Bradley 1992 Green
Pentridge Dorset Cursus	(Gill 2019, p. 30) Geophysics survey of northern ditch of Dorset Cursus in November 2018 reveals two small causeways.	1953 Atkinson 1982 Barrett, Bradley & Green 1984 Barrett, Bradley & Green

The importance of spring meadows to prehistoric cattle, emphasized by Rogers et al.'s (2018, p. 142) research on Mesolithic aurochs, is still stressed within Henton's (2010, pp. 105–14) research on later Bronze Age cattle. It therefore appears that, for around four millennia, the annual cycle of spring meadows played an important role in the lives of cattle. Until artificial feed and root crops became widely available in the nineteenth century, hay was the principal winter feed for farm animals. This resulted in large numbers of domestic stock having to be killed prior to the onset of winter, for without these measures the breeding stock could not have been kept alive until the following spring (Cook et al. 2003, pp. 155–62). While hay can be grown in many fields, the best hay-grass appears to have been that grown in carefully managed waterside meadows, which encouraged early grass growth in the spring months, providing early feed for livestock.

It is highly probable that the natural floodplains adjacent to cursus monument sites acted like later medieval water meadows, where areas of grassland running alongside a river or stream were irrigated to produce a rich hay crop and lush grazing. As with natural floodplains, watering of the medieval water meadows would have begun in late December and continued through to March when sheep and cattle were placed on the refreshed meadows, remaining there until at least May when other pastures could provide the same tall, lush and nutrient-rich grasses. Water meadow operations known as floating or drowning would have deposited nutrient-laden silt, resulting in an oxidation of the soil which would have reduced the effects of frost throughout the winter months. This would have raised soil temperatures, allowing the grass to grow earlier. Modern experiments suggest that the irrigated water within a water meadow raises the temperature of the soil above 5°C, allowing earlier germination and growth to occur (Kerridge 1967, pp. 251–67). This potentially suggests that early spring grasses were naturally available alongside cursus monuments at the time when Rogers et al. (2018, p. 142) indicate that 'cattle were leaving forest cover to commence their eight-month period of feeding on the grasslands'.

Medieval water meadows were so successful that they became an essential component of the agricultural economy of Wiltshire, Dorset and Hampshire, which endured for over 400 years. Prior to the introduction of water meadows, the size of sheep flocks appears to have been limited by the amount of food available to sustain them through the winter, particularly during the months of March and April, when hay supplies were low and grass had not yet grown. The initial simple medieval practice of 'floating upwards', which involved the blocking of a watercourse by controlling the flow at the point of exit (Kerridge 1967, p. 254), causing it to overflow and flood the surrounding farmland, potentially occurred naturally throughout the prehistoric period. Investigations reveal virtually all cursus monuments within the study had some form of association with either floodplains or springlines that would result in similar effects to the floating upwards of water meadows. The use of water meadows was such a successful and long-lasting practice that it only started to decline in the late nineteenth century with the onset of a deep agricultural recession, where imports of cheap foreign grain, the development of new grass strains and artificial fertilizers meant that pastures could be improved by reseeding and nitrogen application.

Figure 13: Springline encouraging additional grass growth (Wolverton Cursus)

The only cursus monuments that do not appear to have any association with cattle husbandry are the Ivinghoe Cursus, discovered by Gover (2000) using geophysical survey and the Fimber Cursus in the Yorkshire Wolds. Unlike in other areas where cursus monuments tend to be sited upon the floodplains, the cursus monument at Ivinghoe Beacon (Gover 2000) was constructed along the chalk hilltop. However, this could be as a result of Holgate's (1995, p. 3) suggestion that 'woodland clearance does not start within the valleys of the Chiltern Hills until the early second millennium BC' and Brown and Field's (2000, p. 5) suggestion that 'Bronze Age/Iron Age society of the Chiltern Hills farm the slopes of the hills avoiding the heavily forested valley bottoms'. Perhaps the natural openness of these hilltops resulted in an inverse form of cursus monument construction. It should be noted that both Loveday (2017, personal communication) and Brophy (2018, personal communication) are showing a growing interest in upland cursus monuments.

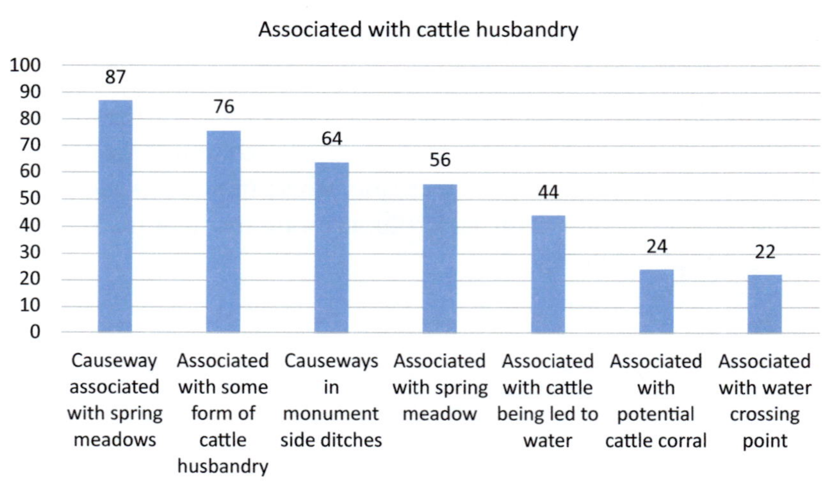

Figure 14: Cursus monument association with cattle husbandry

The Fimber Cursus appears to be the only other cursus monument that does not have a correlation with potential cattle husbandry in the vicinity of water. It is of note, however, that the monument has a dew pond at its western terminal. Dew ponds – artificial ponds usually situated towards the top of a hill – are used for watering livestock in areas where a natural supply of surface water may not be readily available. They are usually shallow, saucer-shaped depressions, which have been lined with clay or chalk upon an insulating straw layer over a bottom layer of chalk or lime, which are slowly filled by the condensation that forms within them. Martin (1915, p. 96) described how Sussex farmers created dew ponds 'using teams of oxen harnessed to broad-wheeled carts to grind the chalk to powder, the result of which would stand for years without letting water through'. Although the two dew ponds on Chanctonbury Hill have potentially been dated to the Neolithic period, there is currently no evidence that the dew pond at the Fimber Cursus dates to this period.

While undertaking a boots-on-the-ground survey around the western end of the Stonehenge Greater and Lesser Cursus monuments, I observed a phenomenon where, although the high ground was clear and bright in the morning sunshine, the valley bottoms remained shrouded in mist. A surrounding field contained around 100 cows and, although they had the choice to graze in the sunshine or in areas still shrouded by mist, about 95 per cent of the cows had chosen to graze in the mist. On speaking to local farmer Richard Crook, who has over fifty years' experience of farming the Stonehenge landscape, he identified that mist naturally increases the nutritional value of the surrounding grass, which is why downland cattle appear to relish eating grass within mist-filled valleys. This could suggest that the dew pond sited at the western end of the Fimber Cursus was potentially aiding with cattle husbandry in an area that did not benefit from the close vicinity of water.

Figure 15: Cattle in mist at western terminal of Stonehenge Greater Cursus

Mist, fog and cloud are all formed when the air cools to its dew point (less than 2.5°C) resulting in water particles in the air condensing when encountering a cold surface such as the ground. The formation of mist, its intensity, duration and area of coverage are governed by the summary of the mesoscale forces of thermodynamics and air circulation, although these can be further modified by an area's local physical conditions (Croft and Ward 2015, p. 180). The Met Office's (2017) official definition for mist is a visibility between 1,000 m and 2,000 m, while fog has a visibility of less than 1,000 m. However, mist does not necessarily affect all areas the same, as microclimates may be subtly different to the general conditions prevailing over the area and from those that might be reasonably expected under certain types of pressure or cloud cover. For example, upland areas have a specific type of climate notably different from the surrounding lower levels. This is because temperature usually falls in comparison to height at a rate of between 5°C and 10°C per 1,000 m, depending on the humidity of the air. This means that even quite modest upland regions, such as the Cotswolds, can be significantly colder on average than somewhere like the nearby Severn Valley in Gloucestershire.

The most common type of mist to occur over land is radiation mist, which forms on clear nights when the wind is light. Radiation from the ground escapes out toward space, causing the ground to cool, which in turn cools the air in contact with it. This causes the air layer nearer the earth to be cooler than the air immediately above it, a condition called an inversion. On a still night, if the air beneath the inversion layer is sufficiently moist and cools to its dew point, condensation occurs on the ground, forming dew. In the case of a calm wind, this cooling by conduction affects only a very shallow layer. However, slight air movement (3–5 knots) causes turbulent currents that are enough to spread the mist through deeper layers, forming condensation nuclei which produce mist or fog. Increased wind speeds (between 5 and 10 knots) will usually result in the mist thickening vertically, although if the wind further increases or a layer of cloud covers the region the fog is likely to clear (Schemenauer and Cereceda 1994).

The type of mist I experienced at the Stonehenge Lesser and Greater Cursus monuments is known as valley mist. Mist in low-lying valleys is a common occurrence across many regions from early autumn through to late spring. As the night in a valley becomes cool, clear and calm, the ground releases energy gained from the sun during the day back into the air, causing the ground to cool. As it cools, the air tries to offset the change in temperature of the ground. However, the ground continues to cool, still radiating energy, including that transferred to it from the air. This results in the air nearer the ground beginning to cool around the valley bottom, a process known as cold air drainage. If there is sufficient moisture in the air, mist will begin to form in layers as the air temperature approaches the dew point. The requirements for widespread valley mists are light winds throughout the lowest few thousand feet above the ground during the overnight hours, a recent rainfall to enhance moisture or soil moisture above normal and clear overnight skies. Rivers and streams that flow in the base of valleys can also enhance the potential of mist forming, due to their relatively warmer water. This is due to water taking longer to cool and warm throughout the year in comparison to the air in the atmosphere. Maximum and minimum temperatures of air in the atmosphere occur, on average, six weeks into the summer and winter respectively, while the maximum and minimum temperatures of water bodies occur on average twelve weeks into the summer and winter respectively. Therefore, during the autumn and winter months, the water at the bottom of a valley will be relatively warmer than the air around it, resulting in water molecules evaporating, putting moisture into the air. The warmer the water body, the more evaporation of water and the more moisture present in the air. On cool, calm and clear nights, the air temperature will approach its dew point and mist will begin to form. Valley mist is essentially radiation mist confined by local topography, which can last for several days in calm conditions. However, during the spring, when the ground can still be very cold, warm, moist air moving over it from the Gulf Stream of the Atlantic is chilled from below. This causes it to condense into tiny water droplets, resulting in advection mist.

It was my intention to investigate whether a correlation could be established between the location of Neolithic cursus monuments and areas readily affected by modern-day incidences of fog and mist. However, more than 75 per cent of the population now lives in towns or cities, which has created an urban microclimate causing double the amount of winter fog occurrence at these locations, which would result in any correlation between these factors being worthless. Nevertheless, as slight air movement is a significant requirement in the process of turning dew into mist, it has been possible to use Met Office mean wind speed data to identify areas of the study group readily affected by light winds, which are most likely to produce valley mist, thereby increasing the nutritional value of the grasslands.

Use of Pearson's correlation coefficient to investigate the correlation of cursus monument locations in association with potential cattle husbandry occurring within the vicinity of water identifies a strong positive relationship (76 per cent) when the association is applied across all aspects of cattle husbandry. This tallies with George et al.'s (2007, p. 5) investigations into factors and practices that influenced American Mid-west range cattle distribution, when he suggested 'the horizontal distance to water was potentially the most critical component to influence grazing capacity'. The requirement for good-quality water in the Neolithic could well have increased the cattle's use of shallow floodplains, where the herd could encroach further into the water without churning up sediments, ensuring a cleaner drinking supply for all members of the herd (Jacques and Phillips 2014, p. 22). The early spring grasses of these shallow floodplains and the increased nutritional value of the grassland due to valley mist appears to have resulted in the landscapes adjacent to what were to become cursus monument sites being excellent riparian zones throughout the late Mesolithic and into the Neolithic period.

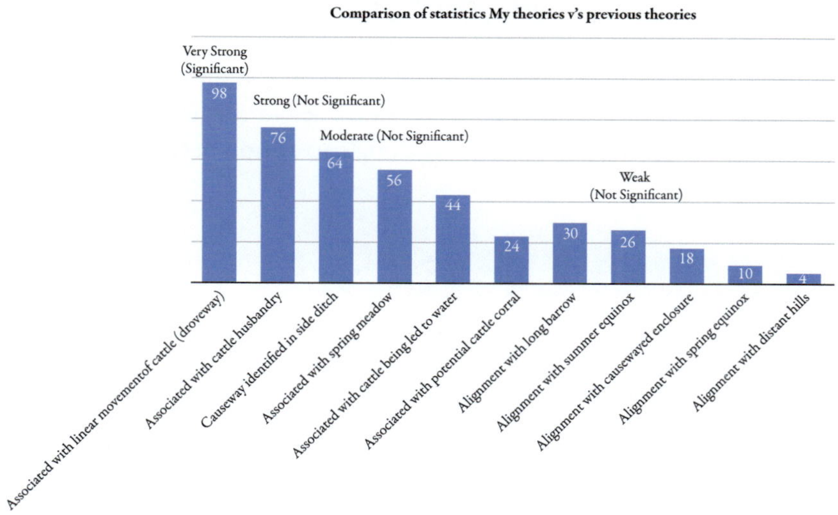

Figure 16: Comparison of my theories versus previous theories

This study, which included fifty cursus monuments situated on or adjacent to the English Chalkland belt, identified that the natural topography of the landscape is a significant factor with regard to the alignment of cursus monuments. All but one of the cursus monuments within the study group align with the potential direction of cattle movement across the landscape in accordance with George et al.'s (2007, p. 5) recommendations, with a very strong and significant relationship identified using Pearson's correlation coefficient. The pit alignment of the potential timber cursus monument at Harlaxton, in Lincolnshire, is the only cursus monument within the study group that does not have a direct alignment with cattle movement along the natural topography of the surrounding landscape.

Although cursus monument alignment has a very strong correlation to the 5-to-10-degree slope gradient, and thus to the probable direction of domestic cattle movement through the valley profile, this alone does not appear to determine the actual location upon which the cursus monument was constructed. The study also identified that cursus monument locations have a strong correlation with potential cattle husbandry taking place in the vicinity of areas affected by the winter flooding of pasture, potentially creating natural water meadows resulting in earlier grass growth. This would correspond to Rogers et al.'s (2018, p. 142) ($\delta^{13}C$) isotopic evidence that 'aurochs started to leave the forest cover during the early spring to commence their eight-month period of living on the grasslands'.

It would also correspond to Harding's (1999, p. 31) theory that 'cursus monuments acted as some form of barrier'. Identification of causeways across the side ditches at 66 per cent of cursus monuments within the study group appears to imply that some form of control was being asserted to the landscape either side of the monument, either blocking the landscape or controlling access to it. While Pearson's correlation coefficient initially appears to identify only a moderate positive relationship, when the association of cursus monument with spring meadows, enabling the herds to feed on early grass growth as indicated by Rogers et al. (2018, p. 142), are introduced into the equation, the Pearson correlation coefficient increases to a strong positive relationship.

The alignment of cursus monuments in accordance with George et al.'s (2007, p. 5) recommendations for the movement of cattle could suggest they commenced life as droveways, perhaps identifying an initial practical function of the landscape prior to its possible later ritual importance as a cursus monument, where 44 per cent of the cursus monuments appear to align in a manner associated with the moving of cattle to water and a further 22 per cent would be associated with assisting cattle to actually cross the water. However, the siting of cursus monuments across bends in the river or the siting of multiple cursus monuments alongside a single river, as in the case of the Eynesbury Cursus complex where three monuments and the River Great Ouse appear to corral the area, identifies only a weak relationship (24 per cent) suggesting this was not a significant factor.

This study has analysed each individual cursus monument using a unique combination of practical and quantifiable methodologies, backed up by boots-on-the-ground field walking across each site. It is the first-time cursus monuments have been analysed in this way and it is also the first time an archaeologist has been able to establish a very strong correlation between cursus monument sites and earlier cattle movement.

CHAPTER 8

A Case Study: The Milfield Basin

Isotopic analysis by Rogers et al. (2018, p. 142) on two pieces of aurochs tooth enamel recovered from Blick Mead, together with Henton's (2010, pp. 105–14) microwear and oxygen isotopes on later Bronze Age cattle across the Fengate and Flag Fen Basin, identify that cattle movement has occurred across millennia at a local level, while Greaney et al.'s (2018, pp. 26–31) research confirms that this movement also occurred on a much larger scale, with distances in excess of 320 km being recorded for cattle journeys to Durrington Walls. The alignment of virtually all cursus monuments in accordance with George et al.'s (2007, p. 5) recommendations for the movement of cattle strongly suggests that cursus monuments commenced life as some form of cattle management system, perhaps acting as droveways, identifying a practical function for their placement and alignment within the landscape prior to their possible later ritual importance. Harding (1999, p. 31) suggests that 'cursus monuments potentially acted as some form of barrier to the landscape', and this idea appears to be supported by the significant number of causeways identified within the side ditches of cursus monuments throughout the study group (66 per cent), apparently implying that some form of control was being asserted over the landscape either side of the monument, perhaps controlling the herd's ability to feed on early grass growth as indicated by Rogers et al. (2018, p. 142).

Interestingly, this study has also identified one site from the Mesolithic/Neolithic transition period that meets each of the criteria outlined above, yet does not develop from a droveway into a cursus monument. That site is the Milfield Basin in Northumberland, where Waddington's (1999, p. 171) investigations into the Mesolithic/Neolithic transition period suggest that 'the core area of late Mesolithic settlement appears to have been located on the raised gravel terraces of the valley floor, providing easy access to a wide diversity of ecological zones'. The Milfield Basin forms a geographically discrete area which is physically demarcated by its surrounding natural features. These include both the encircling Cheviot Hills to the south and west of the plain and the sandstone escarpment to the north and east, which rises abruptly creating a natural amphitheatre. The plain itself lies on a confluence of the River Glen and the River Till that contained some of the most fertile land in prehistoric Northumberland (Waddington 1999, p. 14). Both rivers meander extensively, and as with many of the rivers within my study group, the River Till in particular is prone to severe winter and spring flooding (Gibson 1986, p. 93).

Waddington (1999, p. 21), identifies that the geography of the basin appears to have been conditioned by its glacial inheritance, consisting of three distinct topographical zones. The Cheviot Hills rise sharply from the Milfield plain, their distinct round, flat-topped hills extending for approximately 650 sq km, where the steep slopes contain thin skeletal and acidic soils with little agricultural value. The gentle slopes and areas of low plateau contain deeper, free-draining, brown earths overlying andesitic drift which are suitable for agriculture including cereal cultivation, and the fell sandstone uplands form a sweeping, almost continuous escarpment, averaging a height of approximately 150 m. The fell sandstone region produces acid soils, which are particularly poor in terms of agricultural potential, generally producing open moorland dominated by heather and bracken.

Sandwiched between these two areas of high ground, the Milfield plain contains a complex sedimentary sequence which overlies the cement stone bedrock. Course-grained glacio-deltaic sands and gravels fan out from the Glen Valley providing free-draining terraces which would have provided attractive locations for prehistoric settlement and agriculture. However, poor drainage and the inundation of flood waters upon the alluvial and clay silt soils of the valley floor would have tended to result in most of these areas being unsuitable for early agriculture.

As the starting point for his investigations, Waddington (ibid., p. 30) used Tipping's (1996) earlier research to establish the environmental history of the basin. He then undertook a mixture of field walking and the excavation of test pits to establish the distribution of lithic scatters and identified a system of morphometric mapping to determine the slope gradients within the area. Waddington (1999, p. 45) established the topographical and geomorphological context by creating a detailed slope map, where he divided the study area into component slope units. However, it is noticeable that the slope units he devised were solely identified by his personal field observations. It therefore appears that Waddington's morphometric classification system was purely a visual interpretation of the region, providing coverage of the physiography of the valley floor, rather than incorporating any scientific reasoning behind the selection of his slope differentiations, such as the slope gradients identified by George et al. (2007, p. 5) for cattle movement.

However, Waddington (1999, p. 106) was able to use the palaeo-environmental and lithic data to reconstruct a model of the Mesolithic settlement throughout the valley basin, identifying that 'land-use exploitation in the Milfield Basin was ordered around the core settlement area of the gravel terraces which may have provided a relatively permanent zone of occupation by the Late Mesolithic'. Waddington (ibid., p. 104) believed that the density of material implied the site 'was a place which was revisited on many occasions rather than a site occupied very intensively for just a short period, and probably represents a winter-spring, and possibly even year-round base camp'.

Waddington's (1999, p. 55) model appears to have worked on the principle that 'critical to the understanding of the Milfield Basin data set is the degree to which the lithic population of the Milfield Basin is under-represented in the visible surface samples', perhaps a result of the steep slopes. A significant proportion of the lithics had been buried in colluvial drapes and positive lynchets, no longer detectable by surface survey. A further problem appears to have been that around 40 per cent of the Milfield lithic assemblage consisted of chert, volcanic material and quartz, all of which are difficult to recognize as worked lithics during field investigations. This potentially meant that the Mesolithic and early Neolithic lithic densities from the Milfield Basin were always going to be smaller than in other flint-rich areas.

The distribution of lithic scatters also led Waddington (ibid., p. 171) to suggest that smaller camps appear to have been 'occupied for relatively short periods on a seasonal basis'. These were located close to springs and tributary streams, where the woodland had been managed by burning. Tipping (1996, p. 23) believes that 'this would have potentially assisted with improving the hunting of game such as deer, aurochs and wild boar throughout the area, which appears to have played an important activity throughout the spring and summer months', and this corresponds with Rogers et al.'s (2018, p. 142) ($\delta^{13}C$) isotopic evidence that 'aurochs started to leave the forest cover during the early spring to commence an eight-month period of living on the grasslands'. Waddington (1999, p. 173) notes that the Mesolithic archaeology of the Milfield Basin 'provided few marked departures from other regional syntheses', where the settlement location around watercourses, springs and the raised gravel terraces corresponds with others identified across other parts of northern England.

Neolithic pollen diagrams from the Cheviot Hills suggest that initially in this area early Neolithic groups continued to build upon previous Mesolithic woodland disturbance while undertaking only limited cereal cultivation, while pollen diagrams from the sandstone escarpments identify that after the Mesolithic period no further human activity appears to take place in this area until the Bronze Age. However, the valley floor appears to tell a totally different story, where the Akeld Steads pollen diagrams indicate that major woodland clearance occurred from the earlier half of the fourth millennium, perhaps suggesting the start of landscape clearance for crop production.

Lithic distribution from the Neolithic period identifies that the majority of finds were clustered on the gravel terraces and the low slopes and plateau areas of the Chilton dip slopes. However, the overall lithic distribution for the Neolithic period appears to differ from that of the Mesolithic due to a significant reduction in lithic density on the sandstone slopes. The few finds discovered suggest processing activities rather than hunting or industrial activities were taking place in this area.

Waddington (1999, p. 113) contests the view that the Neolithic period consisted of an economic phenomenon introduced from outside and imposed upon the Mesolithic indigenous population. He (ibid., p. 113) suggests that 'there is currently no evidence for direct contact between the Milfield Basin and the continent during the early Neolithic', nor for the fact that Mesolithic people retreated into the hills while the new farmers occupied the valleys and lowlands. Waddington (1999, p. 113) believes his investigation shows that during the Mesolithic/Neolithic transition period the core area of settlement within the Milfield Basin appears to have remained located upon the raised gravel terrace and the fringe areas of the low Cheviot slopes.

But from the viewpoint of this study, the most interesting factor within the Milfield Basin was the construction of the Coupland complex. The structure has been radiocarbon dated to around 3800 BC, which would place it slightly earlier than the cursus monuments within the study group. It has been defined by Loveday (2006, p. 108) as 'an avenue', yet importantly Waddington (1999, p. 134) defines it as 'an enclosure and droveway'. Emily Mercer's (Bradford University) phosphate analysis of the feature (Waddington 1999, p. 136) identifies the landscape, which has been trampled by stock, includes 'a high incidence of cattle faecal remains', suggesting the droveway may have been used for the movement of stock in and out of the Coupland enclosure, possibly during the seasonal movement of herds thought to have taken place on these uplands at this time.

Waddington (1999, p. 136) has overlaid aerial photographs along the course of the structure on to detailed geomorphological maps, leading him to suggest that 'the structure used a deep naturally incised gully at the northern end, to lead cattle directly to the water's edge, while at the southern end, the structure led to the Galewood Depression, which in turn led to the River Glen'. This has led Waddington (ibid., p. 136) to further suggest that 'the southern course of the feature may have been used for the daily requirement of watering stock'.

The alignment of the Coupland enclosure in accordance with George et al.'s (2007, p. 5) recommendations for the movement of cattle, together with the phosphate analysis undertaken by Emily Mercer (Waddington 1999, p. 136), would strongly suggest that the feature started its life along the same lines as the cursus monuments within the study group, as some form of cattle management system, probably acting as a droveway. However, for some unknown reason, this initial practical function of the landscape did not go on to acquire the format structure of a cursus monument. However, two sets of double pit alignments at Milfield North and Ewart, which appear to have been situated within spur valleys leading to water at either end of the Coupland enclosure, could potentially be identified as timber cursus monuments.

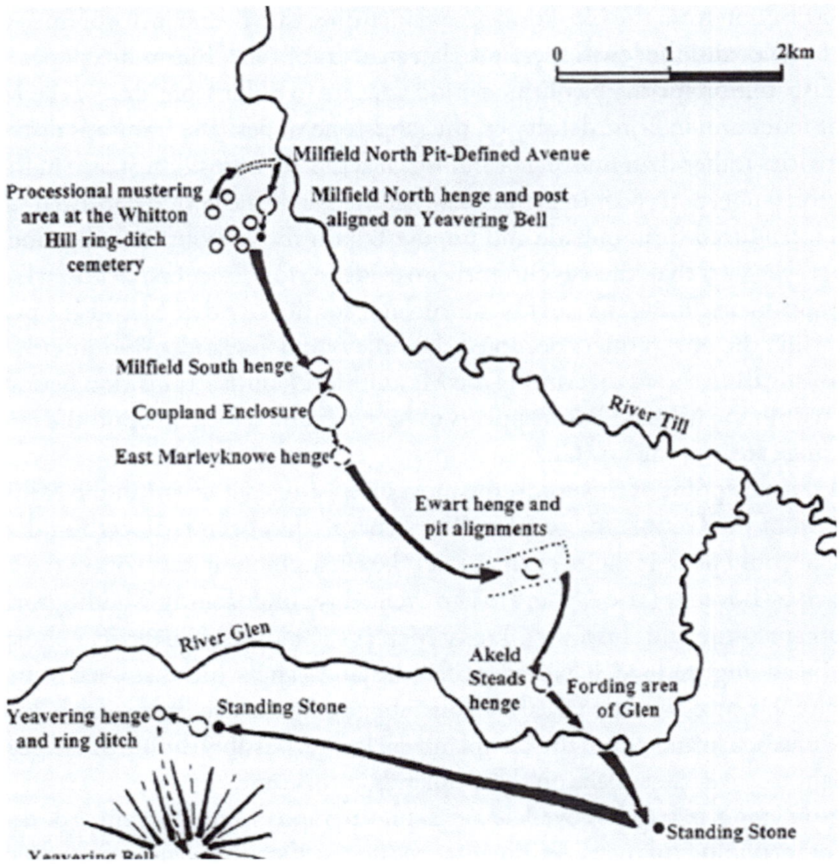

Map 32: Waddington's interpretation of processional routes across the Milfield Plain (After Waddington 1999, p. 160. Reproduced with kind permission of BAR Publishing, <www.barpublishing.com>)

The construction of the Coupland enclosure, its alignment with the topography of the landscape and its association with cattle and water makes the feature extremely like the cursus monuments that have been identified throughout this study. The pit alignments at the Milfield North Pit-defined Avenue and either side of the Ewart henge also appear similar to pit alignments interpreted by Brophy (2000) as timber cursus monuments. Waddington (1999, p. 164) seems to add weight to this argument when he identifies that the northern sector of the Ewart double pit alignment had an almost right-angled turn at its eastern end. Could this be similar to the squared terminal of the timber cursus monument at Douglasmuir? Although Brophy (2016, p. 60) initially believed this type of monument to be 'a genuinely regional phenomenon, with no convincing examples found outside Scotland', discussions regarding English Neolithic pit alignments in Lincolnshire and the north-east (Brophy 2017, personal communication, April) are causing him to reassess this position.

Conclusion

The research outlined in this book incorporates a fresh methodology, previously unexplored in cursus monument investigation, which has enabled me to identify a very strong correlation between the placement and alignment of Neolithic cursus monuments and the movement of large herbivores within that landscape. Because previous cursus monument studies have tended to focus on the construction or post-construction phases of the monument, rather than upon the reasons a Neolithic community decided to locate and align these monuments where they did, this study makes an original contribution to the cursus monument debate.

The study has been able to combine archaeological landscape methodologies to identify a correlation between the initial placement and alignment of Neolithic cursus monuments and the movement of large herbivores. It has also been able to identify the specific species of animal. The fact that red deer hunting tended to occur within closed-canopy woodlands, due to the herd's tendency to feed within or on the edge of forest cover rather than on open grasslands, strongly suggests there is no correlation between the hunting of red deer and the landscapes where many later cursus monuments were to be constructed.

While it is highly probable that herds of aurochs roamed the same landscapes later used by Neolithic pastoralists and their domestic cattle, field observation is unable to place any potential hunting-associated landscape within a specific Mesolithic period. Only ten per cent of monument sites could be directly identified as having a high probability of there being any correlation between previous aurochs movement and the later alignment and placement of cursus monument sites, which suggests only a weak correlation exists between these factors and that they do not appear significant.

This research suggests that a different set of economic practices, other than those of an agriculturalist economic culture, defined the start of the Neolithic period. It appears Early Neolithic migrants undertook a lifestyle that was more pastoral than fully agricultural, moving with their herds along traditional paths and territories previously used by Mesolithic communities. The study shows how the key factors of slope gradient and proximity to water can be combined to identify a very strong correlation between the alignment of cursus monuments and the movement of cattle across the landscape. Quantifiable data identifies that, in order to achieve optimum grazing capacity for the herd and thus maximum milk yield in lactating cows, migrating cattle need to travel parallel to the contour of the 5 to 10-degree slope gradient. The research explored within this book shows that virtually all cursus monument sites align with this slope gradient, and thus with the routes that herds of domestic cattle would have been required to take to follow the valley profile. This could suggest that cursus monuments commenced life as cattle droveways, perhaps identifying an initial practical function within the landscape prior to their possible later ritual importance.

Although cursus monument alignment seems to follow the recommendations for the movement of cattle, appearing to have a very strong correlation with the 5 to 10-degree slope gradient and thus to the direction cattle would have moved through the valley profile, this alone does not appear to determine where a cursus monument was constructed. However, the identification of causeways within the side ditches of two-thirds of the study group strongly suggests that the monument was in some way acting as some form of barrier. This, together with the significant levels of both extreme event and first influx fluvial flooding, appears to imply some form of control was being asserted over the landscape, probably as some form of cattle husbandry. The suggestion that the earliest British Neolithic field systems were laid out for the use of livestock appears to be supported within the Milfield Basin by the prehistoric construction of the Coupland complex, which has been radiocarbon dated to around 3800 BC, just prior to the commencement of cursus monument construction.

Field observation within the Milfield Basin appears to suggest that the Coupland complex, together with the two pit alignments at either end, would have enabled Neolithic pastoralists to assert control over access to spring meadows in the landscape between the complex and the adjacent river. It also appears that, throughout the study area, cursus monument locations have a strong correlation with areas where winter flooding of pasture occurred, potentially creating natural water meadows which would have resulted in earlier grass growth.

It appears that a correlation exists between the immediate landscape upon which cursus monument construction occurred and the movement of domestic cattle. Pearson's correlation coefficient suggests a strong relationship exists between Neolithic communal use of the landscape to herd domestic cattle and the use of the landscape as a cursus monument construction site. Domestic cattle would be required to forage and browse in forest cover to survive the winter months. The cattle would need to be moved in the early spring, this time to specialized feeding grounds created by areas of floodplain where spring meadows resulted in earlier grass growth. Throughout the rest of the year, the cattle would need to be regularly moved to riparian zones with plentiful supplies of water.

This study identifies a strong correlation with cattle husbandry as a whole and more specifically identifies an association with several aspects of cattle husbandry, such as spring meadows (56 per cent), leading cattle to water (44 per cent), cattle crossing water (22 per cent) and the cursus monument together with the river forming a potential cattle corral (24 per cent). Investigation of fauna records throughout the study group area together with recommendations for the movement of cattle have enabled this study to move away from the 'enigmatic' approach to investigating cursus monuments. Instead, each individual monument has been analysed using a unique combination of practical and quantifiable methodologies, backed up by boots-on-the-ground field walking of each site. This is the first time that cursus monuments have been analysed in this way. It is also the first time an archaeologist has established a very strong correlation between cursus monument sites and cattle movement.

Appendices

Appendix A Domestic Cattle Fauna records from locations adjacent to cursus monuments

Cursus monument site	**Domestic Cattle Fauna**	**Excavation**
Rudston A Cursus	(Harding 2006, p. 119) Identifies three pits at Low Caythorpe 1, along the course of the Caythorpe Gas Pipeline, which produced the partial remains of as many as nine domesticated head of cattle.	1877 Greenwell 1958 C & E Grantham (Mentioned in Dymond D 1966) 1988 Manby
Rudston B Cursus	(Harding 2006, p. 119) Identifies three pits at Low Caythorpe 1, along the course of the Caythorpe Gas Pipeline, which produced the partial remains of as many as nine domesticated head of cattle.	
Rudston C Cursus	(Harding 2006, p. 119) Identifies three pits at Low Caythorpe 1, along the course of the Caythorpe Gas Pipeline, which produced the partial remains of as many as nine domesticated head of cattle.	
Rudston D Cursus	(Harding 2006, p. 119) Identifies three pits at Low Caythorpe 1, along the course of the Caythorpe Gas Pipeline, which produced the partial remains of as many as nine domesticated head of cattle.	
Duggleby Cursus	Not excavated	
Fimber Cursus	Not excavated	
Kirby Underdale Cursus	Not excavated	
Bag Enderby Pit Alignment	Not excavated	
Harlaxton	Not excavated	
Stenigot Pit Alignment	Not excavated	
Hanworth Cursus	Not excavated	
Fornham All Saints Cursus	Not excavated	
Stratford St Mary	Not excavated	
Barnack Cursus	Not excavated	
Eynesbury Cursus		Sections were cut across the Northern Cursus (Macaulay 1994; Kemp 1998)
Godmanchester Cursus		Excavations by McAvoy and the Central Excavation Unit in 1988–91 at Rectory Farm, Godmanchester
Brampton Cursus		Mortuary enclosure at eastern end of Cursus excavated by Malim in 1991
Maxey Cursus	(Ainsley. In: Pryor 2005, p. 84) The dominance of domestic cattle is remarkable at the Etton Landscapes site (485 relating to 64 per cent)	1962–63 Simpson 1979–81 Pryor 1982–84 Pryor (et al.) 1985 The Fenland Project, No.1: Archaeology and Environment in the Lower Welland Valley

Cursus monument site	**Domestic Cattle Fauna**	**Excavation**
Etton Cursus	(Ainsley. In: Pryor 2005, p. 84) The dominance of domestic cattle is remarkable at the Etton Landscapes site (485 relating to 64 per cent)	1982–87 Pryor
Springfield Cursus	(Buckley et al. 2001, p. 147) Suggest fragments identifiable to cattle included four sesamoids, part of an ulna, two fragments of first phalanx, three fragments of vertebra and several fragments of unidentified long bone.	1979–84 Hedges and Buckley
Stanwell Cursus 1	Only two Neolithic features produced any animal bone: the HE1 enclosure and both ditches of the C1 Stanwell Cursus. However, the presence of domestic animals at the site during this period is noteworthy (Knight. In: Lewis et al. 2010). Two pieces were identified as cow (Bos Taurus), 34 fragments were identified as cow/red deer and 4 fragments were identified as large mammal.	1979–85 O'Connell 2006–10 Framework Archaeology
Stanwell Cursus 2	Only two Neolithic features produced any animal bone: the HE1 enclosure and both ditches of the C1 Stanwell Cursus. However, the presence of domestic animals at the site during this period is noteworthy (Knight. In: Lewis et al. 2010). Two pieces were identified as cow (Bos Taurus), 34 fragments were identified as cow/red deer and 4 fragments were identified as large mammal.	1979–85 O'Connell 2006–10 Framework Archaeology
Stanwell Cursus 3	Only two Neolithic features produced any animal bone: the HE1 enclosure and both ditches of the C1 Stanwell Cursus. However, the presence of domestic animals at the site during this period is noteworthy (Knight. In: Lewis et al. 2010). Two pieces were identified as cow (Bos Taurus), 34 fragments were identified as cow/red deer and 4 fragments were identified as large mammal.	1979–85 O'Connell 2006–10 Framework Archaeology
Stanwell Cursus 4	Only two Neolithic features produced any animal bone: the HE1 enclosure and both ditches of the C1 Stanwell Cursus. However, the presence of domestic animals at the site during this period is noteworthy (Knight. In: Lewis et al. 2010). Two pieces were identified as cow (Bos Taurus), 34 fragments were identified as cow/red deer and 4 fragments were identified as large mammal.	1979–85 O'Connell 2006–10 Framework Archaeology
Stanwell Cursus 5	Only two Neolithic features produced any animal bone: the HE1 enclosure and both ditches of the C1 Stanwell Cursus. However, the presence of domestic animals at the site during this period is noteworthy (Knight. In: Lewis et al. 2010). Two pieces were identified as cow (Bos Taurus), 34 fragments were identified as cow/red deer and 4 fragments were identified as large mammal.	1979–85 O'Connell 2006–10 Framework Archaeology
Biggleswade Cursus		2004 Albion Archaeology
Cardington Cursus	Not excavated	
Cople Cursus	Not excavated	
Ivinghoe Beacon Cursus	Not excavated	

Cursus monument site	Domestic Cattle Fauna	Excavation
Wolverton Cursus 1	(Rajkovaca. In: Hogan 2013b, p. 38) The fieldwork at Manor Farm resulted in the recovery of 231 assessable fragments of animal bone of which 48 were identifiable to species of which 27 (56 per cent) were from domestic cattle. A further 47 were cattle sized but the specimen could not be further identified.	2008–11 Cambridge Archaeological Unit
Wolverton Cursus 2	(Rajkovaca. In: Hogan 2013b, p. 38) The fieldwork at Manor Farm resulted in the recovery of 231 assessable fragments of animal bone of which 48 were identifiable to species of which 27 (56 per cent) were from domestic cattle. A further 47 were cattle sized but the specimen could not be further identified.	2008–11 Cambridge Archaeological Unit
Wolverton Cursus 3	(Rajkovaca. In: Hogan 2013b, p. 38) The fieldwork at Manor Farm resulted in the recovery of 231 assessable fragments of animal bone of which 48 were identifiable to species of which 27 (56 per cent) were from domestic cattle. A further 47 were cattle sized but the specimen could not be further identified.	2008–11 Cambridge Archaeological Unit
Wolverton Cursus 4	(Rajkovaca. In: Hogan 2013b, p. 38) The fieldwork at Manor Farm resulted in the recovery of 231 assessable fragments of animal bone of which 48 were identifiable to species of which 27 (56 per cent) were from domestic cattle. A further 47 were cattle sized but the specimen could not be further identified.	2008–11 Cambridge Archaeological Unit
Wolverton Cursus 5	(Rajkovaca. In: Hogan 2013b, p. 38) The fieldwork at Manor Farm resulted in the recovery of 231 assessable fragments of animal bone of which 48 were identifiable to species of which 27 (56 per cent) were from domestic cattle. A further 47 were cattle sized but the specimen could not be further identified.	2008–11 Cambridge Archaeological Unit
Benson Cursus	Not excavated	
Dorchester Cursus		1947–52 Atkinson 1981 Chambers 1988 Bradley and Chambers 2010–17 Hey
Drayton St Leonard	Not excavated	
Drayton North Cursus	(Ayres and Powell. In: Barclay et al. 2003, p. 159) A total of 248 bone fragments were discovered from the east cursus ditch. Around half were identifiable the majority being domestic cattle sized (100). The west cursus ditch provided similar results with the predominance of cattle bones in the identifiable material (20).	1977 Parrington 1979–82 Ainslie and Wallis 1985–86 Oxford Archaeological Unit
Drayton South Cursus	(Wilson. In: Barclay et al. 2003, p. 29). Various features yielded scattered bone, but only those from the Neolithic pit 107 were recorded as soil samples from this feature were sieved for bones. Mainly domestic pig but one fragmentary domestic cattle tooth	1921–37 E T Leeds 1994 Oxford Archaeological Unit
Buscot Cursus	Not excavated	

Cursus monument site	Domestic Cattle Fauna	Excavation
Lechlade Cursus	(Ayres and Powell. In: Barclay et al. 2003, p. 207) The Lechlade Cursus produced a total of 148 fragments of animal bone, the majority unidentifiable. Most of the bone was recovered from the upper fills of the east cursus ditch. Of the identifiable fragments the majority were cattle sized (28).	1965 Vatcher & Vatcher 1985 Oxford Archaeological Unit
North Stoke Cursus		Rescue excavations were undertaken in the summer of 1950, again in the spring of 1951 and for a fortnight in the summer of 1952. Among those who worked on them were Hector Catling, Salvatore Puglisi and Isobel Smith. 1982 Case
South Stoke Cursus	Not excavated	
Stadhampton Cursus	Not excavated	
Sonning Cursus	Not excavated	
Stonehenge Greater Cursus	72 per cent of identifiable animal fauna recovered from the Stonehenge Greater Cursus comprised of domestic cattle (Serjeantson 2011, p. 16).	1947 Stone 1963 Christie 1983 Richards 2008 Parker Pearson et al.
Stonehenge Lesser Cursus	(Richards 1990, p. 83) A total of 178 fragments of animal bone were recovered. With the exception of red deer antlers very few bones were recovered from the bottom of the ditches. 38 fragments were identified as large mammal which Richards (1990, p. 83) suggests could possibly be aurochs, while 15 fragments were from domestic cattle.	1983 Richards
Yatesbury Cursus (Avebury)	Not excavated	
Gussage Dorset Cursus	(Legge. In: companion volume of Barrett et al. 1991, p. 20) undertook the bone analysis which identified the Neolithic fauna were completely dominated by species suited to a woodland habitat, particularly cattle and pig.	1953 Atkinson 1986 Bradley 1991 Barrett et al. 1992 Green
Pentridge Dorset Cursus	(Legge. In: companion volume of Barrett et al. 1991, p. 20) undertook the bone analysis which identified the Neolithic fauna were completely dominated by species suited to a woodland habitat, particularly cattle and pig.	1953 Atkinson 1982 Barrett et al. 1984 Barrett et al.

Appendix B Identified causeways in cursus monuments

Cursus monument site	Possible Causeways	Excavation
Rudston A Cursus	(Harding 1999, p. 31) Identifies two causeways in the western ditch and at least two causeways in the eastern ditch of the Rudston A Cursus.	1877 Greenwell 1958 C & E Grantham 1988 Manby
Rudston B Cursus	(Harding 1999, p. 31) Identifies one causeway in the southern ditch of the Rudston B Cursus.	
Rudston C Cursus	(Harding 1999, p. 31) Identifies a possible two causeways in the southern ditch and a possible causeway in the northern ditch of the Rudston C Cursus.	
Rudston D Cursus	(Harding 1999, p. 31) Identifies three causeways in the western ditch and at least one causeway in the eastern ditch of the Rudston D Cursus.	
Hanworth Cursus	(HER 18190) Identifies a possible three causeways in the ditches of the Hanworth Cursus.	
Fornham All Saints Cursus	(Loveday 2006, p. 32) Identifies possible causeways in the side ditches at the eastern terminal.	
Eynesbury Cursus	(Malim 1999, p. 79) Identifies at least two causeways in the western ditch and a possible causeway in the eastern ditch of the Eynesbury Middle Cursus.	
Godmanchester Cursus	(Malim 1999, p. 84) Causeways identified along the north-western ditch.	
Maxey Cursus	(Pryor, F. 1998, p. 2) Identifies a possible causeway in the northern ditch of the Maxey Cursus.	1962–63 Simpson 1979–81 Pryor 1982–84 Pryor (et al.) 1985 The Fenland Project, no.1: Archaeology and Environment in the Lower Welland Valley
Etton Cursus	(Pryor 1998, p. 4) Identifies at least three possible causeways in the northern ditch and a further three possible causeways in the southern ditch of the Etton Cursus.	1982–87 Pryor
Springfield Cursus	(Pastscape National Monument Number 879395) States there are several gaps along the course of both side ditches – potentially causeways.	1979–84 Hedges and Buckley
Stanwell Cursus Complex	(Lewis et al. 2010, p. 31) Identify at least four causeways in the Stanwell C1 Cursus, at least one in the Stanwell C2 Cursus and a further four in the Stanwell C3 Cursus.	1979–85 O'Connell 2006–10 Framework Archaeology
Biggleswade Cursus	No evidence of causeways.	2004 Albion Archaeology
Cardington Cursus	(Loveday 2006, p. 30) Identifies two possible causeways in the side ditches of the Cardington Cursus.	
Wolverton Cursus Complex	(Hogan 2013b) Identifies a possible two causeways in the Wolverton 1 Cursus and at least two, possibly four, causeways in the Wolverton 2 Cursus.	2008–11 Cambridge Archaeological Unit
Benson Cursus	(Barclay and Hey 1999, p. 72) Identify at least two causeways in the western ditch and at least two, possibly three, in the eastern ditch of the Benson Cursus.	

Cursus monument site	Possible Causeways	Excavation
Dorchester Cursus	(Loveday 1999, p. 51) Identifies a possible six causeways in the south-western ditch and a possible three causeways in the north-eastern ditch of the Dorchester-on-Thames Cursus.	1947–52 Atkinson 1981 Chambers 1988 Bradley and Chambers 2010–17 Hey
Drayton North Cursus	(Barclay et al. 2003, p. 9) identifies four causeways in the western ditch and three causeways in the eastern ditch of the Drayton North Cursus.	1977 Parrington 1979–82 Ainslie and Wallis 1985–86 Oxford Archaeological Unit
Drayton South Cursus	(Barclay et al. 2003, p. 9) Identifies one causeway in the eastern ditch of the Drayton South Cursus.	1921–37 E T Leeds 1994 Oxford Archaeological Unit
Buscot Cursus	(Barclay and Hey 1999, p. 72) Identify a causeway in both the southern and northern ditches of the Buscot Cursus.	
Lechlade Cursus	No evidence of causeways	1965 Vatcher & Vatcher 1985 Oxford Archaeological Unit
North Stoke Cursus	(Loveday 2006, p. 94) Identifies two possible causeways in side ditches at southern terminal.	1982 Case
Sonning Cursus	(Ford 1987) The site was described as having a markedly rectangular end with entrance gap – possibly a causeway – at its far eastern end	
Stonehenge Greater Cursus	(Gaffney et al. 2012) Causeways identified during Stonehenge Hidden Landscapes Project.	1947 Stone 1963 Christie
Stonehenge Lesser Cursus	(Richards 1990) Identified causeways during excavations undertaken as part of Stonehenge Environs Project.	1983 Richards
Gussage Dorset Cursus	(Gill 2019, p. 30) Geophysics survey of northern ditch of Dorset Cursus in November 2018 reveals two small causeways.	1953 Atkinson 1986 Bradley 1991 Bradley 1992 Green
Pentridge Dorset Cursus	(Gill 2019, p. 30) Geophysics survey of northern ditch of Dorset Cursus in November 2018 reveals two small causeways.	1953 Atkinson 1982 Barrett, Bradley & Green 1984 Barrett, Bradley & Green

Appendix C Cursus monuments associated with control of cattle movement

Cursus monument site	**Associated with linear movement of cattle (Droveway)**	**Associated with cattle being led to water**	**Associated with water crossing point**	**Associated with (Spring meadows)**	**Associated with potential cattle pen/corral**
Rudston A Cursus	Yes	Yes	Yes		
Rudston B Cursus	Yes	Yes			
Rudston C Cursus	Yes	Yes	Yes		
Rudston D Cursus	Yes	Yes		Yes	
Duggleby Cursus	Yes			Yes	
Fimber Cursus	Yes				
Kirby Underdale Cursus	Yes				
Bag Enderby Pit Alignment	Yes			Yes	
Harlaxton					
Stenigot Pit Alignment	Yes			Yes	
Hanworth Cursus	Yes			Yes	
Fornham All Saints Cursus	Yes			Yes	
Stratford St Mary	Yes				
Barnack Cursus	Yes				
Eynesbury Cursus	Yes	Yes		Yes	Yes
Godmanchester Cursus	Yes	Yes			Yes
Brampton Cursus	Yes			Yes	
Maxey Cursus	Yes			Yes	
Etton Cursus	Yes			Yes	
Springfield Cursus	Yes				
Stanwell Cursus 1	Yes				Yes
Stanwell Cursus 2	Yes	Yes	Yes		Yes
Stanwell Cursus 3	Yes	Yes	Yes		Yes
Stanwell Cursus 4	Yes				
Stanwell Cursus 5	Yes	Yes	Yes		Yes
Biggleswade Cursus	Yes	Yes		Yes	
Cardington Cursus	Yes	Yes		Yes	
Cople Cursus	Yes	Yes		Yes	
Ivinghoe Beacon Cursus	Yes				
Wolverton Cursus 1	Yes	Yes	Yes	Yes	Yes
Wolverton Cursus 2	Yes			Yes	Yes
Wolverton Cursus 3	Yes			Yes	
Wolverton Cursus 4	Yes	Yes	Yes	Yes	Yes
Wolverton Cursus 5	Yes			Yes	
Benson Cursus	Yes	Yes		Yes	Yes
Dorchester Cursus	Yes				
Drayton St Leonard	Yes			Yes	
Drayton North Cursus	Yes				

Cursus monument site	**Associated with linear movement of cattle (Droveway)**	**Associated with cattle being led to water**	**Associated with water crossing point**	**Associated with (Spring meadows)**	**Associated with potential cattle pen/corral**
Drayton South Cursus	Yes				
Buscot Cursus	Yes			Yes	
Lechlade Cursus	Yes	Yes		Yes	
North Stoke Cursus	Yes			Yes	Yes
South Stoke Cursus	Yes			Yes	Yes
Stadhampton Cursus	Yes	Yes		Yes	
Sonning Cursus	Yes	Yes		Yes	
Stonehenge Greater Cursus	Yes	Yes	Yes	Yes	
Stonehenge Lesser Cursus	Yes				
Yatesbury Cursus (Avebury)	Yes	Yes	Yes	Yes	
Gussage Dorset Cursus	Yes	Yes	Yes		
Pentridge Dorset Cursus	Yes	Yes	Yes		
Total (%)	98	44	22	56	24

Appendix D1 Summary of cursus monuments – Construction dates

Cursus monument Site	OS Co-ordinates	Excavated	Date	Dating Method
Rudston A Cursus	TA 099657 to TA 101680	Yes	3750–3250 BC	Radiocarbon
Rudston B Cursus	TA 081669 to TA 094675			
Rudston C Cursus	TA 089680 to TA 099680			
Rudston D Cursus	TA 099717 to TA 096679			
Duggleby Cursus	SE 879669 to SE 892670			
Fimber Cursus	SE 893610 to SE 907610			
Kirby Underdale Cursus	SK 823594 to SK 807586			
Bag Enderby Pit Alignment	TF 351725	Yes		
Harlaxton	SK891339			
Stenigot Pit Alignment	TF 245811 to TF 244812			
Hanworth Cursus	TG 207362			
Fornham All Saints Cursus	TL 829687 to TL 842672			
Stratford St Mary	TM 048343 to TM 046345			
Barnack Cursus	TF 083066 To TF 084067			
Eynesbury Cursus	TL 184584	Yes	4860–3450 BC	OSL
Godmanchester Cursus	TL255709			
Brampton Cursus	TL 203716	Yes		
Maxey Cursus	TF125078 To TF139063	Yes		
Etton Cursus	TF138074	Yes		
Springfield Cursus	TL725067 to TL 735084	Yes	3400–3000 BC	Radiocarbon
Stanwell Cursus 1		Yes	3600–3300 BC	Radiocarbon
Stanwell Cursus 2		Yes	3600–3300 BC	Radiocarbon
Stanwell Cursus 3		Yes	3600–3300 BC	Radiocarbon
Stanwell Cursus 4		Yes	3600–3300 BC	Radiocarbon
Stanwell Cursus 5		Yes	3600–3300 BC	Radiocarbon
Biggleswade Cursus	TL 197466	Yes		
Cardington Cursus	TL 089497 To TL 089499	Yes		
Cople Cursus	TL 093500			
Ivinghoe Beacon Cursus	SP 961168	Geophys		
Wolverton Cursus 1	SP 802423	Yes	3500–3000 BC	Radiocarbon
Wolverton Cursus 2	SP 804421 to SP 807423	Yes	3500–3000 BC	Radiocarbon
Wolverton Cursus 3	SP807423 to SP 808423	Yes	3500–3000 BC	Radiocarbon
Wolverton Cursus 4	SP 803419	Yes	3500–3000 BC	Radiocarbon
Wolverton Cursus 5	SP 808423	Yes	3500–3000 BC	Radiocarbon
Benson Cursus	SU 624910 to SU 629919			
Dorchester Cursus	SU 569958 to SU 581948	Yes	3380–2920 BC	Radiocarbon
Drayton St Leonard	SU 601969 to SU 602973			
Drayton North Cursus	SU 490941 to SU 492950	Yes	3610–3380 BC	Radiocarbon
Drayton South Cursus	SU 486935 To SU 489941	Yes		
Buscot Cursus	SU 217989 to SU 222985			
Lechlade Cursus	SP 212002 to SP 212005	Yes		
North Stoke Cursus	SU 611856		3630–3340 BC	Radiocarbon

Cursus monument Site	OS Co-ordinates	Excavated	Date	Dating Method
South Stoke Cursus	SU 595831			
Stadhampton Cursus	SU 599991 to SU 597985			
Sonning Cursus	SU 767760	Yes		
Stonehenge Greater Cursus	SU 109429 to SU 137431	Yes	3630–3370 BC	Radiocarbon
Stonehenge Lesser Cursus	SU 103434 to SU 107435	Yes	3640–3130 BC	Radiocarbon
Yatesbury Cursus (Avebury)				
Gussage Dorset Cursus	ST 969125 to SU 018160	Yes	3360–3030 BC	Radiocarbon
Pentridge Dorset Cursus	SU 018160 to SU 040193	Yes		

Appendix D2 Summary of cursus monuments – Classification of Cursus Monument

Cursus monument Site	Length (m)	Width (m)	Long Mortuary Enclosure	Minor Cursus	Major Cursus	Mega Cursus
Rudston A Cursus	2,700	70				Yes
Rudston B Cursus	1,550	90			Yes	
Rudston C Cursus	1,480	60			Yes	
Rudston D Cursus	4,000	50–90				Yes
Duggleby Cursus	1,200	160			Yes	
Fimber Cursus	1,300	18–27			Yes	
Kirby Underdale Cursus	800	30			Yes	
Bag Enderby Pit Alignment	129	46	Yes			
Harlaxton	250			Yes		
Stenigot Pit Alignment	148	140	Yes			
Hanworth Cursus	380	55		Yes		
Fornham All Saints Cursus	1,900	42			Yes	
Stratford St Mary	295	65		Yes		
Barnack Cursus	120	20	Yes			
Eynesbury Cursus	200	15		Yes		
Godmanchester Cursus	500	90		Yes		
Brampton Cursus	300	25		Yes		
Maxey Cursus	1,710	58			Yes	
Etton Cursus	2,000	58			Yes	
Springfield Cursus	680	45		Yes		
Stanwell Cursus 1	3,800	20				Yes
Stanwell Cursus 2	480	60		Yes		
Stanwell Cursus 3	230	19		Yes		
Stanwell Cursus 4	82	21	Yes			
Stanwell Cursus 5	230	19		Yes		
Biggleswade Cursus	750	75		Yes		
Cardington Cursus	180			Yes		
Cople Cursus	125	15	Yes			
Ivinghoe Beacon Cursus	140	30	Yes			
Wolverton Cursus 1	Unknown	20		Yes		

Cursus monument Site	Length (m)	Width (m)	Long Mortuary Enclosure	Minor Cursus	Major Cursus	Mega Cursus
Wolverton Cursus 2	400	30		Yes		
Wolverton Cursus 3	300	50		Yes		
Wolverton Cursus 4	100	Unknown		Yes		
Wolverton Cursus 5	80	15	Yes			
Benson Cursus	1,090	65			Yes	
Dorchester Cursus	1,600	64			Yes	
Drayton St Leonard	410	45		Yes		
Drayton North Cursus	650	75			Yes	
Drayton South Cursus	750	70			Yes	
Buscot Cursus	750	50		Yes		
Lechlade Cursus	300	45		Yes		
North Stoke Cursus	240	20		Yes		
South Stoke Cursus	200	30		Yes		
Stadhampton Cursus	400	45		Yes		
Sonning Cursus	250	45		Yes		
Stonehenge Greater Cursus	2,730	100–50				Yes
Stonehenge Lesser Cursus	400	60		Yes		
Yatesbury Cursus (Avebury)					Yes	
Gussage Dorset Cursus	5,640	90				Yes
Pentridge Dorset Cursus	4,290	90				Yes

Appendix D3 Summary of cursus monuments – Terminal types

Cursus monument Site	Terminal Type Rounded	Terminal Type Squared	Terminal Type Bi
Rudston A Cursus		Yes	
Rudston B Cursus		Yes	
Rudston C Cursus		Ends not found	
Rudston D Cursus		Yes	
Duggleby Cursus		Ends not found	
Fimber Cursus		Ends not found	
Kirby Underdale Cursus		Ends not found	
Bag Enderby Pit Alignment		Pit Alignment	
Harlaxton		Pit Alignment	
Stenigot Pit Alignment		Pit Alignment	
Hanworth Cursus		Yes	
Fornham All Saints Cursus	Yes		
Stratford St Mary		Yes	
Barnack Cursus			Yes
Eynesbury Cursus			Yes
Godmanchester Cursus		Ends not found	
Brampton Cursus		Yes	

Cursus monument Site	**Terminal Type Rounded**	**Terminal Type Squared**	**Terminal Type Bi**
Maxey Cursus			
Etton Cursus			Yes
Springfield Cursus			Yes
Stanwell Cursus 1	Yes		
Stanwell Cursus 2	Yes		
Stanwell Cursus 3		Yes	
Stanwell Cursus 4	Yes		
Stanwell Cursus 5	Yes		
Biggleswade Cursus			Yes
Cardington Cursus			Yes
Cople Cursus	Yes		
Ivinghoe Beacon Cursus	Yes		
Wolverton Cursus 1		Yes	
Wolverton Cursus 2		Yes	
Wolverton Cursus 3		Yes	
Wolverton Cursus 4	Yes		
Wolverton Cursus 5			Yes
Benson Cursus			Yes
Dorchester Cursus	Yes		
Drayton St Leonard			Yes
Drayton North Cursus			Yes
Drayton South Cursus			Yes
Buscot Cursus			Yes
Lechlade Cursus			Yes
North Stoke Cursus		Yes	
South Stoke Cursus	Yes		
Stadhampton Cursus			Yes
Sonning Cursus			Yes
Stonehenge Greater Cursus		Yes	
Stonehenge Lesser Cursus		Yes	
Yatesbury Cursus (Avebury)		Ends not found	
Gussage Dorset Cursus		Yes	
Pentridge Dorset Cursus		Yes	

Appendix D4 Summary of cursus monuments – Straightness of cursus monument ditches

Cursus monument Site	**Cursus ditch straight**	**Cursus ditch sinuous**	**Cursus ditch Angular**
Rudston A Cursus		Yes	
Rudston B Cursus	Yes		
Rudston C Cursus	Yes		
Rudston D Cursus			Yes
Duggleby Cursus		Yes	
Fimber Cursus	Yes		
Kirby Underdale Cursus		Unknown	
Bag Enderby Pit Alignment	Yes		
Harlaxton	Yes		
Stenigot Pit Alignment	Yes		
Hanworth Cursus	Yes		
Fornham All Saints Cursus			Yes
Stratford St Mary	Yes		
Barnack Cursus	Yes		
Eynesbury Cursus		Yes	
Godmanchester Cursus		Yes	
Brampton Cursus		Yes	
Maxey Cursus		Yes	
Etton Cursus	Yes		
Springfield Cursus	Yes		
Stanwell Cursus 1	Yes		
Stanwell Cursus 2	Yes		
Stanwell Cursus 3		Yes	
Stanwell Cursus 4	Yes		
Stanwell Cursus 5	Yes		
Biggleswade Cursus	Yes		
Cardington Cursus	Yes		
Cople Cursus	Yes		
Ivinghoe Beacon Cursus	Yes		
Wolverton Cursus 1	Yes		
Wolverton Cursus 2	Yes		
Wolverton Cursus 3			Yes
Wolverton Cursus 4	Yes		
Wolverton Cursus 5	Yes		
Benson Cursus	Yes		
Dorchester Cursus	Yes		
Drayton St Leonard	Yes		
Drayton North Cursus	Yes		
Drayton South Cursus	Yes		
Buscot Cursus	Yes		
Lechlade Cursus	Yes		
North Stoke Cursus	Yes		

Cursus monument Site	Cursus ditch straight	Cursus ditch sinuous	Cursus ditch Angular
South Stoke Cursus	Yes		
Stadhampton Cursus	Yes		
Sonning Cursus	Yes		
Stonehenge Greater Cursus	Yes		
Stonehenge Lesser Cursus	Yes		
Yatesbury Cursus (Avebury)			
Gussage Dorset Cursus		Yes	
Pentridge Dorset Cursus		Yes	

Appendix D5 Summary of cursus monuments – Geology

Cursus monument site	Location	Geology
Rudston A Cursus	Rolling Downland	Chalk
Rudston B Cursus	Rolling Downland	Chalk
Rudston C Cursus	Rolling Downland	Chalk
Rudston D Cursus	Rolling Downland	Chalk
Duggleby Cursus	Rolling Downland	Chalk
Fimber Cursus	Rolling Downland	Chalk
Kirby Underdale Cursus	Rolling Downland	Chalk
Bag Enderby Pit Alignment	Rolling Downland	Chalk
Harlaxton	Raised terrace	Mudstone and muddy limestone
Stenigot Pit Alignment	Raised terrace	Chalk
Hanworth Cursus	Rolling Downland	Chalk
Fornham All Saints Cursus	Raised terrace	Chalk
Stratford St Mary	Gently sloping gravel terrace	Sand, clay and gravel
Barnack Cursus	Raised terrace	Mudstone and muddy limestone
Eynesbury Cursus	Gently sloping gravel terrace	Mudstone and muddy limestone
Godmanchester Cursus	Landscape destroyed by dev	Mudstone and muddy limestone
Brampton Cursus	Gently sloping gravel terrace	Mudstone and muddy limestone
Maxey Cursus	Raised terrace	Mudstone and muddy limestone
Etton Cursus	Raised terrace	Mudstone and muddy limestone
Springfield Cursus	Sloping gravel terrace	London clay
Stanwell Cursus 1	Raised terrace	London clay
Stanwell Cursus 2	Raised terrace	London clay
Stanwell Cursus 3	Raised terrace	London clay
Stanwell Cursus 4	Raised terrace	London clay
Stanwell Cursus 5	Raised terrace	London clay
Biggleswade Cursus	Sloping gravel terrace	Kelaway and Oxford clay
Cardington Cursus	Sloping gravel terrace	Kelaway and Oxford clay
Cople Cursus	Sloping gravel terrace	Kelaway and Oxford clay
Ivinghoe Beacon Cursus	Rolling Downland	Chalk with flint capping
Wolverton Cursus 1	Gravel terrace	Kelaway and Oxford clay

Cursus monument site	Location	Geology
Wolverton Cursus 2	Gravel terrace	Kelaway and Oxford clay
Wolverton Cursus 3	Gravel terrace	Kelaway and Oxford clay
Wolverton Cursus 4	Gravel terrace	Kelaway and Oxford clay
Wolverton Cursus 5	Gravel terrace	Kelaway and Oxford clay
Benson Cursus	Lower slopes of Chiltern Hills	Gault and upper greensand
Dorchester Cursus	2nd floodplain gravel terrace	Gault and upper greensand
Drayton St Leonard	2nd floodplain gravel terrace	Gault and upper greensand
Drayton North Cursus	Floodplain gravel terrace	Gault and upper greensand
Drayton South Cursus	2nd floodplain gravel terrace	Gault and upper greensand
Buscot Cursus	Floodplain gravel terrace	Gault and upper greensand
Lechlade Cursus	2nd floodplain gravel terrace	Gault and upper greensand
North Stoke Cursus	2nd floodplain gravel terrace	Gault and upper greensand
South Stoke Cursus	2nd floodplain gravel terrace	Gault and upper greensand
Stadhampton Cursus	2nd floodplain gravel terrace	Gault and upper greensand
Sonning Cursus	2nd floodplain gravel terrace	Gault and upper greensand
Stonehenge Greater Cursus	Rolling Downland	Chalk
Stonehenge Lesser Cursus	Rolling Downland	Chalk
Yatesbury Cursus (Avebury)	Rolling Downland	Chalk
Gussage Dorset Cursus	Rolling Downland	Chalk
Pentridge Dorset Cursus	Rolling Downland	Chalk

Appendix D6 Summary of cursus monuments – Association with rivers and streams

Cursus monument Site	**Associated with rivers and streams**			
	Crosses river at right angle	**Crosses river at angle**	**Runs parallel to river**	**River runs at end of Cursus**
Rudston A Cursus	Yes			
Rudston B Cursus	Yes			
Rudston C Cursus		Yes		
Rudston D Cursus			Yes	
Duggleby Cursus				Yes
Fimber Cursus				
Kirby Underdale Cursus			Yes	
Bag Enderby Pit Alignment			Yes	
Harlaxton				Yes
Stenigot Pit Alignment			Yes	
Hanworth Cursus				Yes
Fornham All Saints Cursus			Yes	
Stratford St Mary			Yes	
Barnack Cursus				
Eynesbury Cursus			Yes	
Godmanchester Cursus			Yes	
Brampton Cursus			Yes	
Maxey Cursus			Yes	
Etton Cursus			Yes	
Springfield Cursus			Yes	
Stanwell Cursus 1			Yes	
Stanwell Cursus 2				Yes
Stanwell Cursus 3				Yes
Stanwell Cursus 4				Yes
Stanwell Cursus 5				Yes
Biggleswade Cursus				Yes
Cardington Cursus			Yes	
Cople Cursus			Yes	
Ivinghoe Beacon Cursus				
Wolverton Cursus 1	Yes			
Wolverton Cursus 2			Yes	
Wolverton Cursus 3			Yes	
Wolverton Cursus 4	Yes			
Wolverton Cursus 5			Yes	
Benson Cursus				Yes
Dorchester Cursus			Yes	
Drayton St Leonard				Yes
Drayton North Cursus			Yes	
Drayton South Cursus				Yes
Buscot Cursus			Yes	Yes
Lechlade Cursus			Yes	Yes

Cursus monument Site	**Associated with rivers and streams**			
North Stoke Cursus			Yes	
South Stoke Cursus			Yes	
Stadhampton Cursus			Yes	Yes
Sonning Cursus			Yes	
Stonehenge Greater Cursus	Yes			
Stonehenge Lesser Cursus				
Yatesbury Cursus (Avebury)				Yes
Gussage Dorset Cursus	Yes			
Pentridge Dorset Cursus	Yes			

Appendix D7 Summary of cursus monuments – Alignment

Cursus monument site	**Alignment**			
	Long barrow	**Causewayed enclosure**	**Distant hills**	**Animal movement**
Rudston A Cursus	Yes			Yes
Rudston B Cursus	Yes			Yes
Rudston C Cursus	Yes			Yes
Rudston D Cursus				Yes
Duggleby Cursus		Yes		Yes
Fimber Cursus	Yes		Yes	Yes
Kirby Underdale Cursus				Yes
Bag Enderby Pit Alignment				Yes
Harlaxton				90 degrees
Stenigot Pit Alignment	Yes			Yes
Hanworth Cursus		Yes		Yes
Fornham All Saints Cursus		Yes		Yes
Stratford St Mary				Yes
Barnack Cursus		Yes		Yes
Eynesbury Cursus	Yes			Yes
Godmanchester Cursus				Yes
Brampton Cursus				Yes
Maxey Cursus		Yes		Yes
Etton Cursus	Yes	Yes		Yes
Springfield Cursus		Yes		Yes
Stanwell Cursus 1				Yes
Stanwell Cursus 2				Yes
Stanwell Cursus 3				Yes
Stanwell Cursus 4				Yes
Stanwell Cursus 5				Yes
Biggleswade Cursus				Yes
Cardington Cursus		Yes		Yes
Cople Cursus				Yes
Ivinghoe Beacon Cursus				Yes

Cursus monument site	Alignment			
Wolverton Cursus 1				Yes
Wolverton Cursus 2				Yes
Wolverton Cursus 3				Yes
Wolverton Cursus 4				Yes
Wolverton Cursus 5				Yes
Benson Cursus	Yes			Yes
Dorchester Cursus	Yes			Yes
Drayton St Leonard	Yes			Yes
Drayton North Cursus	Yes			Yes
Drayton South Cursus				Yes
Buscot Cursus				Yes
Lechlade Cursus		Yes		Yes
North Stoke Cursus	Yes			Yes
South Stoke Cursus				Yes
Stadhampton Cursus				Yes
Sonning Cursus				Yes
Stonehenge Greater Cursus	Yes		Yes	Yes
Stonehenge Lesser Cursus		Yes		Potential
Yatesbury Cursus (Avebury)				Yes
Gussage Dorset Cursus	Yes			Yes
Pentridge Dorset Cursus	Yes			Yes

Appendix D8 Summary of cursus monuments – Celestial Alignment

Cursus monument site	Alignment		
	Summer solstice	**Winter solstice**	**Moon**
Rudston A Cursus			
Rudston B Cursus	Yes		
Rudston C Cursus			
Rudston D Cursus			
Duggleby Cursus			
Fimber Cursus			
Kirby Underdale Cursus			
Bag Enderby Pit Alignment			
Harlaxton			
Stenigot Pit Alignment			
Hanworth Cursus	Yes	Yes	
Fornham All Saints Cursus			
Stratford St Mary			
Barnack Cursus	Yes		
Eynesbury Cursus			
Godmanchester Cursus	Yes		Yes
Brampton Cursus			
Maxey Cursus			

Cursus monument site	**Alignment**		
Etton Cursus			
Springfield Cursus	Yes		
Stanwell Cursus 1			
Stanwell Cursus 2			
Stanwell Cursus 3			
Stanwell Cursus 4			
Stanwell Cursus 5			
Biggleswade Cursus			
Cardington Cursus			
Cople Cursus	Yes		
Ivinghoe Beacon Cursus	Yes		
Wolverton Cursus 1			
Wolverton Cursus 2	Yes		
Wolverton Cursus 3			
Wolverton Cursus 4			
Wolverton Cursus 5			
Benson Cursus	Yes		
Dorchester Cursus		Yes	
Drayton St Leonard			
Drayton North Cursus	Yes		
Drayton South Cursus	Yes		
Buscot Cursus	Yes		
Lechlade Cursus			
North Stoke Cursus			
South Stoke Cursus			
Stadhampton Cursus			
Sonning Cursus			
Stonehenge Greater Cursus			
Stonehenge Lesser Cursus			
Yatesbury Cursus (Avebury)			
Gussage Dorset Cursus	Yes		
Pentridge Dorset Cursus			

Appendix D9 Summary of cursus monuments – Neolithic arrowheads

Cursus monument Site	Neolithic arrowhead type
Rudston A Cursus	Arrowhead at South Side Mound
Rudston B Cursus	
Rudston C Cursus	
Rudston D Cursus	16 leaf-shaped arrowheads from North Burton
Duggleby Cursus	
Fimber Cursus	Arrowheads
Kirby Underdale Cursus	Leaf arrowhead
Bag Enderby Pit Alignment	
Harlaxton	Leaf and barbed and tanged arrowheads
Stenigot Pit Alignment	
Hanworth Cursus	
Fornham All Saints Cursus	
Stratford St Mary	
Barnack Cursus	
Eynesbury Cursus	Barbed and tanged arrowhead
Godmanchester Cursus	
Brampton Cursus	
Maxey Cursus	Arrowheads
Etton Cursus	
Springfield Cursus	
Stanwell Cursus 1	Barbed and tanged arrowhead
Stanwell Cursus 2	Barbed and tanged arrowhead
Stanwell Cursus 3	Barbed and tanged arrowhead
Stanwell Cursus 4	Barbed and tanged arrowhead
Stanwell Cursus 5	Barbed and tanged arrowhead
Biggleswade Cursus	
Cardington Cursus	
Cople Cursus	
Ivinghoe Beacon Cursus	
Wolverton Cursus 1	A mixed flint industry of Neolithic date were found
Wolverton Cursus 2	A mixed flint industry of Neolithic date were found
Wolverton Cursus 3	A mixed flint industry of Neolithic date were found
Wolverton Cursus 4	A mixed flint industry of Neolithic date were found
Wolverton Cursus 5	A mixed flint industry of Neolithic date were found
Benson Cursus	Barbed and tanged arrowheads found at Benson Hill
Dorchester Cursus	Wessex-type and lozenge-type arrowhead
Drayton St Leonard	Lozenge-type arrowhead
Drayton North Cursus	22 arrowheads
Drayton South Cursus	Barbed and tanged arrowhead
Buscot Cursus	
Lechlade Cursus	Leaf-shaped arrowhead
North Stoke Cursus	Flint arrowhead

Cursus monument Site	**Neolithic arrowhead type**
South Stoke Cursus	
Stadhampton Cursus	
Sonning Cursus	
Stonehenge Greater Cursus	Numerous arrowheads
Stonehenge Lesser Cursus	Numerous arrowheads
Yatesbury Cursus (Avebury)	
Gussage Dorset Cursus	Numerous arrowheads at terminal ends
Pentridge Dorset Cursus	Numerous arrowheads at terminal ends

Appendix D10 Mesolithic finds in the vicinity of cursus monuments

Cursus monument	**National Monument Number**	**Finds**
Rudston A Cursus	81218 910812	Approximate site of two long barrows, one with a possible round barrow at the western end, recorded by Greenwell circa 1870. The long barrow contained numerous animal bones, flint chippings, charcoal and sherds of plain, dark-coloured pottery, principally at the level of the old ground surface.
		A curved flint sickle blade fragment from Rudston is in the British Museum.
Rudston B Cursus	910837	A tranchet axe was found in 1928 on ploughed land to the north of Kilham Lane between Kilham and Rudston.
Rudston C Cursus	Manby (1976, pp. 133–37)	Mesolithic flint flakes beneath the Neolithic soil levels at Octon Long Barrow.
Rudston D Cursus	910808	Sixteen flint leaf-shaped arrowheads from North Burton.
Duggleby Cursus	Hurst (1983, pp. 77–78)	Mesolithic flints found near source of Gypsey Race.
Fimber Cursus	910846 910845	Three Mesolithic pebble maceheads and one tranchet axe were found from Fimber. A Mesolithic tranchet axe was found from Towthorpe.
Kirby Underdale Cursus	Painsthorpe 99	Two groups of flints which included 21 flakes, leaf arrowheads, serrated flakes and a knife.
Bag Enderby Pit Alignment	354419 354461	Flint chippings found on the field around the edge of the quarry.
		A Neolithic flint axe head.
Harlaxton	323811 323697 323799	Prehistoric implements including microliths (Mesolithic or possibly Upper Palaeolithic). Flints – including leaf and barbed and tanged arrowheads
		Prehistoric implements including scrapers, broken Neolithic axes, re-used flakes of polished axes; transverse, leaf-shaped and barbed and tanged arrowheads.
Stenigot Pit Alignment	1375375	A Neolithic long barrow is visible as a cropmark on air photographs taken on 25th May 2012.
Hanworth Cursus		Currently no record of Mesolithic activity.
Fornham All Saints Cursus Bury St Edmunds	382839	Mesolithic tranchet axe and flints including a platform core and 2 blades.
Stratford St Mary	386604	A scatter of flints was found in 1958 on fields between Higham Road and the river west of the A12 at Stratford St Mary.

Cursus monument	**National Monument Number**	**Finds**
Barnack Cursus	350238 348124	Scatter of Mesolithic flints were found just north of Maxey.
	361487	Mesolithic flint scatter was found at Southorpe.
	871244	Mesolithic implements including two cores, eight blades or flakes and four scrapers were found at Wansford.
Eynesbury Cursus	362837	Exploratory excavation of part of a ring ditch on the line of the St Neots bypass revealed a quarry pit of Mesolithic date, Neolithic flint debris and microliths.
	363429	Barbed and tanged flint arrowhead found in this field in 1957. Mesolithic tranchet axe found in St Neots.
	362868	Mesolithic flints, comprising 16 cores, 29 blades and flakes, 5 scrapers and 3 other flints.
Godmanchester Cursus	366807	Neolithic flakes found on rising land above river.
	366701	Mesolithic flint core and flake found at Hartford.
	366702	A Mesolithic tranchet axe found at Hartford.
Brampton Cursus	366807	Neolithic flakes found on rising land above river.
	366701	Mesolithic flint core and flake found at Hartford.
	366702	A Mesolithic tranchet axe found at Hartford.
Maxey Cursus	350238	Scatter of Mesolithic flints were found just north of Maxey.
	1217290	Several Mesolithic sites were found while undertaking watching briefs and a geophysical survey at Dyke 1, 2, 3 and 4, part of the south-west dyke survey.
	1215165	A Mesolithic flint scatter was discovered during an excavation at Crowtree Farm.
Etton Cursus	350238	Scatter of Mesolithic flints were found just north of Maxey.
	1217290	Several Mesolithic sites were found while undertaking watching briefs and a geophysical survey at Dyke 1, 2, 3 and 4, part of the south-west dyke survey.
	1215165	A Mesolithic flint scatter was discovered during an excavation at Crowtree Farm.
Springfield Cursus	375534	Several worked flints, including an arrowhead, a 'Neolithic knife' and a spearhead have been found in the gravel pit near Admiral's Park (TL 695074).
	879221	Collection of about 400 flints of Mesolithic to Neolithic date found in residual contexts.
	879425	Wymer lists one Mesolithic tranchet axe, twelve cores, 25 unretouched blades and flakes, two scrapers and one Microlith from Chelmsford.
	879189	Mesolithic, Neolithic and Bronze Age flints.
	879436	Mesolithic implements including 1 pebble macehead, 1 tranchet axe, 1 axe or adze, 40 cores, 52 unretouched flakes and blades, 40 scrapers, 3 microliths and 1 microburin from Great Baddow.
Stanwell Cursus 1	Landscape evolution in the Middle Thames Valley: Heathrow Terminal 5 Excavations	41 Mesolithic Flints from Cursus Ditches or tree throws. Pit Complex c7000–6500 BC. Posthole complex c6000 BC.

Cursus monument	National Monument Number	Finds
Stanwell Cursus 2	Landscape evolution in the Middle Thames Valley: Heathrow Terminal 5 Excavations	41 Mesolithic Flints from Cursus Ditches or tree throws. Pit Complex c7000–6500 BC. Posthole complex c6000 BC.
Stanwell Cursus 3	Landscape evolution in the Middle Thames Valley: Heathrow Terminal 5 Excavations	41 Mesolithic Flints from Cursus Ditches or tree throws. Pit Complex c7000–6500 BC. Posthole complex c6000 BC.
Stanwell Cursus 4	Landscape evolution in the Middle Thames Valley: Heathrow Terminal 5 Excavations	41 Mesolithic Flints from Cursus Ditches or tree throws. Pit Complex c7000–6500 BC. Posthole complex c6000 BC.
Stanwell Cursus 5	Landscape evolution in the Middle Thames Valley: Heathrow Terminal 5 Excavations	41 Mesolithic Flints from Cursus Ditches or tree throws. Pit Complex c7000–6500 BC. Posthole complex c6000 BC.
Biggleswade Cursus	Archi UK Database	Mesolithic flint microlith core found in Stratford
Cardington Cursus	1090270	Early Neolithic ground and polished axes
Cople Cursus	362728	Small barbed flint arrowhead with the tip missing was found in 1972 on a farm at Cople TL 105485.
Ivinghoe Beacon Cursus	346367 346405	A flint core, probably Mesolithic, was found on the surface of a ploughed field SP 972159.
		A Neolithic flint axe was found in the garden of 54 Glebe Close, Ivinghoe.
Wolverton Cursus 1	345117 Hogan 2011	A late Neolithic Settlement site at Stacey Bushes, Wolverton. A mixed flint industry of Mesolithic and Neolithic date were found.
		Manor Farm faunal remains of aurochs in Mesolithic tree throws.
Wolverton Cursus 2	345117 Hogan 2011	A late Neolithic Settlement site at Stacey Bushes, Wolverton. A mixed flint industry of Mesolithic and Neolithic date were found.
		Manor Farm faunal remains of aurochs in Mesolithic tree throws.
Wolverton Cursus 3	345117 Hogan 2011	A late Neolithic Settlement site at Stacey Bushes, Wolverton. A mixed flint industry of Mesolithic and Neolithic date were found.
		Manor Farm faunal remains of aurochs in Mesolithic tree throws.
Wolverton Cursus 4	345117 Hogan 2011	A late Neolithic Settlement site at Stacey Bushes, Wolverton. A mixed flint industry of Mesolithic and Neolithic date were found.
		Manor Farm faunal remains of aurochs in Mesolithic tree throws.
Wolverton Cursus 5	345117 Hogan 2011	A late Neolithic Settlement site at Stacey Bushes, Wolverton. A mixed flint industry of Mesolithic and Neolithic date were found.
		Manor Farm faunal remains of aurochs in Mesolithic tree throws.
Benson Cursus	242051	A flint pick found in the Thames near Benson Weir in 1912, is now in the Ashmolean Museum.

Cursus monument	National Monument Number	Finds
Dorchester Cursus		Currently no record of Mesolithic activity, however Wessex-type and lozenge-type arrowheads suggest hunting continues into the Neolithic period.
Drayton St Leonard		Currently no record of Mesolithic activity, however a lozenge-type arrowhead suggests hunting continues into the Neolithic period.
Drayton North Cursus	Holgate 2003	A high proportion of flintwork from the ground surface of the cursus monument is identified as Mesolithic indicating a seasonal occupation
Drayton South Cursus	Holgate 2003	A high proportion of flintwork from the ground surface of the Drayton North cursus monument is identified as Mesolithic indicating a seasonal occupation
Buscot Wick Cursus		Currently no record of Mesolithic activity.
Lechlade Cursus		Currently no record of Mesolithic activity.
North Stoke Cursus	1207006 241737	A Mesolithic flint scatter was discovered during an excavation at South Stoke.
		A Mesolithic tranchet axe from Little Stoke is in the Ashmolean Museum.
South Stoke Cursus	1207006 241737	A Mesolithic flint scatter was discovered during an excavation at South Stoke.
	241943	A Mesolithic tranchet axe from Little Stoke is in the Ashmolean Museum.
		Flint implements found at Watch Folley
Stadhampton Cursus		Currently no record of Mesolithic activity.
Sonning Cursus	244705	Ten Mesolithic tranchet axes were listed as being found at Sonning. The find spots were: four from the Thames, one from Sonning Golf course, one from Sonning Bridge, two from Sonning Cutting and two located to Sonning only.
Stonehenge Greater Cursus	Vatcher 1963 Wessex Archaeology 2015 Parker Pearson 2012 Parker Pearson 2012 Jacques 2014 959525 109688 Coady 2004	Mesolithic postholes, Stonehenge carpark. Mesolithic posthole Boscombe Down. Mesolithic long flint blades and microliths adjacent to spring at Blue Stonehenge. Mesolithic hunting camp 400 m south of Mesolithic postholes. Blick Mead, Mesolithic home base, 35,000 pieces of worked and burnt flint. A Mesolithic tranchet axe found south-west of Stonehenge. A pit of Mesolithic date encountered during excavations in the car park at Stonehenge in 1988–89. Mesolithic flint scatter west from Countess Farm.
Stonehenge Lesser Cursus	Vatcher 1963 Wessex Archaeology 2015 Parker Pearson 2012	Mesolithic postholes, Stonehenge carpark. Mesolithic posthole Boscombe Down. Mesolithic long flint blades and microliths adjacent to spring at Blue Stonehenge.
	Parker Pearson 2012 Jacques 2014	Mesolithic hunting camp 400 m south of Mesolithic postholes.
	959525 109688	Blick Mead, Mesolithic home base, 35,000 pieces of worked and burnt flint.
	Coady 2004	A Mesolithic tranchet axe found south-west of Stonehenge.

Cursus monument	National Monument Number	Finds
		A pit of Mesolithic date encountered during excavations in the car park at Stonehenge in 1988–89.
		Mesolithic flint scatter west from Countess Farm.
Yatesbury Cursus (Avebury)	Evans & Smith 1967 Kendall 1916	Mesolithic flints at home base at Cherhill. Mesolithic assemblages at Hackpen Hill.
	Evans 1993	Mesolithic flintwork indicating a short-stay camp.
	Pollard 2002 969847	8 Mesolithic find spots around the Galteemore Springs relating to hunting activity.
	969848 1011278	Mesolithic implements including axes and tranchet axes found at the foot of Avebury Down.
	969761	Mesolithic flint implements including a tranchet axe, pick and cores found on Four Barrow Hill. A Mesolithic pick was found near Beckhampton. A Mesolithic tranchet axe and a microlith both reportedly found along the course of the West
		Kennet Avenue.
	216364	A single microlith found on Windmill Hill.
Gussage Dorset Cursus	Green 2000 Higgs 888699	4 sites that contain a high percentage of microliths in the area of Down Farm are likely to represent hunting sites.
	213833 888922 888929	Mesolithic site at Downton. A Mesolithic core was found at Bowersmain Farm. A Mesolithic axe and two micro-scrapers have been found on Brockington Down, by Martin Green and Barry Lewis. Mesolithic flint implements including a pick, 3
		cores and 2 scrapers were found on East Chase Farm.
		A Mesolithic site at St Giles Field, Down Farm. A surface concentration of Mesolithic flints among a broader scatter was sampled by excavation in 1976.
Pentridge Dorset Cursus	213607 213623 888848	A scatter of Mesolithic cores and blades has been found on Pentridge Hill, Cranborne. Mesolithic and Neolithic flints from Penbury
	888856 213507	Knoll include axeheads, scrapers, microliths and other implement types.
	888853	Mesolithic and Neolithic flints, including an arrowhead, have been found on Pentridge Hill.
		Mesolithic flints including a tranchet axe, an axe, 12 blades and 4 scrapers were found in a field South of Penbury Knoll Camp.
		A Mesolithic tranchet axe and borer were found on Garston Down, Pentridge.
		Mesolithic flints including a tranchet axe and some unretouched blades and flakes have been found on Garston Down.

Appendix D11 Summary of cursus monuments – Openness of landscape

Cursus monument Site	Openness of landscape
Rudston A Cursus	Open
Rudston B Cursus	Open
Rudston C Cursus	Open
Rudston D Cursus	Open
Duggleby Cursus	
Fimber Cursus	
Kirby Underdale Cursus	
Bag Enderby Pit Alignment	
Harlaxton	
Stenigot Pit Alignment	
Hanworth Cursus	
Fornham All Saints Cursus	
Stratford St Mary	
Barnack Cursus	
Eynesbury Cursus	Open
Godmanchester Cursus	
Brampton Cursus	
Maxey Cursus	Open
Etton Cursus	Open
Springfield Cursus	
Stanwell Cursus 1	Cleared prior to Construction
Stanwell Cursus 2	Cleared prior to Construction
Stanwell Cursus 3	Cleared prior to Construction
Stanwell Cursus 4	Cleared prior to Construction
Stanwell Cursus 5	Cleared prior to Construction
Biggleswade Cursus	Open
Cardington Cursus	
Cople Cursus	
Ivinghoe Beacon Cursus	Open
Wolverton Cursus 1	
Wolverton Cursus 2	
Wolverton Cursus 3	
Wolverton Cursus 4	
Wolverton Cursus 5	
Benson Cursus	
Dorchester Cursus	Partly wooded
Drayton St Leonard	
Drayton North Cursus	Partly wooded
Drayton South Cursus	
Buscot Cursus	
Lechlade Cursus	
North Stoke Cursus	

Cursus monument Site	Openness of landscape
South Stoke Cursus	
Stadhampton Cursus	
Sonning Cursus	
Stonehenge Greater Cursus	Open
Stonehenge Lesser Cursus	Open
Yatesbury Cursus (Avebury)	
Gussage Dorset Cursus	Open
Pentridge Dorset Cursus	Open

Appendix D12 Summary of cursus monuments – Associated with river, floodplain or mist

Cursus monument site	Associated with river or floodplain	Associated with springs	Associated with mist	Area restricted to animal movement due to floodplains or marshland
Rudston A Cursus	Yes	Yes	Yes	Yes
Rudston B Cursus	Yes	Yes	Yes	Yes
Rudston C Cursus	Yes	Yes	Yes	Yes
Rudston D Cursus	Yes	Yes	Yes	Yes
Duggleby Cursus	Yes	Yes	Yes	Yes
Fimber Cursus			Yes	
Kirby Underdale Cursus	Yes	Yes	Yes	Yes
Bag Enderby Pit Alignment	Yes	Yes	Yes	Yes
Harlaxton	Yes	Yes	Yes	Yes
Stenigot Pit Alignment	Yes	Yes	Yes	Yes
Hanworth Cursus	Yes			Yes
Fornham All Saints Cursus	Yes		Yes	Yes
Stratford St Mary	Yes		Yes	Yes
Barnack Cursus	Yes		Yes	Yes
Eynesbury Cursus	Yes		Yes	Yes
Godmanchester Cursus	Yes		Yes	Yes
Brampton Cursus	Yes		Yes	Yes
Maxey Cursus	Yes		Yes	Yes
Etton Cursus	Yes		Yes	Yes
Springfield Cursus	Yes	Yes	Yes	Yes
Stanwell Cursus 1	Yes		Yes	Yes
Stanwell Cursus 2	Yes		Yes	Yes
Stanwell Cursus 3	Yes		Yes	Yes
Stanwell Cursus 4	Yes		Yes	Yes
Stanwell Cursus 5	Yes		Yes	Yes
Biggleswade Cursus	Yes	Yes	Yes	Yes
Cardington Cursus	Yes		Yes	Yes
Cople Cursus	Yes		Yes	Yes
Ivinghoe Beacon Cursus				
Wolverton Cursus 1	Yes	Yes	Yes	Yes

Cursus monument site	**Associated with river or floodplain**	**Associated with springs**	**Associated with mist**	**Area restricted to animal movement due to floodplains or marshland**
Wolverton Cursus 2	Yes	Yes	Yes	Yes
Wolverton Cursus 3	Yes	Yes	Yes	Yes
Wolverton Cursus 4	Yes	Yes	Yes	Yes
Wolverton Cursus 5	Yes	Yes	Yes	Yes
Benson Cursus	Yes	Yes	Yes	Yes
Dorchester Cursus	Yes		Yes	Yes
Drayton St Leonard	Yes		Yes	Yes
Drayton North Cursus	Yes		Yes	Yes
Drayton South Cursus	Yes		Yes	Yes
Buscot Cursus	Yes		Yes	Yes
Lechlade Cursus	Yes		Yes	Yes
North Stoke Cursus	Yes		Yes	Yes
South Stoke Cursus	Yes		Yes	Yes
Stadhampton Cursus	Yes		Yes	Yes
Sonning Cursus	Yes	Yes	Yes	Yes
Stonehenge Greater Cursus	Yes	Yes	Yes	Yes
Stonehenge Lesser Cursus		Yes	Yes	Yes
Yatesbury Cursus (Avebury)		Yes	Yes	Yes
Gussage Dorset Cursus	Yes	Yes	Yes	Yes
Pentridge Dorset Cursus	Yes	Yes	Yes	Yes

Appendix E Location of aurochs' bones with respect to cursus monuments

Cursus monument Site	**Excavated**	**Contains Aurochs' bones**	**Distance from Cursus monument**
Rudston A Cursus	Yes	Manby's (1976) excavations of the Kilham long barrow discovered 2 horn fragments and the proximal end of a mesial phalanx of a large auroch in pit 'B', the right astragalus of a large auroch in pit 'F' and an auroch's tooth in the secondary silt of the southern quarry ditch of trench 19.	10 kilometres from the Rudston Cursus complex
Rudston B Cursus		Manby's (1976) excavations of the Kilham long barrow discovered 2 horn fragments and the proximal end of a mesial phalanx of a large auroch in pit 'B', the right astragalus of a large auroch in pit 'F' and an auroch's tooth in the secondary silt of the southern quarry ditch of trench 19.	10 kilometres from the Rudston Cursus complex
Rudston C Cursus		Manby's (1976) excavations of the Kilham long barrow discovered 2 horn fragments and the proximal end of a mesial phalanx of a large auroch in pit 'B', the right astragalus of a large auroch in pit 'F' and an auroch's tooth in the secondary silt of the southern quarry ditch of trench 19.	10 kilometres from the Rudston Cursus complex

Cursus monument Site	Excavated	Contains Aurochs' bones	Distance from Cursus monument
Rudston D Cursus		Manby's (1976) excavations of the Kilham long barrow discovered 2 horn fragments and the proximal end of a mesial phalanx of a large auroch in pit 'B', the right astragalus of a large auroch in pit 'F' and an auroch's tooth in the secondary silt of the southern quarry ditch of trench 19.	10 kilometres from the Rudston Cursus complex
Duggleby Cursus			
Fimber Cursus			
Kirby Underdale Cursus			
Bag Enderby Pit Alignment	Yes		
Harlaxton			
Stenigot Pit Alignment			
Hanworth Cursus			
Fornham All Saints Cursus			
Stratford St Mary			
Barnack Cursus		Pryor (1998) discovered two aurochs' skulls, including their horns buried upon an oak plank within ditch segment No 12 of the Etton causewayed enclosure	5 Kilometres from the Cursus monument
Eynesbury Cursus	Yes		
Godmanchester Cursus			
Brampton Cursus	Yes		
Maxey Cursus	Yes	Pryor (1998) discovered two aurochs' skulls, including their horns buried upon an oak plank within ditch segment No 12 of the Etton causewayed enclosure	Immediate vicinity
Etton Cursus	Yes	Pryor (1998) discovered two aurochs' skulls, including their horns buried upon an oak plank within ditch segment No 12 of the Etton causewayed enclosure	Immediate vicinity
Springfield Cursus	Yes		
Stanwell Cursus 1	Yes	Grigson (1989) identified nine bones and teeth which were assumed to be from Mesolithic aurochs at Stratford's Yard, Chesham, in Buckinghamshire,	35 kilometres north-east of the Stanwell Cursus complex,
		Eton Rowing Lake, Dorney, a Mesolithic auroch's bone	Twelve kilometres north-east of the Stanwell Cursus complex,
		Holloway Lane, Cotton (1991) discovered a pit which contained the complete skeleton of an auroch which had been killed by six Conygar type barbed and tanged arrowheads, then butchered and placed in a large pit.	Two kilometres north-east of the Stanwell Cursus complex,

Cursus monument Site	Excavated	Contains Aurochs' bones	Distance from Cursus monument
Stanwell Cursus 2	Yes	Grigson (1989) identified nine bones and teeth which were assumed to be from Mesolithic aurochs at Stratford's Yard, Chesham, in Buckinghamshire,	35 kilometres north-east of the Stanwell Cursus complex,
		Eton Rowing Lake, Dorney, a Mesolithic auroch's bone	Twelve kilometres north-east of the Stanwell Cursus complex,
		Holloway Lane, Cotton (1991) discovered a pit which contained the complete skeleton of an auroch which had been killed by six Conygar type barbed and tanged arrowheads, then butchered and placed in a large pit.	Two kilometres north-east of the Stanwell Cursus complex,
Stanwell Cursus 3	Yes	Grigson (1989) identified a nine bones and teeth which were assumed to be from Mesolithic aurochs at Stratford's Yard, Chesham, in Buckinghamshire,	35 kilometres north-east of the Stanwell Cursus complex,
		Eton Rowing Lake, Dorney, a Mesolithic auroch's bone	Twelve kilometres north-east of the Stanwell Cursus complex,
		Holloway Lane, Cotton (1991) discovered a pit which contained the complete skeleton of an auroch which had been killed by six Conygar type barbed and tanged arrowheads, then butchered and placed in a large pit.	Two kilometres north-east of the Stanwell Cursus complex,
Stanwell Cursus 4	Yes	Grigson (1989) identified a nine bones and teeth which were assumed to be from Mesolithic aurochs at Stratford's Yard, Chesham, in Buckinghamshire,	35 kilometres north-east of the Stanwell Cursus complex,
		Eton Rowing Lake, Dorney, a Mesolithic auroch's bone	Twelve kilometres north-east of the Stanwell Cursus complex,
		Holloway Lane, Cotton (1991) discovered a pit which contained the complete skeleton of an auroch which had been killed by six Conygar type barbed and tanged arrowheads, then butchered and placed in a large pit.	Two kilometres north-east of the Stanwell Cursus complex,
Stanwell Cursus 5	Yes	Grigson (1989) identified a nine bones and teeth which were assumed to be from Mesolithic aurochs at Stratford's Yard, Chesham, in Buckinghamshire,	35 kilometres north-east of the Stanwell Cursus complex,
		Eton Rowing Lake, Dorney, a Mesolithic auroch's bone	Twelve kilometres north-east of the Stanwell Cursus complex,
		Holloway Lane, Cotton (1991) discovered a pit which contained the complete skeleton of an auroch which had been killed by six Conygar type barbed and tanged arrowheads, then butchered and placed in a large pit.	Two kilometres north-east of the Stanwell Cursus complex,
Biggleswade Cursus			
Cardington Cursus	Yes		

Cursus monument Site	**Excavated**	**Contains Aurochs' bones**	**Distance from Cursus monument**
Cople Cursus			
Ivinghoe Beacon Cursus			
Wolverton Cursus 1	Yes	Rajkovaca (2013) identified a near complete auroch's first phalanx from the Wolverton 1 Cursus, while two tree throws (F.96/97 and F.98) produced fragmentary remains identified as aurochs and further tree throws (F.106, F.111, F.112, F.114 and F.131) yielded a number of limb fragments which were identified as probably belonging to aurochs.	Immediate vicinity
Wolverton Cursus 2	Yes	Rajkovaca (2013) identified a near complete auroch's first phalanx from the Wolverton 1 Cursus, while two tree throws (F.96/97 and F.98) produced fragmentary remains identified as aurochs and further tree throws (F.106, F.111, F.112, F.114 and F.131) yielded a number of limb fragments which were identified as probably belonging to aurochs.	Immediate vicinity
Wolverton Cursus 3	Yes	Rajkovaca (2013) identified a near complete auroch's first phalanx from the Wolverton 1 Cursus, while two tree throws (F.96/97 and F.98) produced fragmentary remains identified as aurochs and further tree throws (F.106, F.111, F.112, F.114 and F.131) yielded a number of limb fragments which were identified as probably belonging to aurochs.	Immediate vicinity
Wolverton Cursus 4	Yes	Rajkovaca (2013) identified a near complete auroch's first phalanx from the Wolverton 1 Cursus, while two tree throws (F.96/97 and F.98) produced fragmentary remains identified as aurochs and further tree throws (F.106, F.111, F.112, F.114 and F.131) yielded a number of limb fragments which were identified as probably belonging to aurochs.	Immediate vicinity
Wolverton Cursus 5	Yes	Rajkovaca (2013) identified a near complete auroch's first phalanx from the Wolverton 1 Cursus, while two tree throws (F.96/97 and F.98) produced fragmentary remains identified as aurochs and further tree throws (F.106, F.111, F.112, F.114 and F.131) yielded a number of limb fragments which were identified as probably belonging to aurochs.	Immediate vicinity
Benson Cursus			
Dorchester Cursus	Yes	Auroch's skull discovered at Site II at Dorchester-on-Thames Cursus (Hey 2011)	Immediate vicinity
Drayton St Leonard			
Drayton North Cursus	Yes	Ayres and Powell (2003), who undertook the animal fauna investigation of the Drayton North Cursus identified two aurochs bones in the Cursus ditch	Immediate vicinity
Drayton South Cursus	Yes	Ayres and Powell (2003), who undertook the animal fauna investigation of the Drayton North Cursus identified two aurochs bones in the Drayton North Cursus ditch	One kilometre from Drayton South Cursus
Buscot Cursus			

Cursus monument Site	**Excavated**	**Contains Aurochs' bones**	**Distance from Cursus monument**
Lechlade Cursus	Yes		
North Stoke Cursus			
South Stoke Cursus			
Stadhampton Cursus			
Sonning Cursus		35 bones of aurochs skeleton discovered at Thames Valley Park, Reading	One kilometre from Sonning Cursus
Stonehenge Greater Cursus	Yes	Jacques (2014) identifies 155 bone fragments at Blick Mead that belong to aurochs, which made up 59 per cent of the total fauna assemblage.	Two kilometres from Stonehenge Greater Cursus Immediate vicinity
		Richards' (1990) suggests that 38 fragments of unidentified large mammal found within the ditches of the Stonehenge Lesser Cursus were potentially from aurochs	Five kilometres from Stonehenge Greater Cursus
		Discovery of an auroch's skeleton within a pit at Boscombe Down (Wessex Archaeology 2015)	
Stonehenge Lesser Cursus	Yes	Jacques (2014) identifies 155 bone fragments at Blick Mead that belong to aurochs, which made up 59 per cent of the total fauna assemblage.	Two kilometres from Stonehenge Greater Cursus Immediate vicinity
		Richards' (1990) suggests that 38 fragments of unidentified large mammal found within the ditches of the Stonehenge Lesser Cursus were potentially from aurochs	Five kilometres from Stonehenge Greater Cursus
		Discovery of an auroch's skeleton within a pit at Boscombe Down (Wessex Archaeology 2015)	
Yatesbury Cursus (Avebury)		Evens (1983) discovers aurochs' bones at Cherhill Mesolithic site	Five kilometres from Yatesbury Cursus
Gussage Dorset Cursus	Yes	Bradley (1991) discovers an auroch bone from the Down Farm Shaft and two aurochs bones from the western ditch of the Dorset Cursus	Immediate vicinity
Pentridge Dorset Cursus	Yes	Bradley (1991) discovers an auroch bone from the Down Farm Shaft and two aurochs bones from the western ditch of the Dorset Cursus	Maximum of five kilometres from the Pentridge section of the Dorset Cursus Cursus

Appendix F Cursus monument aerial photography and excavation records

Cursus monument site	Aerial photography	Excavation
Rudston A Cursus	APs (RAF 541/546/4066-7 1946) APs (NMR TA 0865/2 95 97 and 100 17.7.70.)	1877 William Greenwell 1958 C & E Grantham 1988 T G Manby
Rudston B Cursus	Aerial photographs taken by Dr J K St Joseph 1961	
Rudston C Cursus	Aerial photographs taken by Dr J K St Joseph 1961	
Rudston D Cursus		
Duggleby Cursus	RCHME/EH/HE Aerial Photographers comment, MacLeod, 1997	
Fimber Cursus	Aerial photograph taken by John Dent	
	Catherine Stoertz 1997 Ancient landscapes of the Yorkshire Wolds: aerial photographic transcription and analysis [Map 1: The North-western Wolds] Page(s)27–29	
Kirby Underdale Cursus	Catherine Stoertz 1997 Ancient landscapes of the Yorkshire Wolds: aerial photographic transcription and analysis [Map 3] Page(s)17, 27–30, 85	
Bag Enderby Pit Alignment	RCHME 1992–1996. National Mapping Programme, Lincolnshire. TF3572:LI.17.8.1	
Harlaxton	RCHM, 1995, National Mapping Programme SK 83NE aerial photo overlay	
Stenigot Pit Alignment	RCHME. 1992–1996. National Mapping Programme. Lincolnshire. TF2481:LI.216.8.1	
Hanworth Cursus	RCHME/EH/HE Aerial Photographers comment Carolyn Dyer/18-DEC-1996/RCHME: Roughton Causewayed Enclosure Project	
Fornham All Saints Cursus	Photographic survey carried out by the RCHME Air Photography Unit on this site, between November 1995 and February 1996.	
Stratford St Mary	RCHME/EH/HE Aerial Photographers comment Andrew Miller/05-AUG-	
	1996/RCHME: AP Primary Recording Project. Aerial photograph NMR, TM 0433/11/235–236; TM 0434/1–3, 6–10, 12–14; TM 0534/12/438–439, TM 0534/13/442–443, TM 0534/15/448–450, TM 0534/16/462–463, TM 0534/17–18	
Barnack Cursus	Prepared by the Royal Commission on Historical Monuments (England) 1960 *A matter of time: an archaeological survey of the river gravels of England*. Page(s)33FF RCHME/EH/HE Aerial Photographers comment	
	Helen Winton/05-DEC-1995/Lincolnshire NMP	
Eynesbury Cursus	Aerial photographs taken by Dr J K St Joseph 1959	
Godmanchester Cursus	Oblique aerial photograph reference number TL 2571 30-JUL-1984 NMR 2173/1299–1316	
Brampton Cursus	Aerial photographs (J K St Joseph) 1962 Air photographs BCS 69, 70, 71, 72, 73 (Dept. of Aerial photography Cambridge) Field Investigators Comments F1 BHS 11-FEB-71	

Cursus monument site	Aerial photography	Excavation
Maxey Cursus	The Maxey cursus is clearly visible from Dr St Joseph's photograph 1956.	1962–63 WG Simpson 1979–81 Francis Pryor 1982–84 Francis Pryor, Charles French ... (et al.) 1985 The Fenland Project, no. 1: Archaeology and Environment in the Lower Welland Valley
	Royal Commission on Historical Monuments (England) 1960 A matter of time: an archaeological survey of the river gravels of England. Page(s)24–25	
	In 1997 an air photograph interpretation and transcription, undertaken as a part of the RCHME: Etton Causewayed Enclosure project, included the course of the Maxey Cursus	
Etton Cursus	First recorded as cropmarks in by Crawford 1927 RCHME/EH/HE Aerial Photographers comment	1982–87 Francis Pryor
	Ann Carter/07-JUL-1997/RCHME: Etton Causewayed Enclosure	
Springfield Cursus		1979–84 Hedges and Buckley
Stanwell Cursus 1	Poulton R, 1978, Cropmarks at Stanwell Fiona Small/02-SEP-1995/RCHME: Heathrow Mapping Project	1979–85 M O'Connell 2006–10 Framework Archaeology
Stanwell Cursus 2	Poulton R, 1978, Cropmarks at Stanwell Fiona Small/02-SEP-1995/RCHME: Heathrow Mapping Project	1979–85 M O'Connell 2006–10 Framework Archaeology
Stanwell Cursus 3	Poulton R, 1978, Cropmarks at Stanwell Fiona Small/02-SEP-1995/RCHME: Heathrow Mapping Project	1979–85 M O'Connell 2006–10 Framework Archaeology
Stanwell Cursus 4	Poulton R, 1978, Cropmarks at Stanwell Fiona Small/02-SEP-1995/RCHME: Heathrow Mapping Project	1979–85 M O'Connell 2006–10 Framework Archaeology
Stanwell Cursus 5	Poulton R, 1978, Cropmarks at Stanwell Fiona Small/02-SEP-1995/RCHME: Heathrow Mapping Project	1979–85 M O'Connell 2006–10 Framework Archaeology
Biggleswade Cursus	RCHME/EH/HE Aerial Photographers comment Andrew Miller/24-FEB-1995/RCHME: AP Primary Recording Project	2004 Albion Archaeology
	Oblique aerial photograph reference number NMR, TL 1946/1/321–2, TL 1946/2/325–7, TL 1946/3/448–51, TL 1946/4/452–4, TL 1946/5/124–7, TL 1946/6/128–9, TL 1946/8/2227–8	
Cardington Cursus	Oblique aerial photograph reference number CUCAP ADO70 06-JUL-1961	
	Oblique aerial photograph reference number CUCAP BXU100 22-JUN-1976	

Cursus monument site	Aerial photography	Excavation
Cople Cursus	Vertical aerial photograph reference number EARTH.GOOGLE.COM 01-JAN-2006 ACCESSED 31-AUG-2016	
Ivinghoe Beacon Cursus	Geophysical Survey J Gover 2000	
Wolverton Cursus 1		2008–11 Cambridge Archaeological Unit
Wolverton Cursus 2		2008–11 Cambridge Archaeological Unit
Wolverton Cursus 3		2008–11 Cambridge Archaeological Unit
Wolverton Cursus 4		2008–11 Cambridge Archaeological Unit
Wolverton Cursus 5		2008–11 Cambridge Archaeological Unit
Benson Cursus	1933 Major Allen 1934 Leeds ET, *Rectangular Enclosures of the Bronze Age in the Upper Thames Valley*, Antiquity Journal Vol. 14 No. 4 1934 pp.	
	414–416 Victoria Fenner/07-JAN-1993/RCHME: Thames Valley NMP	
Dorchester Cursus	1927 O G S Crawford 1933 Major Allen RCHME/EH/HE Aerial Photographers comment	1947–52 Atkinson 1981 Chambers 1988 Bradley and Chambers 2010–17 Gill Hey
	Fiona Small/12-MAY-1994/RCHME: Thames Valley NMP	
Drayton St Leonard	RCHME/EH/HE Aerial Photographers comment	
	Fiona Small/21-OCT-1996/RCHME: Drayton St Leonard Enclosure Project	
Drayton North Cursus	RCHME/EH/HE Aerial Photographers comment Moraig Brown/11-MAR-1993/RCHME: Thames Valley NMP	1977 Michael Parrington 1979–82 Ainslie and Wallis 1985–86 Oxford Archaeological Unit
Drayton South Cursus	1933 Major Allen Aerial photographs taken by Dr St Joseph 1962	1921–37 E T Leeds 1994 Oxford Archaeological Unit
Buscot Cursus	1969 Crop marks – triple ring and apparent cursus, seen on Air Photo taken July 1969 at Buscot	
	RCHME/EH/HE Aerial Photographers comment	
	Carolyn Dyer/11-FEB-1993/RCHME: Thames Valley NMP	
Lechlade Cursus	1947 D N Riley 1961 Aerial photography as part of Fairey Survey	1965 Vatcher & Vatcher 1985 Oxford Archaeological Unit
	Aerial photographs taken by Dr St Joseph 1962	
North Stoke Cursus	1933 Major Allen	1982 Case

Cursus monument site	Aerial photography	Excavation
South Stoke Cursus	Aerial photograph RCHME SU5983/1–2	
Stadhampton Cursus	1986 Discovered by RCHME Air Photography Unit during routine reconnaissance	
Sonning Cursus	Aerial photographs taken by Dr St Joseph 1959	
	RCHME/EH/HE Aerial Photographers comment	
	Helen Winton/21-JAN-1993/RCHME: Thames Valley NMP	
Stonehenge Greater Cursus	Vertical aerial photograph reference number NMR SU 1142/4 (CCC 11752/OS1734)	1947 J Stone 1963 Christie
	Vertical aerial photograph reference number NMR SU 1143/18 (CCC 5203/05666) 09-FEB-1934	
	Vertical aerial photograph reference number USAAF US/7PH/GP/LOC122 1060–65 24-DEC-1943	
	Oblique aerial photograph reference number NMR SU1142/73–4 (18559/22–3) 09-JAN-2000	
Stonehenge Lesser Cursus Stonehenge Lesser Cursus Cont.	Vertical aerial photograph reference number NMR SU 1043/3 (CCC 8533/128) 28-APR-1921 Vertical aerial photograph reference number	1983 J Richards
	NMR SU 1043/13 (CCC 5203/05667) 09-FEB-1934	
	Vertical aerial photograph reference number USAAF US/7PH/GP/LOC122 1065–66 24-DEC-1943	
	Oblique aerial photograph reference number NMR SU 1043/6 (CAP GF 47) 30-JUN-1951	
	Oblique aerial photograph reference number NMR SU 1043/364–5 (1352/364–5) 11-OCT-1978	
Yatesbury Cursus (Avebury)	1935 Major G W G Allen (5/76 26.6.35 and 5/96 1.6.35)	
Gussage Dorset Cursus	O G S Crawford and Alexander Keiller 1928 Wessex from the air	1953 Atkinson 1986 Bradley
	Page(s)232	1991 Bradley 1992 Green
Pentridge Dorset Cursus	O G S Crawford and Alexander Keiller 1928 Wessex from the air Page(s)232	1953 Atkinson 1982 Barrett Bradley & Green 1984 Barrett Bradley & Green

Bibliography

Ainsley, C. (2005). The faunal remains. In: Pryor, F. and French, C. (eds). *Archaeology and environment of the Etton landscape*. East Anglian Archaeology Report No. 109. Fenland Archaeological Trust.

Ainslie, R., Wallis, J. and Wilson, B. (1987). Excavations on the cursus at Drayton, Oxon. *Oxoniensia* Vol. 52: pp. 1–10.

Allen, M. (1991). Analysing the landscape: A geographical approach to archaeological problems. In: Schofield, A. (ed.). *Interpreting artefact scatters. Contributions to ploughzone archaeology*. Oxford: Oxbow Books, pp. 39–57.

Allen, M. (1997). Environment and land-use; the economic development of the communities who built Stonehenge (an economy to support the stones). In: Cunliffe, B. and Refrew, C. (eds). *Science and Stonehenge*. Oxford: Oxford University Press, pp. 115–44.

Allen, M. (2000). Soils, pollen and lots of snails. In: Green, M. (ed). *A landscape revealed 10,000 years on a chalkland farm*. Stroud: The History Press, pp. 36–49.

Allen, M. (2002). The chalkland landscape of cranborne chase: A prehistoric human ecology. *Landscapes*. Vol. 3: pp. 54–68.

Allen, M. (2007). Land use and land development; the molluscan evidence. In: French, C., Lewis, H., Allen, M., Green, M., Scaife, R. and Gardiner, J. (eds). *Prehistoric landscape development and human impact in the upper Allen Valley, Cranborne Chase, Dorset*. Cambridge: McDonald Institute for Archaeological Research, pp. 151–89.

Allen, M. (2017). *Molluscs in archaeology: Methods, approaches and applications*. Studying Scientific Archaeology 3. Oxford: Oxbow Books.

Allen, M. and Gardiner, J. (2004). Neolithic of the Wylye valley 1: Millennium re-investigation of the Corton long barrow ST 9308 4034. *Wiltshire Archaeological Magazine*. Vol. 97: pp. 63–77.

Allen, M. and Gardiner, J. (2006). Codford Circle: Iron Age pits and feasting. *PAST*. Vol. 53: pp. 1–3.

Allen, M. and Gardiner, J. (2009). If you go down to the woods today. A re-evaluation of the chalkland post-glacial woodland: Implications for prehistoric communities. In: Allen, M., Sharples, N. and O'Connor, T. (eds). *Land and people: Papers in memory of John G Evans*. Prehistoric Society Research Paper 2. Oxford: Oxbow Books, pp. 49–66.

Allen, M. and Gardiner, J. (2012). Not out of the woods yet: Some reflections on Neolithic ecological relationships with woodland. In: Jones, A., Pollard, J., Allen, M. and Gardiner, J. (eds). *Image, memory and monumentality: Archaeological engagements with the material world*. Prehistoric Society Research Paper 5. Oxford: Oxbow Books, pp. 93–107.

Allen, T., Barclay, A. and Lamdin-Whymark, H. (2004). Opening the wood, making the land. The study of a Neolithic landscape in the Dorney area of the Middle Thames Valley. In: Cotton, J. and Field, D. (eds). *Towards a New Stone Age: Aspects of the Neolithic in south-east England*. CBA Research Report Series 137. York: Council for British Archaeology, pp. 82–98.

Allen, T., Barclay, A., Cromarty, A., Anderson-Whymark, H., Parker, A., Robinson, M. and Jones, G. (2013). *Opening the wood, making the land: The archaeology of a middle Thames landscape. The Eton rowing course at Dorney and the Maidenhead, Eton and Winsor flood alleviation channel.* Vol. 1: *Mesolithic to Early Bronze Age*. Oxford: Oxford Archaeology Thames Valley Landscapes Monograph. Vol. 38.

Andersen, S. (2013). *Tybrind Vig submerged mesolithic settlements in Denmark. Jutland*. Archaeological Society Publications. Vol. 77. Moesgard: Hojbjerg.

Ashbee, P. (1970). *The earthen long barrow in Britain*. London: Dent.

Ashbee, P. (2004). Early ditches: Their forms and infills. In: Cleal, R. and Pollard, J. (eds). *Monuments and material culture. Papers in honour of Avebury Archaeologist: Isobel Smith*. Salisbury: Hobnob Press, pp. 1–14.

Ashbee, P., Bell, M. and Proudfoot, E. (1989). *Wilsford shaft: Excavations 1960–62*. Swindon: English Heritage.

Ashmore, W. and Knapp, B. (1999). *Archaeologies of landscape: Contemporary perspectives*. Oxford: Blackwell.

Aston, M. (1985). *Interpreting the landscape: Landscape archaeology and local history*. London and New York: Routledge.

Atkinson, R. (1951). The excavations at Dorchester, Oxfordshire 1946–1951. *Archaeological Newsletter*. Vol. 4: pp. 56–59.

Atkinson, R. (1955). The Dorset Cursus. *Antiquity*. Vol. 29(113): pp. 4–9.

Atkinson, R. (1956). *Stonehenge*. London: Hamish Hamilton.

Atkinson, R. (1979). *Stonehenge: Archaeology and interpretation*. London: Penguin.

Atkinson, R., Piggott, S. and Sandars, N. (1951). *Excavations at Dorchester, Oxon*. Oxford: Department of Antiquities Ashmolean Museum.

Ayres, K. and Powell, A. (2003). Animal bone. In: Barclay, A., Lambrick, G., Moore, J. and Robinson, M. (eds). *Lines in the landscape. Cursus monuments in the Upper Thames Valley: Excavation at the Drayton and Lechlade Cursuses*. Oxford University School of Archaeology (Oxford Archaeology, Thames Valley Landscapes Monograph 15). Oxford: Holywell Press.

Bailey, G. (2007). Time perspectives, palimpsests and the archaeology of time. *Journal of Anthropological Archaeology*. Vol. 26: pp. 198–223.

Barber, M. (2011). *Introductions to heritage assets: Prehistoric avenues and alignments*. Swindon: Historic England.

Barclay, A. (2014). Re-dating the Coneybury Anomaly and its implications for understanding the earliest Neolithic pottery from southern England. *PAST*. Vol. 77: pp. 11–13.

Barclay, A. and Bayliss, A. (1999). Cursus monuments and the radiocarbon problem. In: Barclay, A. and Harding, J. (eds). *Pathways and ceremonies. The cursus monuments of Britain and Ireland*. Neolithic Studies Group Seminar Papers 4. Oxford: Oxbow Books, pp. 11–29.

Barclay, A. and Harding, J. (1999). An introduction to the cursus monuments of Neolithic Britain and Ireland. In: Barclay, A. and Harding, J. (eds). *Pathways and ceremonies. The cursus monuments of Britain and Ireland*. Neolithic Studies Group Seminar Papers 4. Oxford: Oxbow Books, pp. 1–10.

Barclay, A. and Hey, G. (1999). Cattle, cursus monuments and the river: The development of ritual and domestic landscapes in the Upper Thames Valley. In: Barclay, A. and Harding, J. (eds). *Pathways and ceremonies. The cursus monuments of Britain and Ireland*. Neolithic Studies Group Seminar Papers 4. Oxford: Oxbow Books, pp. 67–76.

Barclay, A., Lambrick, G., Moore, J. and Robinson, M. (2003). *Lines in the landscape*. Oxford Archaeological Unit. Oxford: Holywell Press.

Barclay, G. and Maxwell, G. (1998). *The Cleaven Dyke and Littleour: Monuments in the Neolithic of Tayside*. Society of Antiquities of Scotland: Monograph Series No. 13. Kings Stanley: Past Historic.

Barker, G. and Webley, D. (1978). Causewayed camps and early Neolithic economies in central southern England. *Proceedings of the Prehistoric Society*. Vol. 44: pp. 161–86.

Barnatt, J.,Beswick, P., Chambers, F. and Evans, J. (1994). Excavation of a Bronze Age unenclosed cemetery, cairns and field boundaries at Eaglestone Flat, Cubar, Derbyshire 1989–90. *Proceedings of the Prehistoric Society*. Vol. 60: pp. 287–370.

Barrett, J., Bradley, R. and Green, M. (1991). *Landscape, monuments and society: The pre-history of Cranbourne Chase*. Cambridge: Cambridge University Press.

Bayes, T. and Price, R. (1763). An essay towards solving a problem in the doctrine of chance. By the late Rev. Mr. Bayes, communicated by Mr. Price, in a letter to John Canton, MA. and F.R.S. *Philosophical Transactions of the Royal Society of London*. Vol. 53: pp. 370–418.

Bayliss, A., Whittle, A. and Healy F. (2008). Timing, tempo and temporalities in the early Neolithic of southern Britain. In: Fokkens, H., Coles, B., van Gijn, A., Kleijne, J., Ponjee, H. and Slappendel, C. (eds). *Between foraging and farming: An extended broad spectrum of papers presented to Leendert Louwe Kooijmans*. Analecta Praehistorica Leidensia 40. Leiden: Faculty of Archaeology, Leiden University, pp. 26–42.

Bell, M. (1983). Valley sediments as evidence of prehistoric land-use on the South Downs. *Proceedings of the Prehistoric Society*. Vol. 49: pp. 119–50.

Bell, M. and Nobel, G. (2012). Prehistoric woodland ecology. In: Jones, A., Pollard, J., Allen, M. and Gardiner, J. (eds). *Image, memory and monumentality: Archaeological engagements with the material world*. Prehistoric Society Research Paper 5. Oxford: Oxbow Books, pp. 80–92.

Benson, D. and Whittle, A. (2007). *Building memories: The Neolithic Cotswold long barrow at Ascott-under-Wychwood, Oxfordshire*. Oxford: Oxbow Books.

Bergman, A. (1993). The development of the Bow in Western Europe. A technological and functional perspective. *Archaeological Papers of the American Anthropological Association*. Vol. 4: pp. 95–105.

Binford, L. (1978). *Nunamiut ethnoarchaeology*. New York: Academic Press.

Binford, L. (1981). *Bones ancient men and modern myths*. New York. Academic Press.

Birks, H. (1975). Studies in the vegetational history of Scotland IV. Pine stumps in Scottish blanket peats. *Philosophical Transactions of the Royal Society of London* Biology. Vol. 270: pp. 181–226.

Bloch, M. and Parry, J. (1981). Introduction: Death and the regeneration of life. In: Bloch, M. and Parry, J. (eds). *Death and the regeneration of life*. Cambridge: Cambridge University Press, pp. 1–44.
Boivin, N. and Owoc, M. (eds). (2004). *Soils, stones and symbols: Cultural perceptions of the mineral world*. London: UCL Press.
Boscombe Down Phase II Excavations, Amesbury (1995). *Boscombe sports field, assessment by Wessex Archaeology*. Report Reference 36875.1.
Boscombe Down Phase V Excavations, Amesbury (2004). *Post-excavation assessment report and proposals for analysis and final publication by Wessex Archaeology*. Report Reference 56240.
Bowden, M., Soutar, S., Field, D. and Barber, M. (2015). *The Stonehenge landscape: Analysing the Stonehenge World Heritage Site*. Swindon: Historic England.
Boyd, M. (1994). A Mesolithic tranchet axe from the Yorkshire Wolds. *Yorkshire Archaeological Journal*. Vol. 66: pp. 226–29.
Bradley, R. (1978). *The prehistoric settlement of Britain*. London: Routledge, p. 103.
Bradley, R. (1983). The bank barrows and related monuments of Dorset in the light of recent fieldwork. *Proceedings of the Dorset Natural History and Archaeological Society*. Vol. 105: pp. 15–20.
Bradley, R. (1984a). Studying monuments. In: Bradley, R. and Gardiner, J. (eds). *Neolithic Studies*. Oxford: British Archaeological Reports Vol. 133: pp. 61–66.
Bradley, R. (1984b). *The social foundations of prehistoric Britain*. London: Longman.
Bradley, R. (1984c). Regional systems in Neolithic Britain. In: Bradley, R. and Gardiner, J. (eds). *Neolithic Studies*. Oxford: British Archaeological Reports. Vol. 133: pp. 5–14.
Bradley, R. (1985). *Consumption, Change and the archaeological record*. Edinburgh: Department of Archaeology.
Bradley, R. (1986). *The Dorset Cursus: The archaeology of the enigmatic (Wessex Lecture)*. Council of British Archaeology Group 12.
Bradley, R. (1987). Flint technology and the character of Neolithic settlement. In: Brown, A. and Edmonds, M. (eds). *Lithic analysis and later British Prehistory*. Oxford: British Archaeological Reports No. 162: p. 182.
Bradley, R. (1991). Ritual, time and history. *World Archaeology*. Vol. 23: pp. 209–19.
Bradley, R. (1993). *Altering the earth*. Edinburgh: Society of Antiquaries of Scotland.
Bradley, R. (2002). *The past in prehistoric societies*. London: Routledge.
Bradley, R. and Chambers, R. (1988). A new study of the cursus complex at Dorchester on Thames. *Oxford Journal of Archaeology*. Vol. 7: pp. 271–89.
Bradley, R., Bowden, M., Gaffney, V. and Mepham, L. (1983). The date of the Dorset Cursus two sections at Woodcutts: Earlier Neo to Beaker? *Proceedings of the Prehistoric Society*. Vol. 49: pp. 376–79.
Bradshaw, R. and Mitchell, F. (1999). The palaeoecological approach to reconstructing former grazing-vegetation interactions. *Forest Ecology and Management*. Vol. 120: pp. 3–12.
Branch, N. (2015). Changing environments in the Stonehenge Area from the late Glacial to the Iron Age. *University of Buckingham Lecture*. Society of Antiquaries.
Brasier, C. (1979). Dutch elm disease as an analogue of Neolithic elm decline. *Nature*. Vol. 281: pp. 78–80.
Brophy, K. (1999). Seeing the Cursus as a symbolic river. *British Archaeology*. Vol. 44: pp. 6–7.
Brophy, K. (2009). The map trap: The depiction of regional geographies of the Neolithic. In: Brophy, K. and Barclay, G. (eds). *Defining a regional Neolithic from Britain and Ireland*. Neolithic Studies Group. Seminar Papers 9. Oxford: Oxbow Books, pp. 5–25.
Brophy, K. (2016). *Reading between the lines. The Neolithic cursus monuments of Scotland*. London: Routledge.
Brophy, K. and Cowley, D. (2005). *From the air. Understanding aerial archaeology*. Stroud: Tempus.
Brophy, K. and Millican, K. (2015). Wood and fire: Scotland's timber cursus monuments. *Archaeological Journal*. Vol. 172(2): pp. 1–28.
Brown, A. (1995). The Mesolithic and later flint artefacts. In: Allen, T. (ed.). *Lithics and landscape: Archaeological discoveries on the Thames Water pipeline at Gatehampton Farm, Goring, Oxfordshire 1985–92*. Oxford: Thames Valley Landscapes No. 7, pp. 65–84.
Brown, A. (1997). Clearances and clearings; deforestation in Mesolithic / Neolithic Britain. *Oxford Journal of Archaeology*. Vol. 16: p. 140.
Brown, A. (2000). Floodplain vegetation history: Clearings as potential ritual spaces. In: Fairburn, A. (ed.). *Plants in Neolithic Britain and beyond*. Neolithic Studies Group Papers 5. Oxford: Oxbow Books, pp. 19–31.
Brown, M. (2001). *English Heritage: Ivinghoe Beacon, Ivinghoe, Buckinghamshire*. Archaeological Investigation Report Series 15/2001. Swindon: English Heritage.
Brown, M. and Field, D. (2000). Ivinghoe Beacon, Ivinghoe, Buckinghamshire. *Survey Report Archaeological Investigation Series 15/2001*. Swindon: English Heritage.

Brown, A., Bell, M., Timpany, S. and Nayling, N. (2005). Mesolithic to Neolithic and Medieval coastal environmental change: Intertidal survey at Woolaston, Gloucestershire. *Archaeology in the Severn Estuary*. Vol. 16: pp. 67–83.

Brown, G., Field, D. and McOmish, D. (2005). *The Avebury landscape: Aspects of the field archaeology of the Marlborough Downs*. Oxford: Oxbow Books.

Buckley, D., Hedges, J. and Brown, N. (2001). Excavations at a Neolithic Cursus, Springfield, Essex, 1979–1985. *Proceedings of the Prehistoric Society*. Vol. 67: pp. 101–62.

Bunn, H. and Gurtov, A. (2014). Prey mortality profiles indicate that early Pleistocene homo at Olduvai was an ambush predator. *Quaternary International*, Vol. 322–323: pp. 44–53.

Burl, H. (1987). *The Stonehenge people*. London: Dent.

Bush, M. (1988). Early Mesolithic disturbance; a force on the landscape. *Journal of Archaeological Science*. Vol. 15: pp. 453–62.

Bush, M. (1989). On the antiquity of British grasslands; a response to Thomas. *Journal of Archaeological Science*. Vol. 16: pp. 555–60.

Bush, M. and Ellis, S. (1987). The sedimental and vegetational history of Willow Garth. In: Ellis, S. (ed.). *East Yorkshire field guide*. Cambridge: Quaternary Research Association.

Bush, M. and Flenley, J. (1987). The age of the British chalk grassland. *Nature*. Vol. 329: pp. 434–36.

Canada National Defence Landforce Infantry. (2001). *Ambush and counter-ambush*. (B-GL-392-008/FP-001. Issued on authority of Chief of the Land Staff.

Case, H. (1969). Neolithic explanations. *Antiquity*. Vol. 43: pp. 176–86.

Case, H. (1982). The linear ditches and southern enclosure, North Stoke. In: Case, H. and Whittle, A. (eds). *Settlement patterns in the Oxford region; excavations at the Abingdon causewayed enclosure and other sites*. Research Report 44. Council for British Archaeology and Department of Antiquities, Ashmolean Museum, pp. 60–74.

Catt, J. (1978). The contribution of loess to soils in lowland Britain. In: Limbrey, S. and Evans, J. (eds). *The effect of man on the landscape – The lowland zone*. London: Council for British Archaeology, pp. 12–20.

Chambers, R. (1992). Excavations in the Neolithic and Bronze Age complex at Dorchester-on-Thames, Oxfordshire. In: Whittle, A., Atkinson, R., Chambers, R. and Thomas, N. (eds). Excavations in the Neolithic and Bronze Age complex at Dorchester-on-Thames, Oxfordshire. 1947–52 and 1981. *Proceedings of the Prehistoric Society*. Vol. 58: pp. 143–201.

Champion, J. (2005). *The Enigmatic Cursus*. Available at <http://www.megalithic.co.uk/article.php?sid=2146412148> [Megalithic Portal accessed 18 October 2017].

Chapman, H. (2005). Rethinking the cursus problem – Investigating the Neolithic landscape archaeology of Rudston, East Yorkshire, UK using GIS. *Proceedings of the Prehistoric Society*. Vol. 71: pp. 159–70.

Chapman, H. (2006). *Landscape archaeology and GIS*. Stroud: Tempus.

Chapman, B. and Trani, M. (2007). The Nature Conservancy and the US Forest Service, southern region. In: Trani, M., Ford, W. and Chapman, B. (eds). *The land manager's guide to mammals of the South*. Durham, NC: The Nature Conservancy, South-eastern Region, pp. 540–44.

Charlton, S. (2018, In press). *Fragmentary skeletal material from Vespasian's Camp*. ZooMS Analysis.

Chenery, C., Pashley, V., Lamb, A., Sloane, H. and Evans, J. (2012). The oxygen isotope relationship between phosphate and structural carbonate fractions of human bioapatite. *Rapid Communications in Mass Spectrometry*. Vol. 26: pp. 309–19.

Cherry, J. (1978). Generalisation and the archaeology of the state. In: Green, D., Haselgrove, C. and Spriggs, M. (eds). *Social organisation and settlement*. Oxford: British Archaeological Reports S47, pp. 411–37.

Child, G. (1931). The forest cultures of northern Europe: A study in evolution and diffusion. *Journal of the Royal Anthropological Institute*. Vol. 61: pp. 325–48.

Chisham, C. (2004). *Mesolithic human activity and environmental change: A case study of the Kennet Valley*. Unpublished PhD Thesis. Department of Archaeology. University of Reading.

Christie, P. (1963). The Stonehenge Cursus. *Wiltshire Archaeological and Natural History Magazine*. Vol. 58: pp. 370–82.

Clark, G. (1954). *Excavations at Star Carr*. Cambridge: Cambridge University Press.

Clark, G. (1972). *Star Carr: A case study in bioarchaeology*. Module in Anthropology 10. Addison-Wesley.

Clarke, R. (2013). *Watching brief at the rear of building 98, The Officers Mess, MOD Boscombe Down, Salisbury, Wiltshire 11–12 November 2013*. For Boscombe Down Conservation Group. BDCG 2013-11-(1). Farnborough, Hampshire: QinetiQ plc.

Clay, P. (2001). An archaeological resource assessment and research agenda for the Neolithic and Early-Middle Bronze Age of the East Midlands. *East Midlands Archaeological Research Framework,* pp. 1–40.

Clay, P. (2009). Core or periphery? The case of the Neolithic of the East Midlands. In: Brophy, K. and Barclay, G. (eds). *Defining a regional Neolithic from Britain and Ireland.* Neolithic Studies Group. Seminar Papers 9. Oxford: Oxbow Books, pp. 92–105.

Cleal, R., Allen, M., Harding, P. and Newman, C. (2004). An Archaeological and environmental study of the Neolithic and later prehistoric landscape of the Avon Valley and Durrington Walls environs. *Wiltshire Archaeological and Natural History Magazine*. Vol. 97: pp. 218–48.

Coady, I. (2004). *What is the nature and extent of early Holocene activity within the Stonehenge environs?* Unpublished undergraduate thesis. School of Conservation Sciences. Bournemouth.

Colt Hoare, R. (1812). *The history of ancient Wiltshire*. Vol. 1. London: William Miller.

Colt Hoare, R. (1819). *The history of ancient Wiltshire*. Vol. 2. London: Hughes, Harding, Maver and Lepard.

Conneller, C. (2011). *An archaeology of materials. Substantial transformations in early prehistoric Europe*. New York and Abingdon: Routledge.

Conneller, C., Milner, N., Taylor, B. and Taylor, M. (2012). Substantial settlement in the European Early Mesolithic: New research at Star Carr. *Antiquity*. Vol. 86(334): pp. 1004–20.

Cook, H., Stearne, K. and Williamson, T. (2003). The origins of water meadows in England. *Agricultural History Review*. Vol. 51(2): pp. 155–62.

Coope, G. (1998). Insects. In: Preece, R. and Bridgland, D. (eds). *Late quaternary environmental change in North-West Europe: Excavations at Hollywell Coobe, South-East England.* London: Chapman and Hall, pp. 213–33.

Cotton, J. (1990). Finds from the cursus ditches. In: O'Connell, M. (ed.). Excavations during 1979–1985 of a multi-period site at Stanwell. *Surrey Archaeological Collection*. Vol. 80: pp. 28–29.

Cotton, J. (1991). Prehistory in Greater London. *Current Archaeology*. Vol. 124: pp. 151–54.

Cotton, J., Elsden, N., Pipe, A. and Rayner, L. (2006). Taming the wild: A final Neolithic / Earlier Bronze Age aurochs deposit in west London. In: Serjeantson, D. and Fields, D. (eds). *Animals in the Neolithic of Britain and Europe*. Oxford: Oxbow Books, pp. 149–67.

Crane, N. (2016). *The making of the British landscape: From the Ice Age to the present.* London: Weidenfeld and Nicolson. Orion Publishing.

Crawford, O. (1935). Rectangular enclosures: A note on Mr Leeds paper. *Antiquaries Journal.* Vol. 15: pp. 77–78.

Croft, R. and Mynard, D. (1993). The changing landscape of Milton Keynes. *Buckinghamshire Archaeology Society Monograph*. Vol. 5: Aylesbury.

Croft, P. and Ward, B. (2015). Fog. In: North, G., Pyle, J. and Zhang, F. (eds). *Encyclopaedia of Atmospheric Sciences.* Vols 1–6: pp. 180–88.

Cruise, G. (1993). Soil pollen. In: Whittle, A., Rouse, A. and Evans, J. (eds). A Neolithic downland monument in its environment: Excavations at the Easton Down long barrow, Bishops Cannings, North Wiltshire. *Proceedings of the Prehistoric Society*. Vol. 59: pp. 219–21.

Crutchley, S. (2005). Recent aerial survey work in the Marlborough Downs region. In: Brown, G., Field, D. and McOmish, D. (eds). *The Avebury landscape: Aspects of the field archaeology of the Marlborough Downs.* Oxford: Oxbow Books, pp. 34–42.

Cunliffe, B. (2012). *Britain begins*. Oxford: Oxford University Press.

Darvill, T. (1987). *Prehistoric Britain*. London: Batsford.

Darvill, T. (2006). *Stonehenge, The Biography of a Landscape*. Stroud: Tempus.

Darvill, T. (2008). *Oxford concise dictionary of archaeology*. Oxford. Oxford University Press.

Darvill, T. (2013). *A research framework for the Stonehenge, Avebury and associated sites World Heritage Site. Research activity in the Stonehenge landscape 2005–2012*. Swindon: Historic England.

David, B. and Thomas, J. (2008). *Handbook of landscape archaeology*. London and New York: Routledge.

Day, S. (1991). Post-glacial vegetation history of the Oxford Region. *New Phytologist*. Vol. 119: pp. 445–70.

Devereux, P. and Pennick, N. (1989). *Lines on the landscape. Leys and other linear enigmas*. London: Robert Hale.

Dimbleby, G. (1965). Post-glacial changes in soil samples. *Proceedings of the Royal Society Biological.* Vol. 161: pp. 355–62.

Dimbleby, G. (1979) Pollen analysis. In Ashbee, P., Smith, I. Evans, J (eds). Excavations of three long barrows near Avebury. *Proceedings of the Prehistoric Society*. Vol. 45: pp. 207–300 (Horslip pp. 276–78, Beckhampton pp. 280–81 and South Street pp. 284–88).

Dincauze, D. and Hasenstab, R. (1989). Explaining the Iroquois: Tribalisation on a prehistoric periphery. In: Champion, T. (ed.). *Centre and periphery: Comparative studies in archaeology*. London: Unwin Hyman, pp. 67–87.

Dinnin, M. and Sadler, J. (1999). 10,000 years of change: The Holocene entomofauna of the British Isles. *Quaternary Proceedings*. Vol. 7: pp. 545–62.

Drennan, R. (1996). *Statistics for archaeologists: A common sense approach*. New York: Springer US.

Drew, C. and Piggott, S. (1936). The excavation of long barrow 163a on Thickthorn Down Dorset. *Proceedings of the Prehistoric Society*. Vol. 41: pp. 77–96.

Dymond, D. (1966). Ritual monuments at Rudston, East Yorkshire, England. *Proceedings of the Prehistoric Society*. Vol. 32: pp. 86–95.

Earnshaw, J. (1973). The site of a medieval post mill and prehistoric site at Bridlington. *Yorkshire Archaeological Journal*. Vol. 45: pp. 19–40.

Edmonds, M. (1999). *Ancestral geographies of the Neolithic. Landscape, monuments and memory*. London: Routledge.

Edwards, K. and Hirons, K. (1984). Cereal pollen grains in pre-elm decline deposits: Implications for the earliest agriculture in Britain and Ireland. *Journal of Archaeological Science*. Vol. 11: pp. 71–80.

Ellis, S. and Newsome, D. (1991). *Chalkland soil formation and erosion on the Yorkshire Wolds, northern England*. Amsterdam: Elsevier.

Entwistle, R. (1985). Molluscan analysis. In: Bradley, R and Entwistle, R. (eds). Thickthorn Down long barrow – a new assessment. *Dorset Natural History and Archaeological Society*. Vol. 107: pp. 174–76.

Entwistle, R. and Bowden, M. (1991). Cranborne Chase; the molluscan evidence. In: Barrett, J., Bradley, R. and Hall, M. (eds). *Papers on the prehistoric archaeology of Cranborne Chase*. Monograph Series 11. Oxford: Oxbow, pp. 20–48.

Evans, J. (1966). Late-Glacial and Post-Glacial subaerial deposits at Pitstone, Buckinghamshire. *Proceedings of the Geologists Association*. Vol. 77(3): pp. 347–64.

Evans, J. (1967). *The stratification of mollusca in chalk soils and their relation to archaeology*. Unpublished PhD thesis. University of London, Institute of Archaeology.

Evans, J. (1971). Habitat change on the calcareous soils of Britain: The impact of Neolithic man. In: Simpson, D. (ed.). *Economy and settlement in Neolithic and Early Bronze Age Britain and Europe*. Leicester: Leicester University Press, pp. 11–16.

Evans, J. (1972). *Land snails in archaeology*. London: Seminar Press.

Evans, J. (1975). *The environment of early man in the British Isles*. London: Paul Elek.

Evans, J. (1976). Land snails. In: Everson, P. (ed.). Iron Age enclosures at the Queensway Health Centre site, Hardwick Park, Wellingborough. *Northamptonshire Archaeology*. Vol. 11: pp. 97–98.

Evans, J. (2000). Discussion of the molluscan evidence. In: Needham, S. (ed.). *The passage of the Thames; Holocene environment and settlement at Runnymead*. London: British Museum Press, pp. 138–45.

Evans, J. (2003). *Environmental Archaeology and the social order*. London: Routledge.

Evans, J. and Dimbleby, G. (1976). In: Manby, T. (ed.). Excavation of the Kilham long barrow, East Riding of Yorkshire. *Proceedings of the Prehistoric Society*. Vol. 42: pp. 150–59.

Evans, J. and Jones, H. (1979). Mount Pleasant and Woodhenge: The land Mollusca. In: Wainwright, G. (ed.). *Mount Pleasant, Dorset: Excavations 1970–1971*. Research Report of the Society of Antiquaries, pp. 190–213.

Evans, J. and Rouse, A. (1991). The land Mollusca. In: Sharples, N. (ed.). *Maiden Castle: Excavations and field survey 1985–6*. Report No. 19. London: English Heritage.

Evans, J., Smith, I., Darvill, T. and Grigson, C. (1983). Excavations at Cherhill, North Wiltshire, 1967. *Proceedings of the Prehistoric Society*. Vol. 49: pp. 43–117.

Evans, J., Rouse, A. and Sharples, N. (1988). The landscape setting of causewayed camps: Some recent work on Maiden Castle enclosure. In: Barrett, J. and Kinnes, I. (eds). *The archaeology of context in the Neolithic and Bronze Age: Recent trends*. Sheffield: J. Collis, pp. 73–84.

Evans, J., Davies, P., Mount, R. and Williams, D. (1992). Molluscan taxocenes from Holocene overbank alluvium in central southern England. In: Needham, S. and Macklin, G. (eds). *Alluvial archaeology in Britain*. Monograph Vol. 27. Oxford: Oxbow, pp. 65–74.

Evans, J., Limbery, S., Mate, I. and Mount, R. (1993). An environmental history of the upper Kennet Valley, Wiltshire, for the past 10,000 years. *Proceedings of the Prehistoric Society*. Vol. 59: pp. 188.

Farley, M. (1983). Mirror burial at Dorton. *Proceedings of the Prehistoric Society*. Vol. 49: pp. 269–302.

Farley, M. (2010). *An illustrated history of early Buckinghamshire*. Buckingham Archaeological Society.

Fergusson, J. (1872). *Rude Stone monuments in all countries. Their age and uses*. London: John Murray.

Fisher, P. (1996). Extending the applicability of viewsheds in landscape planning. *Photogrammetric Engineering & Remote Sensing*. Vol. 62(11): pp. 1297–302.

Fishpool, M. (1999). Land Mollusca. In: Whittle, A., Pollard, J. and Grigson, C. (eds). *The harmony of symbols: The Windmill Hill causewayed enclosure*. Oxford: Oxford Books.

Fleming, A. (1999). Phenomenology and the megaliths of Wales: A dream too far? *Oxford Journal of Archaeology*. Vol. 18(2): pp. 119–25.

Ford, S. (1987). *East Berkshire Archaeological Survey. Occasional paper*. Berkshire County Council.

Fraser, D. (1983). *Land and society in Neolithic Orkney*. BAR British Series 117. Oxford: BAR.

Fraser, F. and King, J. (1954). Faunal remains. In: Clark, G. (ed.). *Excavations at Star Carr*. Cambridge: Cambridge University Press, pp. 70–95.

French, C. (2002). *Geoarchaeology in action: Studies in soil micromorphology and landscape evolution*. London: Routledge.

French, C. (2009). A landscape tail of two soil histories in lowland zones of England. In: Allen, M., Sharples, N. and O'Connor, T. (eds). *Land and people, papers in memory of John G Evans*. Prehistoric Society Research Paper 2. Oxford: Oxbow Books, pp. 89–104.

French, C. and Pryor, F. (2005). Archaeology and environment of the Etton landscape. *Norwich Fenland Archaeological Trust*. East Anglian Archaeology Report No. 9.

French, C., Macklin, M. and Passmore, D. (1992). Archaeology and palaeochannels in the Lower Welland and Nene Valleys: Alluvial archaeology at the fen-edge, Eastern England. In: Needham, S. and Macklin, M. (ed.). *Alluvial archaeology in Britain: Proceeding of a conference sponsored by the RMC Group plc. 3–5 January 1991*. Oxbow Monograph 27. Oxbow: British Museum.

French, C., Lewis, H., Scaife, R. and Allen, M. (2005). New perspectives on Holocene landscape development in the southern English chalklands: The Upper Allen Valley, Cranborne Chase, Dorset. *Geoarchaeology*. Vol. 20: pp. 109–34.

French, C., Scaife, R., Allen, M., Parker-Pearson, M., Pollard, J., Richards, C., Thomas, J. and Welham, K. (2012). Durrington Walls to West Amesbury by way of Stonehenge: A major transformation of the Holocene landscape. *Antiquaries Journal*. Vol. 92: pp. 1–36.

Frison, G. (2004). *Survival by hunting: Prehistoric human predators and animal prey*. Los Angeles: University of California Press.

Gaddis, J. (2002). *The landscape of history: How historians map the past*. Oxford: Oxford University Press.

Gaffney, V., Bintliff, J. and Slapsak, B. (1991). Site formation processes and the Hvar survey project, Yugoslavia. In: Schofield, A. (ed.). *Interpreting artefact scatters. Contributions to ploughzone archaeology*. Oxford: Oxbow Books, pp. 59–77.

Gaffney, V., Gaffney, C., Neubauer, W., Baldwin, E., Chapman, H., Garwood, P., Moulden, H., Sparrow, T., Bates, R., Locker, K., Hinterleitner, A., Trinks, I., Nau, E., Zitz, T., Floery, S., Verhoeven, G. and Doneus, M. (2012). The Stonehenge hidden landscapes project. *Archaeological Prospection*. Vol. 19(2): pp. 147–55.

Gardiner, J. (1984). Lithic distributions and settlement patterns in central southern England. In: Bradley, R. and Gardiner, J. (eds). *Neolithic studies*. British Archaeological Report No 133: pp. 15–40.

Gardiner, J. and Allen, M. (2009). Peopling the landscape; prehistory of the Wylye Valley, Wiltshire. In: Allen, M., Sharples, N. and O'Connor, T. (eds). *Land and people: Papers in memory of John G Evans*. Prehistoric Society Research Paper 2. Oxford: Oxbow Books, pp. 77–88.

Gell, A. (1995). The language of the forest: Landscape and phonological iconism in Umeda. In: Hirsch, E. and O'Hanlon, M. (eds). *The Anthropology of landscape. Perspectives on place and space*. Oxford: Oxford University Press, pp. 232–54.

Geneste, J., Castel, J. and Chadelle, J. (2008). From physical to social landscapes: Multidimensional approaches to the archaeology of social place in the European Upper Palaeolithic. In: David, B. and Thomas, J. (eds). *Handbook of landscape archaeology*. London and New York: Routledge, pp. 228–36.

George, M., Bailey, D., Borman, M., Ganskopp, D., Surber, G. and Harris, N. (2007). *Factors and practices that influence livestock distribution*. Rangeland Management Series. Publication No. 8217. University of California.

Gibson, A. (1986). Diatom analysis of clays and late Neolithic pottery from the Milfield Basin, Northumberland. *Proceedings of the Prehistoric Society*. Vol. 52: pp. 89–103.

Gibson, A., Bayliss, A., Heard, H., Mainland, I., Ogden, A. and Bronk Ramsey, C. (2009) Recent research at Duggleby Howe, North Yorkshire. *Archaeological Journal*. Vol. 166: pp. 39–78.

Giddens, A. (1984). *The constitution of society: Outline of the theory of structuration*. Cambridge: Polity Press.

Gill, M. (2019). Putting old kit to good use: The Locate Geophysics Project. *British Archaeology*. Vol. 165: pp. 30–35.

Godwin, H. (1975). *The history of the British flora; a factual basis for photogeography*. Cambridge: Cambridge University Press.

Gosden, C. (1994). *Social being and time*. Oxford: Blackwell.

Gover, J. (2000). *A geophysical investigation of Ivinghoe Beacon, Chiltern Hills*. Unpublished document.

Greaney, S., Madgwick, R. and Parker Pearson, M. (2018). Neolithic food miles: Feeding the builders of Stonehenge. *Current Archaeology*. Vol. 334: pp. 26–31.

Green, C., et al. (2017). Understanding the Spatial Patterning of English Archaeology: Modelling Mass Data, 1500 BC to AD 1086. *Archaeological Journal*. Vol. 174(1): pp. 244–80.

Green, M. (2000). *A landscape revealed, 10,000 years on a chalkland farm*. Stroud: Tempus.

Greenwell, W. (1877). *British barrows: A record of the examination of sepulchral mounds in various parts of England*. Cambridge University Press (2014), pp. 233–37.

Griffiths, S. (2014). Points in time: The Mesolithic-Neolithic transition and the chronology of late rod microliths in Britain. *Oxford Journal of Archaeology*. Vol. 33: pp. 221–43.

Harding, J. (1995). Social Histories and regional perspectives in the Neolithic of lowland England. *Proceedings of the Prehistoric Society*. Vol. 61: pp. 117–36.

Harding, J. (1999). Pathways to new realms: Cursus monuments and symbolic territories. In: Barclay, A. and Harding, J. (eds). *Pathways and ceremonies. The cursus monuments of Britain and Ireland.* Neolithic Studies Group Seminar Papers 4. Oxford: Oxbow Books, pp. 30–38.

Harding, J. (2006). Pit-digging, occupation and structured deposition on Rudston Wold, Eastern Yorkshire. *Oxford Journal of Archaeology*. Vol. 25(2): pp. 109–26.

Harding, A. and Lee, G. (1987). *Henge monuments and related sites of Great Britain: Air photographic evidence and catalogue*. British Archaeological Report (BAR) British Series 1 (1974) – Site 007 175, pp. 74–81.

Harding, J. and Healy, F. (2007). *The Raunds area project. A Neolithic and Bronze Age landscape in Northamptonshire*. Swindon: English Heritage.

Harding, J. and Healy, F. (2008). *A Neolithic and Bronze Age landscape in Northamptonshire: Volume 1 – The Raunds area project*. Swindon: English Heritage.

Harding, P. and Rose, F. (1986). Pasture-woodlands in lowland Britain. *Peterborough: Institute of Terrestrial Ecology*, Natural Environment Research Council.

Harris, J. and Evans, J. (1994). Molluscan analysis. In: Whittle, A. (ed.). Excavations at Millbarrow Neolithic chambered tomb, Winterbourne Monkton, North Wiltshire. *Wiltshire Archaeological and Natural History Magazine*. Vol. 87: pp. 26–32.

Hartz, N. and Winge, H. (1906). Om uroksen fra Vig. Saaret og dræbt med flintvaaben. *Aarbøger for Nordisk Oldkyndighed og Historie*. pp. 225–36.

Healy, F. and Harding, J. (2007). A thousand and one things to do with a round barrow. In: Last, J. (ed.). *Beyond the Grave: New Perspectives on Barrows*. Oxford: Oxbow Books, pp. 53–71.

Hedges, J. and Buckley, D. (1981). *The Springfield cursus and the cursus problem*. Chelmsford: Essex County Council.

Henton, E. (2010). The application of oxygen isotopes and microwear from cattle tooth enamel at Fengate and the Flag Fen Basin. In: Pryor, F. and Bamforth, M. (eds). *Flag Fen, Peterborough excavation and research 1995–2007*. Oxford: Oxbow Books, pp. 105–14.

Herne, A. (1988). A time and a place for the Grimston bowl. In: Barrett, J. and Kinnes, I. (eds). *The archaeology of context in the Neolithic and Bronze Age: Recent trends*. Sheffield: Department of Archaeology, p. 25.

Hey, G. (2010). Late Upper Palaeolithic and Mesolithic period. In: Hey, G. and Hind, J. (eds). *Solent-Thames research framework for the historic environment. Resource assessments and research agendas*. Oxford Wessex Monograph No. 6. Oxford: Oxford Archaeology, pp. 83–86.

Hey, G. with Robinson, M. (2011). Mesolithic communities in the Thames Valley: Living in the natural landscape. In: Hey, G., Garwood, P., Robinson, M., Barclay, A., Bradley, P. et al. (eds). *The Thames through time. The archaeology of the gravel terraces of the Upper and Middle Thames. Early prehistory to 1500 BC*. Monograph 32. Oxford: Oxford School of Archaeology.

Hey, G. (16 March 2016). *Walking into the sunset: New work on the Dorchester-on-Thames cursus and the Thames Valley cursus phenomenon*. Winter Lecture Series. Aylesbury: Buckinghamshire Historical Association.

Hey, G. and Hind, J. (2014). *Solent-Thames research framework for the historic environment: Resource assessments and research agendas*. Oxford Wessex Monograph No. 6. Oxford: Oxford Archaeology.

Hey, G., Dennis, C., Mayes, A., et al. (2007). Archaeological investigations on Whiteleaf Hill, Princess Risborough, Buckinghamshire 2002–06. *Records of Bucks*. Vol. 47, part 2: pp. 1–80.

Higgs, E. (1959). The excavation of a late Mesolithic site at Downton near Salisbury Wiltshire. *Proceedings of the Prehistoric Society*. Vol. 25: pp. 209–32.

Higgs, E. (1961). Some Pleistocene faunas of the Mediterranean coastal areas. *Proceedings of the Prehistoric Society*. Vol. 27: pp. 144–54.

Hinchliffe, J. and Schadla-Hall, R. (1980). *The past under the plough*. London: HMSO.

Hodder, I. (1982). *Symbols in action. Ethnoarchaeological studies of material culture*. Cambridge: Cambridge University Press.

Hodder, I. (1990). *The domestication of Europe*. Oxford: Blackwell.

Hodder, I. (2006). *The leopard's tale. Revealing the mysteries of Çatalhöyük*. London: Thames and Hudson.

Hogan, S. (April 2013a). Manor Farm cursus complex: Floodplain investigations on the River Great Ouse, Milton Keynes. *PAST.* Vol. 73: pp. 2–4.

Hogan, S. (2013b). *Manor Farm, Old Wolverton, Milton Keynes, Buckinghamshire. 2008–2010 Excavation Report*. Cambridge Archaeological Unit. University of Cambridge.

Holgate, R. (1986). Mesolithic, Neolithic and earlier Bronze Age settlement patterns south-west of Oxford. *Oxeniensia*. Vol. 51: pp. 1–14.

Holgate, R. (1995). Early prehistoric settlement of the Chilterns. In: Holgate, R. (ed.). *Chiltern archaeology: Recent work*. Dunstable: The Book Castle, pp. 3–16.

Holm, L. (1991). *The use of stone and hunting of reindeer: A study of stone tool manufacture and hunting of large mammals in the Central Scandes c. 6 000–1 BC* Unpublished PhD Thesis. Department of Archaeology. University of Umea.

Honnor, J. and Lane, T. (2002). *Archaeology, arable landscapes and drainage in the Fenland of Eastern England*. Archaeological Project Services. Report No: 27/02. Oxford Archaeology.

Howell, J. (1982). Neolithic settlement and economy in northern France. *Oxford Journal of Archaeology*. Vol. 1: p. 116.

Hubbard, A. and Hubbard, G. (1907). *Neolithic Dew-ponds and Cattle-ways*. Second edition, London: Longmans, Green.

Huntley, B. and Birks, H. (1983). *An atlas of past and present pollen maps for Europe*. Cambridge: Cambridge University Press.

Hurst, J. (1983). The Wharram Research Project: Results to 1983. *Medieval Archaeology*. Vol. 28: pp. 77–111.

Ingold, T. (2000). *The perception of the environment. Essays in livelihood, dwelling and skill*. London: Routledge.

Ingrouille, M. (1995). *Historical ecology of the British flora*. London: Chapman & Hall.

Isaaks, E. and Srivastava, R. (1990). *An introduction to applied geostatistics*. Oxford: Oxford University Press.

Iversen, J. (1973). The development of Denmark's nature since the last glacial. *Denmark's Geology Undersoils*. Vol. 7: pp. 1–126.

Jacques, D. (2016). Blick Mead's Mesolithic home comforts. *Current Archaeology*. Vol. 310: p. 11.

Jacques, D. and Phillips, T. with contributions by Hoare, P., Bishop, B., Legge, A. and Parfitt, S. (2014). Mesolithic settlement near Stonehenge: Excavations at Blick Mead, Vespasians Camp, Amesbury. *The Wiltshire Archaeological and Natural History Magazine*. Vol. 107: pp. 7–27.

Jacques, D., Phillips, T. and Lyons, T. (2014). Return to Blick Mead: Exploring the Mesolithic origins of Stonehenge's ritual landscape. *Current Archaeology*. Vol. 293: pp. 24–29.

Jacques, D., Lyons, T. and Phillips, T. (2017). Blick Mead: Exploring the first place in the Stonehenge landscape. *Current Archaeology*. Vol. 324: pp. 18–23.

Jacques, D., Phillips, T., Lyons, T., et al. (2018). *Blick Mead: Exploring the first place in the Stonehenge landscape*. Oxford: Peter Lang.

Jedrzejewska, B. and Jedrzejewski, W. (1998). *Predation in vertebrate communities. The Biakowieza primeval forest as a case study*. Berlin: Springer.

Jefferies, R. (1892). *Red deer*. London: Longmans, Green.

Jochim, M. (1976). *Hunter-gatherer subsistence and settlement: A predictive model. Studies in archaeology*. New York: Academic Press.

Jochim, M. (1991). Archaeology as long-term ethnography. *American Anthropologist*. Vol. 93: pp. 308–21.

Johnson, R. (1999). An empty path, processions memories and the Dorset Cursus. In: Barclay, A. and Harding, J. (eds). *Pathways and ceremonies. The cursus monuments of Britain and Ireland*. Neolithic Studies Group Seminar Papers 4. Oxford: Oxbow Books, pp. 39–48.

Jones, A. (1998). Where eagles dare: Landscape, animals and the Neolithic of Orkney. *Journal of Material Culture*. Vol. 3: pp. 301–24.

Jones, D. (1998). Long barrows and elongated enclosures in Lincolnshire. An analysis of the photographic evidence. *Proceedings of the Prehistoric Society*. Vol. 64: pp. 83–114.

Jonsson, L. (1988). The vertebrate faunal remains from the Late Atlantic settlement skateholm in Scania, South Sweden. *Acta Regiae Societatia Humaniorum Litterarum Lundensia*. Vol. 79: pp. 56–88.

Keiller, A. (1965). *Windmill Hill and Avebury: Excavations by Alexander Keiller 1925–1929*. Oxford: Clarenden Press.

Keith-Lucas, D. (2002). Pollen analysis. In: Ford, S. (ed.). *Charnham Lane, Hungerford. Archaeological investigations 1988–1997*. Monograph No. 1. Reading, MA: Thames Valley Archaeological Society.

Kemp, S. (1998). Neolithic and Bronze Age ritual activity at Barford Road, Eynesbury, St Neots. *Cambridgeshire County Council Archaeology Report 148*.

Kendall, H. (1916). Excavations at Hackpen Hill, Wilts. *Proceedings of the Society of Antiquaries of London*. Second Series. Vol. 28: pp. 26–48.

Kennard, A. (1936). Report on non-marine Mollusca. In: Drew, C. and Piggott, S. (eds). Excavation of Long Barrow 163a on Thickthorne Down, Dorset. *Proceedings of the Prehistoric Society*. Vol. 2: pp. 04–95.

Kerridge, E. (1967). *The Agricultural revolution*. London: George Allen & Unwin.

Kidd, S. (2010). Prehistoric farmers. In: Farley, M. (ed.). *An illustrated history of early Buckinghamshire*. Aylesbury: Buckinghamshire Archaeological Society, pp. 27–74.

Kinnes, I. (1985). Circumstances not context: The Neolithic of Scotland as seen from outside. *Proceedings of the Society of Antiquaries of Scotland*, p. 26.

Knight, M. (2009). Excavating a Bronze Age timber platform at Must Farm, Whittlesey, near Peterborough. *PAST*. Vol. 63: pp. 1–4.

Knight, M. (2013). Prehistoric pottery. In: Hogan, S. (ed.). *Manor Farm, Old Wolverton, Milton Keynes, Buckinghamshire. 2008–2010 Excavation Report*. Cambridge Archaeological Unit. University of Cambridge.

Knight, S. (2010). Animal bone. In: Lewis, J., Leivers, M., Brown, L., Smith, A., Cramp, K., Mephan, L. and Phillpotts, C. (eds). *Landscape evolution in the Middle Thames Valley: Heathrow Terminal 5 Excavations*. Vol. 2: Framework Archaeology Monograph No. 3. Framework Archaeology.

Kohler, T. and Parker, S. (1986). Predictive models for archaeological resource location. In: Schiffer, M. (ed.). *Advances in archaeological method and theory*. Vol. 9: New York: Academic Press, pp. 397–452.

Kugler, W. (2011). *The ecological value of Feral populations in Europe: Project to review the current situation and develop a network for the management of feral and semi-feral animal populations*. Interim report, end of 2011: Initial recording and assessment of feral and semi-feral livestock populations in Europe. Safeguard for Agricultural Varieties in Europe.

Lacaille, A. (1963). Mesolithic industries beside Colne waters in Iver and Denham, Buckinghamshire. *Records of Buckinghamshire*. Vol. 17(3): pp. 143–81.

Lake, M., Woodman, P. and Mithen, S. (1998). Tailoring GIS software for archaeological applications: An example concerning viewshed analysis. *Journal of Archaeological Science*. Vol. 25: pp. 27–38.

Lambrick, G. and Moore, J. (1987). Drayton Cursus. *South Midlands Archaeology*. Vol. 17: pp. 85–87.

Lambrick, G. and Robinson, M. (2009). *The Thames through time. The archaeology of the gravel terraces of the upper and middle Thames. Late prehistory 1500 BC–AD 50*. Thames Valley Landscapes Monograph. Vol. 29. Oxford: Oxford Archaeology.

Land Use Consultants. (2005). *South Downs integrated landscape character assessment*. Prepared for the South Downs Joint Committee, in partnership with the Countryside Agency, English Heritage, Hampshire CC, West Sussex CC & East Hampshire CC. London.

Leary, J., Field, D., and Campbell, G. (2013). *Silbury Hill: The largest prehistoric mound in Europe*. Swindon: English Heritage.

Leeds, E. (1934). Rectangular enclosures of the Bronze Age in the Upper Thames Valley. *Antiquaries Journal*. Vol. 14: pp. 414–46.

Legge, A. (1981a). Aspects of cattle husbandry. In: Mercer, J. (ed.). *Farming practice in British prehistory*. Edinburgh: Edinburgh University Press, pp. 169–81.

Legge, A. (1981b). The agricultural economy. In: Mercer, R. (ed.). Grimes Graves, Norfolk. Excavations 1971–72: Vol. 1. *Department of the Environment Archaeological Report 11*. London: Department of the Environment, pp. 79–103.

Legge, A. and Rowley-Conwy, P. (1988). *Star Carr revisited. A re-analysis of the large mammals*. London: Centre for Extra-Mural Studies. Birkbeck College. University of London.

Lewis, J. (1991). Excavation of a late glacial and Early Flandrian site at Three Ways Wharf, Uxbridge. Interim Report In: Barton, R., Roberts, A. and Rowe, D. (eds). *Late Glacial settlement in north-west Europe*. Council for British Archaeology Research Report. No. 77.

Lewis, J. and Rackham, J. (2011). *Three Ways Wharf, Uxbridge: A late glacial and early Holocene hunter-gatherer site in the Colne Valley*. Monograph Series 1. Museum of London Archaeology.

Lewis, J., Leivers, M., Brown, L., Smith, A., Cramp, K., Mephan, L. and Phillpotts, C. (2010). *Landscape evolution in the Middle Thames Valley: Heathrow Terminal 5 Excavations*. Vol. 2: Framework Archaeology Monograph No. 3. Framework Archaeology.

Lockyer, N. (1906). *Stonehenge and other British stone monuments: Astronomically considered*. London: Macmillan.

Lombard, M. (2004). Distribution patterns of organic residues on Middle Stone Age points from Sibudu Cave, KwaZulu-Natal, South Africa. *South African Archaeological Bulletin*. Vol. 59: pp. 37–44.

Lord, T., Thorp, J. and Wilson, P. (2015). *A wild boar dominated ungulate assemblage from an early Holocene natural pit fall trap: Cave shaft sediments in northwest England associated with the 9.3 ka BP cold event*. The Holocene. California: Sage Publications. Vol. 26: pp. 147–53.

Loveday, R. (1985). *Cursuses and related monuments of the British Neolithic*. Unpublished PhD. University of Leicester.

Loveday, R. (1999). Dorchester-on-Thames – ritual complex or ritual landscape. In: Barclay, A. and Harding, J. (eds). *Pathways and ceremonies. The cursus monuments of Britain and Ireland*. Neolithic Studies Group Seminar Papers 4. Oxford: Oxbow Books, pp. 49–68.

Loveday, R. (2006). *Inscribed across the landscape: The Cursus enigma*. Stroud: Tempus Publishing.

Loveday, R. (2009). From ritual to riches – the route to individual power in later Neolithic Eastern Yorkshire. In: Brophy, K. and Barclay, G. (eds). *Defining a regional Neolithic: The evidence from Britain and Ireland*. Oxford: Oxbow Books, pp. 35–52.

Loveday, R. (2016). Monuments to mobility? Investigating cursus patterning in southern Britain. In: Leary, J. and Kador, K. (eds). *Moving on in Neolithic studies: Understanding mobile lives*. Neolithic Studies Group Seminar Paper 14. Oxford: Oxbow Books, pp. 67–109.

Lubbock, J. (1865). *Pre-historic times, As illustrated by ancient remains, and the manners and customs of modern savages*. London and Edinburgh: Williams and Norgate.

Lucas, G. (1995). *Genealogies: Classification, narrative and time: An archaeological study of Eastern Yorkshire, 3700–1300 BC*. Unpublished PhD thesis. University of Cambridge.

Lyons, R. and Machen, R. (2001) Livestock Grazing Distribution: Considerations and Management. *AgriLife Extension Service, The Texas A&M University System*. Vol. 12: pp. 1–6.

Macaulay, S. (1994). Archaeological investigations on a proposed synthetic pitch at Ernulf School, Eynesbury. *Cambridgeshire County Council Archaeology Report A41*.

Mainland, I. (1998). Dental microwear and diet in domestic sheep (Ovis aries) and goats (Capra hircus): Distinguishing grazing and fodder fed ovicaprids using a quantitive analytical approach. *Journal of Archaeological Science*. Vol. 25: pp. 1259–71.

Malim, T. (1993). Neolithic ditches and Iron Age settlement at Thrapston Road, Brampton 1992. *Cambridgeshire Archaeological Reports*. Vol. 81.

Malim, T. (1999). Cursuses and related monuments of the Cambridgeshire Ouse. In: Barclay, A. and Harding, J. (eds). *Pathways and Ceremonies. The cursus monuments of Britain and Ireland*. Neolithic Studies Group Seminar Papers 4. Oxford: Oxbow Books, pp. 77–87.

Manby, T. (1958) A Neolithic site at Driffield, East Riding of Yorkshire. *The Yorkshire Archaeological Journal*. Vol. 39: pp. 169–78.

Manby, T. (1963). The excavation of the Willerby Wold long barrow, East Riding of Yorkshire. *Proceedings of the Prehistoric Society*. Vol. 29: pp. 173–205.

Manby, T. (1974). *Grooved Ware sites in the North of England*. Oxford: British Archaeological Reports. British Series 133.

Manby, T. (1976). The excavation of the Kilham long barrow, East Riding of Yorkshire. *Proceedings of the Prehistoric Society*. Vol. 42: pp. 111–59.

Manby, T. (1988). *Archaeology in eastern Yorkshire: Essays in honour of T C M Brewster*. J R Collis Publications. Department of Archaeology and Prehistory, University of Sheffield, pp. 65–66.

Maroo, S. and Yalden, D. (2000). The Mesolithic animal fauna of Great Britain. *Mammal Society. Mammal Review*. Vol. 30: pp. 243–48.

Marshall, S. (2016). *Exploring Avebury, the essential guide*. Stroud: The History Press.

Marshall, P., Hamilton, W., Van der Plicht, J., Bronks Ramsay, C., Cook, G. and Goslar, T. (2009). Scientific dating. In: Beamish, M. (ed.) Island visits: Neolithic and Bronze Age activity on the Trent valley floor. Excavations at Eggington and Willington, Derbyshire 1998–1999. *Derbyshire Archaeological Journal*. Vol. 132: pp. 80–128.

Martin, E. (1915). *Dew-ponds: History, observation, and experiment*. London: Werner Laurie.

Martin, P. (1978). Scatter of Mesolithic flints found. Council for British Archaeology Group 9. *South Midlands archaeology newsletter*.

Matthews, C. (1976). *Occupation sites on a Chiltern ridge Part 1: Neolithic, Bronze Age and Early Iron Age*. British Archaeological Reports. Vol. 29. Oxford: BAR.

Maxwell, G. (1979). Air photography and the work of the Royal Commission on the Ancient and Historic Monuments of Scotland. *Aerial Archaeology*. Vol. 2: pp. 37–43.

May, J. (1976). *Prehistoric Lincolnshire*. The History of Lincolnshire Committee: Cox & Wyman.

McAvoy, F. (1991). Godmanchester, Cambridgeshire. *English Heritage Conservation Bulletin*. Vol. 14: pp. 16–18.

McOmish, D. (2003). Cursus: Solving a 6,000-year-old puzzle. *British Archaeology*. Vol. 69: pp. 8–13.

McOrmish, D., Field, D. and Brown, G. (2002). *The field archaeology of the Salisbury Plain training area*. Swindon: English Heritage.

Megaw, J. and Simpson, D. (1979). *An introduction to British prehistory*. Leicester: Leicester University Press.

Mellars, P. (1976). Fire ecology, animal populations and man. A study of some ecological relationships in prehistory. *Proceedings of the Prehistoric Society*. Vol. 42: pp. 15–45.

Microclimates. (2011). National Meteorological Library and Archive. Fact sheet 14 (version 01). © Crown copyright 2011 11/0510.

Millican, K. (2016). *The timber monuments of Neolithic Scotland*. BAR British Series 623. Oxford: British Archaeological Reports.

Mithen, S. (1998). *The prehistory of the mind*. London: Phoenix.

Mithen, S. (1990). *Thoughtful foragers: A study of prehistoric decision making*. Cambridge: Cambridge University Press.

Moore, J. (1996). Damp squib. How to fire a major deciduous forest in an inclement climate. In: Pollard, T. and Morrison, A. (eds). *The early prehistory of Scotland*. Edinburgh: Edinburgh University Press, pp. 62–73.

Moore, J. (2003). Enculturation through fire: Beyond hazelnuts and into the forest. In: Larsson, L., Kindgren, H., Knutsson, K., Loeffler, D. and Akerlund, A. (eds). *Mesolithic on the move*. Oxford: Oxbow Books, pp. 139–44.

Moss-Eccardt, J. (1988). Archaeological investigations in the Letchworth area, 1958–1974. *Proceedings of the Cambridge Antiquarian Society*. Vol. 77: pp. 35–103.

Murrell, K. (2012). *Must Farm, Whittlesey 2011–2012: Palaeochannel Investigations*. Interim Statement. University of Cambridge. Cambridge Archaeological Unit. Report No. 1136.

Myers, A. (1989). Reliable and maintainable technology in the Mesolithic of mainland Britain. In: Torrence, R. (ed.). *Time, energy and stone tools*. Cambridge: Cambridge University Press, pp. 78–91.

O'Connell, M. (1990). Excavations during 1979–1985 of a multi-period site at Stanwell, Surrey. *Archaeological Collections*. Vol. 80: pp. 1–62.

Opitz, R. and Cowley, D. (2013). *Interpreting archaeological topography: Lasers, 3D data, observation, visualisation and applications*. Occasional Publication of the Aerial Archaeology Research Group No 5. Oxford: Oxbow Books.

Oswald, A., Dyer, C. and Barber, M. (2001). *The creation of monuments: Neolithic causewayed enclosures in the British Isles*. Swindon: English Heritage.

Palmer, R. (1979). *Aerial photography report*. <https://www.scambs.gov.uk/media/4249/18b2_ns-phase-1-es-tech-appx-b_aerial_photo_report.pdf>.

Parker, A. and Robinson, M. (2003). Palaeoenvironmental investigations on the Middle Thames at Dorney, UK. In: Howard, A., Macklin, M., Balkema, A. and Passmore, D. (eds). *Proceedings of the Alluvial Archaeology of North West Europe and Mediterranean*. Oxford: Oxbow Books, pp. 18–19.

Parker-Pearson, M. (2012). *Stonehenge: Exploring the greatest Stone Age mystery*. London. Simon and Schuster.

Parker Pearson, M. and Ramilisonina. (1998). Stonehenge for the ancestors: Part two. *Antiquity*. Vol. 72(278): pp. 855–56.

Parker-Pearson, M., Richards, C., Allen, M., Payne, A. and Welham, K. (2003). The Stonehenge riverside project: New approaches to Durrington Walls. *PAST*. Vol. 45: pp. 6–8.

Parker Pearson, M., Pollard, J., Richards, C., Thomas, J., Tilley, C. and Welham, K. (2008). *The Stonehenge Riverside Project 2007 Interim Report*.

Parker-Pearson, M., Thomas, J., Marshall, P., Pollard, J., Richards, C., Tilley, C. and Welham, K. (2009). The date of the Greater Stonehenge Cursus. *Antiquity* Vol. 83: p. 319.

Parker Pearson, M., Allen, M., Bayer, O., Casswell, C., Chan, B., French, C., Garwood, P., Nunn, B., Pitts, M., Pollard, J., Pullen, B., Richards, C., Richards, J., Robinson, D., Rylatt, J., Shaw, D., Teather, A. and Thomas, J. (2011). *The Stonehenge Riverside Project 2008 Interim Report*.

Parrington, M. (1982). Review of Archaeology Bedfordshire, Buckinghamshire, Northamptonshire and Oxfordshire. *Council of British Archaeology Newsletter*. Vol. 12: p. 188.

Pastick, J. (2012). The biology of native and invasive Wild Boar (Sus scrofa) and the effect it is having in its invasive range. *Eukaryon*. Vol. 8: pp. 61–63.

Pearson, T. and Field, D. (2011). *Stonehenge cursus: Investigation of earthworks*. Research Department Report Series 103–2011. Swindon: English Heritage.

Pennick, N. and Devereux, P. (1989). *Lines on the landscape. Leys and other linear enigmas*. London: Robert Hale.

Pennington, W. (1974). *The history of British vegetation*. London: Hodder and Stoughton.

Penny, A. and Wood, J. (1973). The Dorset Cursus: A Neolithic astronomical observatory. *Archaeology Journal*. Vol. 130: pp. 44–76.

Phillips, P. (1989). *Archaeology and landscape studies in North Lincolnshire*. BAR British Series 208 i & ii. Oxford: British Archaeological Reports.

Piggott, S. (1954). *The Neolithic cultures of the British Isles; a study of the stone-using agricultural communities of Britain in the second millennium BC*. Cambridge: Cambridge University Press.

Pollard, J (2005). Memory, monuments and middens. In: Brown, G., Field, D. and McOmish, D. (eds). *The Avebury landscape: Aspects of the field archaeology of the Marlborough Downs*. Oxford: Oxbow Books, pp. 103–14.

Pollard, J. and Reynolds, A. (2002). *Avebury: The biography of a landscape*. Stroud: The History Press.

Proudfoot, E. (1965). Bishops Cannings: Roughridge Hill. *Wiltshire Archaeological Magazine*. Vol. 60: pp. 132–33.

Prummel, W. and Niekus, M. (2011). Late Mesolithic hunting of a small female aurochs in the valley of the River Tjonger (the Netherlands) in the light of Mesolithic Aurochs hunting in NW Europe. *Journal of Archaeological Science*. Vol. 38: pp. 1456–67.

Pryor, F. (1982). Problems of survival: Later prehistoric settlement in the southern East Anglian Fenlands. In: *Prehistoric settlement patterns around the southern North Sea*. Papers presented at a colloquium, held in honour of Professor Dr. P.J.R. Modderman, Leiden, 3–7 May 1982. University Press, pp. 125–43.

Pryor, F. (1998). *Etton, excavations at a Neolithic Causewayed enclosure near Maxey, Cambridgeshire, 1982–7*. Swindon: English Heritage Archaeological Report.

Pryor, F. (2015a). *Stonehenge: The story of a sacred landscape*. London: Head of Zeus.

Pryor, F. (2015b). *Home: A time traveller's tales from British prehistory*. London: Penguin Books.

Pryor, F. and French, C. (2005). *Archaeology and environment of the Etton landscape*. East Anglian Archaeology Report No. 109. Fenland Archaeological Trust.

Rackham, O. (1980). *Ancient Woodland. Its history, vegetation and uses in England*. London: Edward Arnold.

Rackham, O. (1986). *The history of the countryside*. London: Dent.

Rackham, O. (1996). *Trees and woodland in the British landscape. The complete history of Britain's trees, woods and hedgerows*. London: Phoenix Giant.

Rajkovaca, V. (2013). Faunal remains. In: Hogan, S. (2013). *Manor Farm, Old Wolverton, Milton Keynes, Buckinghamshire. 2008–2010 Excavation Report*. Cambridge Archaeological Unit. University of Cambridge.

Ramm, H. (1974). Yorkshire Archaeology Register: Aerial reconnaissance and interpretation. *Yorkshire Archaeological Journal*. Vol. 46: pp. 141–57.

Rattray, P. (2009). *Secrets of the reds: A complete guide to hunting red deer*. Available at <https://www.slideshare.net/paulrattray/red-deer-hunting-a-complete-guide-excerpt-2015741> [Accessed 12 October 2016]

Ray, K. and Thomas, J. (2003). In the Kinship of cows: The social centrality of cattle in the earlier Neolithic of southern Britain. In: Parker-Pearson, M. (ed.) *Food, culture and identity in the Neolithic and early Bronze Age*. British Archaeological Report International Series. Vol. 1117. Oxford: Archaeopress, pp. 37–44.

Ray, K. and Thomas, J. (2018). *Neolithic Britain: The transformation of social worlds*. Oxford University Press.

Raymond, A. (1986). Experiments in the function and performance of the weighted atlati. *World Archaeology*. Vol. 18(2): pp. 153–77.

Raymond, F., Bradley, R. and Entwistle, R. (1994). *Prehistoric Land Divisions on Salisbury Plain: The work of the Wessex Linear Ditches Project*. Swindon. English Heritage.

RCHME (1979). *Stonehenge and its environs, monuments and land use*. London: HMSO.

Renfrew, C. (1973). Monuments, mobilisation and social organisation in Neolithic Wessex. In: Ucko, P., Tringham, R. and Dimbleby, R. (eds). *Man, settlement and urbanism*. London: Gerald Duckworth, pp. 539–58.

Renfrew, C. (2007). *Prehistory, the making of the human mind*. London: Phoenix.

Renfrew, C. and Bahn, P. (2012). *Archaeology. Theories, methods and practice*. Sixth edition. London: Thames & Hudson.

Rhodes, E. (2007). *Barford Road, St Neots, Cambridgeshire. Optically stimulated luminescence dating of single grains of quartz from sedimentary fills of two Cursus monuments*. Archaeological Science. Research Department Report Series 32/2007. English Heritage.

Richards, C. (2004). Labouring with monuments: Constructing the dolmen at Carreg Samson, south-west Wales. In: Cummings, V. and Fowler, C. (eds). *The Neolithic of the Irish Sea. Materiality and transitions of practice*. Oxford: Oxbow Books, pp. 72–80.

Richards, J. (1990). *The Stonehenge Environs Project*. Arch Report Vol. 16. Swindon: English Heritage.

Richards, J. (2017). *Stonehenge: The story so far*. Swindon: Historic England.

Richmond, I. (1940). Excavations on the estate of Meikleour, Perthshire, 1939. *Proceedings of the Society for Antiquities in Scotland*. Vol. 74: pp. 37–48.

Riley, D. (1980). Recent air photographs of Duggleby Howe and the Ferrybridge henge. *Yorkshire Archaeological Journal*. Vol. 52: pp. 174–78.

Robertson-Mackay, M. (1980). A "head and hooves" burial beneath a round barrow with other Neolithic and Bronze Age sites, on Hemp Knoll, near Avebury, Wiltshire. *Proceedings of the Prehistoric Society*. Vol. 46: pp. 123–76.

Robinson, M. (2000). Middle Mesolithic to late Bronze Age insect assemblages and an early Neolithic assemblage of waterlogged macroscopic plant remains. In: Needham, S. (ed.). *The passage of the Thames: Holocene environment and settlement at Runnymede (Runnymede Bridge Research Excavations 1)*. London: British Museums Press, pp. 146–67.

Robinson, M. (2005). Neolithic and Bronze Age insect assemblages. In: French, C. and Pryor, F. (eds). *Archaeology and environment of the Etton landscape*. East Anglian Archaeology. No. 109, pp. 153–62.

Robinson, M. (2017). Molluscs from the floodplain alluvial sediments in the Thames Valley. In: Allen, M. (ed.). *Molluscs in archaeology. Methods, approaches and applications*. Oxford: Oxbow Books, pp. 112–26.

Rogers, B. (2015). *Where the wild things are: Analysis of the faunal assemblage from the Mesolithic site at Blick Mead, Wiltshire. Including a study on the diet and migration of the aurochs.* Unpublished thesis. Durham University.

Rogers, B., Grocke, D., Gron, K., Montgomery, J., Rowley-Conwy, P. and Jacques, D. (2019). Isotopic analysis of the Blick Mead dog: A proxy for the dietary reconstruction and mobility of Mesolithic British hunter-gatherers. Journal of Archaeological Science: Reports. Vol. 24: pp. 712–720.

Rogers, B., Gron, K., Montgomery, J., Grocke, D., and Rowley-Conwy, P. with a contribution by Charlton, S. (2018). Aurochs hunters: The large animal bones from Blick Mead. In: Jacques, D., Phillips, T. and Lyons, T. (eds). *Blick Mead exploring the "first place" in the Stonehenge landscape*. Oxford: Peter Lang, pp. 127–52.

Rouse, A. and Evans, J. (1993). Land Molluscs. In: Whittle, A., Rouse, A. and Evans, J. (eds). A Neolithic downland monument in its environment: Excavations at Easton Down long barrow. Bishops Cannings, North Wiltshire. *Proceedings of the Prehistoric Society*. Vol. 59: pp. 197–239.

Rowley-Conwy, P. (1982). Forest grazing and clearance in temperate Europe with special reference to Denmark; an archaeological view. In: Bell, M. and Limbrey, S. (eds). *Archaeological aspects of woodland ecology*. Oxford: British Archaeological Reports, S146, pp. 199–215.

Rowley-Conwy, P. (2004). How the west was lost – A reconsideration of agricultural origins in Britain, Ireland and Southern Scandinavia. *Current Anthropology*. Vol. 45: pp. S83–S113.

Rowley-Conwy, P. (2014). *Animals in the archaeology of southern Britain Mesolithic and Neolithic periods*: University of Buckingham Lecture. Society of Antiquaries.

Rowley-Conwy, P. (2017, In press). To the Upper Lake: Star Carr revisited – by birchbark canoe. In: Rowley-Conwy, P., Serjeantson, D and Halstead, P. (eds). *Economic Zooarchaeology: Studies in hunting, herding and early agriculture*. Oxford: Oxbow Books, pp. 197–207.

Ruggles, C. (1999). *Astronomy in prehistoric Britain and Ireland*. London: Yale University Press.

Ryan, K. (2005). Facilitating milk let-down in traditional cattle herding systems: East Africa and beyond. In: Mulville, J. and Outram, A. (eds). *The zooarchaeology of fats, oils, milk and dairying*. Oxford: Oxbow, pp. 96–107.

Sahlins, M. (1974). *Stone Age economics*. London: Tavistock.

Saunders, D. (2015). *An assessment of the evidence for large herbivore movement and hunting strategies within the Stonehenge landscape during the Mesolithic*. Unpublished MA thesis. Degree of MA in Archaeology (by research): Stonehenge: A Landscape Through Time. The University of Buckingham.

Saunders, D. (2019). An assessment of the evidence for large herbivore movement and hunting strategies within the Stonehenge landscape during the Mesolithic. In: Jacques, D. and Davis, G. (eds). *Stonehenge a landscape through time*. Oxford: Peter Lang, pp. 173–200.

Scaife, R. (1980). *Late Devensian and Flandrian palaeocological studies in the Isle of Wright*. Unpublished PhD thesis, Kings College, University of London.

Scaife, R. (1987). A review of the later Quaternary microfossil and macrofossil research in southern England with special reference to environmental archaeology. In: Keeley, H. (ed.). *Environmental archaeology; A regional review*. Vol. 2: pp. 125–203. London: HBMC.

Scaife, R. (1995). A Boreal and sub-boreal chalk landscape: Pollen evidence. In: Cleal, R., Walker, K. and Montague, R. (eds). *Stonehenge in its landscape: Twentieth-century excavations*. Swindon: English Heritage, pp. 51–55.

Scaife, R. (2004). Soil pollen. In: Allen, M. and Gardiner, J. (eds). Neolithic of the Wylye valley 1: Millennium re-investigation of the Corton long barrow ST 9308 4034. *Wiltshire Archaeological Magazine*. Vol. 97: pp. 63–77.

Scaife, R. (2007). Palynological analysis. In: French, C., Lewis, H., Allen, M., Green, M., Scaife, R. and Gardiner, J. (eds). *Prehistoric landscape development and human impact in the upper Allen Valley, Cranborne Chase, Dorset*. Cambridge: McDonald Institute for Archaeological Research, pp. 43–63.

Schemenauer, R. and Cereceda, P. (1994). The role of wind in rainwater catchment and fog collection. *Water International Journal*. Vol. 19(2): pp. 70–76.

Schlichtherle, H. (2006). Kulhauser in neolithischen Pfahlbausiedlungen des Bodensees. In: Hafner, A. Niffeler, U. and Ruoff, U. (eds). *Die neue Sicht. Unterwasserarchaologie und Geschichtsbild*. Internationalen Kongresses fur Unterwasserarchaologie October 2004 Zurich.

Schofield, A. (1991). *Interpreting artefact scatters. Contributions to ploughzone archaeology*. Oxford. Oxbow Books.

Schulting, R. (2008). Footways and social ecologies from the Early Mesolithic to the Bronze Age. In: Pollard, J. (ed.). *Prehistoric Britain*. Oxford: Blackwell, pp. 90–119.

Serjeantson, D. (1995). Animal bones. [In Chapter 9. Finds assemblages]. In: Cleal, R., Walker, K. and Montague, R. (eds). *Stonehenge in its landscape: Twentieth-century excavations*. London: English Heritage, pp. 51–55.

Serjeantson, D. (2011). *Review of animal remains from the Neolithic and Early Bronze Age of Southern Britain*. Research Department Report Series 29. Swindon: Historic England.

Servet, J. (1982). Primitive order and archaic trade, part ii. *Economy and Society*. Vol. 11: pp. 22–59.

Shearer, I. and Mcllan, K. (2008). Tracing time: Excavations at Knowes and Eweford East (3370–2230 BC). In: Lelong, E. and MacGregor, G. (eds). *The lands of ancient Lothian: Interpreting the archaeology of the A1*. Society of Antiquaries of Scotland, pp. 47–68.

Simmons, I. (1975). Towards an ecology of Mesolithic man in the uplands of Great Britain. *Journal of Archaeological Science*. Vol. 2: pp. 1–15.

Simmons, I. (1996). *The environmental impact of the later Mesolithic cultures: The creation of a moorland landscape in England and Wales*. Edinburgh: Edinburgh University Press.

Simmons, I., Dimbleby, G. and Grigson, C. (1981). The Mesolithic. In: Simmons, I. and Tooley, M. (eds). *The environment in British prehistory*. London: Duckworth, pp. 82–124.

Simpson, W. (1966). Romano-British settlement on the Weiland Gravels. In: Thomas, C. (ed.) *Rural settlement in Roman Britain*. Report No. 7. London: Council for British Archaeology Research Report, pp. 15–25.

Smith, A. (1885). A guide to the British and Roman antiquities of the North Wiltshire Downs. *Wiltshire Archaeological and Natural History Society*. Vol. 23: pp. 59–61.

Smith, I. (1965). *Windmill Hill and Avebury: Excavations by Alexander Keiller, 1925–1939*. Oxford: Clarendon Press.

Smith, R. (1984). The ecology of Neolithic farming systems as exemplified by the Avebury region of Wiltshire. *Proceedings of the Prehistoric Society*. Vol. 50: pp. 99–120.

Smith, I. and Simpson, D. (1966). Excavation of a round barrow on Overton Hill, north Wiltshire. *Proceedings of the Prehistoric Society*. Vol. 32: pp. 122–55.

Smith, A., Whittle, A., Cloutman, E. and Morgan, L. (1989). Mesolithic and Neolithic activity and environmental impact on the South-east Fen-edge in Cambridgeshire. *Proceedings of the Prehistoric Society* Vol. 55: pp. 207–49.

Solounias, N. and Moelleken, S. (1992). Toothwear microwear analysis of Eotrgus sansaniensis (Mammalia Ruminantia), one of the oldest known bovids. *Journal of Vertebrate Palaeontology*. Vol. 12(1): pp. 113–21.

Sorensen, K. and Petersen, E. (1986). The Prejlerup aurochs: An archaeological discovery from boreal Denmark. In: Konigsson, L. (ed.). *Nordic late Quaternary. Biology and Ecology*. Vol. 24: pp. 111–17.

Speth, J. (2010). *The paleoanthropology and archaeology of big-game hunting, interdisciplinary contributions to archaeology*. Springer.

Spitkins, P. (2000). GIS models of past vegetation: An example from northern England, 10,000–5,000 BP. *Journal of Archaeological Science*. Vol. 27: pp. 219–34.

Stainton, B. (1989). Excavation of an early prehistoric site at Stratfords Yard, Chesham. *Records of Buckinghamshire*. Vol. 31: pp. 49–74.

Stiner, M. (1990). The use of mortality patterns in archaeological studies of hominid predatory adaptations. *Journal of Anthropological Archaeology*. Vol. 9(4): pp. 305–51.

Stiner, M. (1994). *Honour among thieves: A zooarchaeological study of Neandertal ecology*. American Antiquity. Vol. 61(4). Princeton: Princeton University Press.

St Joseph, J. (1961). Air Reconnaissance in Britain, 1958–60. *Journal of Roman Studies*. Vol. 51: pp. 119–35.

St Joseph, J. (1965). Air Reconnaissance in Britain, 1961–64. *Journal of Roman Studies*. Vol. 55: pp. 74–89.

Stoertz, C. (1997). Ancient landscapes of the Yorkshire Wolds. *Aerial photographic transcription and analysis*. Map 3: pp. 17, 27–30, 85.

Stone, J. (1947). The Stonehenge Cursus and its affinities. *Archaeological Journal*. Vol. 104: pp. 7–19.

Stukeley, W. (1740). *Stonehenge: A temple restored to the British Druids*. London: Innys and Manby.

Sturt, F. (2006). Local knowledge is required: A rhythm analytical approach to the late Mesolithic and early Neolithic of East Anglian Fenland, UK. *Journal of Maritime Archaeology*. Vol. 1: p. 124.

Surovell, T. and Waguespack, N. (2009). Human prey choice in the late Pleistocene and Its relation to megafaunal extinctions. In: Haynes, G. (ed.) *American megafaunal extinctions at the end of the Pleistocene*. Vertebrate Paleobiology and Paleoanthropology. Dordrecht: Springer, pp. 77–105.

Svenning, J. (2002). A review of natural vegetation openness in north-western Europe. *Biological Conservation*. Vol. 104: pp. 133–48.

Tansley, A. (1939). *The British Islands and their vegetation*. Cambridge: Cambridge University Press.

Taylor, C. (1979). *Roads and tracks of Britain*. London: Dent.

Taylor, C. (1983). *Village and farmstead: A history of rural settlement in England*. London: George Phillip.

Taylor, E. (1865). *Research into the early history of mankind and the development of civilisation*. London: John Murray.

Ten Hove, H. (1968). The Ulmus fall and the transition Antlanticum / Sub-boreal in pollen diagrams. *Palaeogeography, Palaeoclimatology, Palaeoecology*. Vol. 5: pp. 359–69.

Thomas, A. (1960). Further changes in vegetation since the advent of myxomatosis. *Journal of Ecology*. Vol. 48: pp. 287–305.

Thomas, J. (1988). Neolithic explanations revisited: The Mesolithic–Neolithic transition in Britain and south Scandinavia. *Proceedings of the Prehistoric Society*. Vol. 54: pp. 59–66.

Thomas, J. (1991). *Rethinking the Neolithic*. Cambridge: Cambridge University Press.

Thomas, J. (1996a). *Time, culture and identity*. London: Routledge.

Thomas, J. (1996b). Neolithic houses in mainland Britain and Ireland – a sceptical view. In: Thomas, J. and Darvill, T. (eds). *Neolithic houses in northwest Europe and beyond*. Oxford: Oxbow Books.

Thomas, J. (1999). *Understanding the Neolithic*. London: Routledge.

Thomas, J. (2006). On the origins and development of cursus monuments in Britain. *Proceedings of the Prehistoric Society*. Vol. 72: pp. 229–41.

Thomas, J. (2007) Mesolithic – Neolithic transactions in Britain: From essence to inhabitation. *Proceedings of the British Academy*. Vol. 144: pp. 423–39.

Thomas, K. (1977). The land Mollusca from the enclosure on Offham Hill. In Drewett, P. (ed.). The excavation of a Neolithic causewayed enclosure on Offham Hill, East Sussex 1976. *Proceedings of the Prehistoric Society*. Vol. 43: pp. 234–39.

Thomas, K. (1981). Land snail assemblages. In: Bedwin, O. and Aldsworth, F. (eds). Excavations at the Trundle 1980. *Sussex Archaeological Collections*. Vol. 119: pp. 211–14.

Thomas, K. (1982). Neolithic enclosures and woodland habitats on the South Downs in Sussex, England. In: Bell, M. and Limbrey, S. (eds). *Archaeological aspects of woodland ecology*. Oxford: British Archaeology Report S146: pp. 147–70.

Thomas, K. (1983). Mollusc analysis of samples from the ditch fill of trench II. In: Leach, P (ed.). The excavation of a Neolithic causewayed enclosure on Barkhale Down, Bignor Hill, West Sussex. *Sussex Archaeological Collections*. Vol. 121: pp. 28–30.

Thomas, K. (1994). Evidence of the environmental setting of the Neolithic enclosure at Combe Hill, East Sussex. In: Drewett, P (ed.). Dr V. Seton Williams excavations at Combe Hill, 1962 and the role of Neolithic causewayed enclosures in Sussex. *Sussex Archaeological Collections*. Vol. 132: pp. 17–19.

Thomas, K. (1996). A contribution to the environmental history of Whitehawk Neolithic enclosure. In: Russell, M. and Rudling, D. (eds). Excavations at Whitehawk Neolithic enclosure, Brighton, East Sussex: 1991–93. *Sussex Archaeological Collections*. Vol. 134: pp.51–56.

Thomas, N. (1955). A Neolithic pit on Waden Hill, Avebury. *Wiltshire Archaeological and Natural History Magazine*. Vol. 56: pp. 167–71.

Thomas, J., Marshall, P., Parker Pearson, M., Pollard, J., Richards, C., Tilley, C. and Welham, K. (2009). The date of the Greater Stonehenge Cursus. *Antiquity*. Vol. 83: pp. 40–53.

Thurnam, J. (1868). On ancient British barrows. Part 1 – long barrows. *Archaeologia*. Vol. 42: pp. 161–244.

Tilley, C. (1994). *A phenomenology of landscape*. Oxford: Berg.

Tilley, C. (1996). *An ethnography of the Neolithic*. Cambridge: Cambridge University Press.

Tilley, C. (2010). *Interpreting landscapes: Geologies, topographies, identities: Explorations in landscape, Phenomenology 3*. California: Left Coast Press.

Timeline Maps Historical Map Series (2005). *Timeline Historical Map 1811–1817*. Salisbury and the Plain. Number 184.

Tipping, R. (1994). The form and fate of Scotland's woodlands. *Proceedings of the Society of Antiquaries of Scotland*. Vol. 124: p. 9.

Tipping, R. (1996). The Neolithic landscapes of the Cheviot Hills and hinterland: Palaeo-environmental research. *Northern Archaeology*. Special edition. Frodsham, P. (ed.). Vol. 13/14: pp. 17–33.

Topping, P. (1982). Excavations at the cursus at Scorton, North Yorkshire 1978. *Yorkshire Archaeological Journal*. Vol. 54: pp. 7–21.

Trantalidou, K. and Masseti, M. (2015). Archaeozoology of the red deer in the southern Balkan Peninsula and the Aegean region during antiquity: Confronting bones and paintings. In: Baker, K., Carden, R. and Madgwick, R. (eds). *Deer and People*. Oxford: Oxbow Books, pp. 59–77.

Tuan, Y. (1978). Space, time, place: A humanistic frame. In: Carlstein, T., Parks, D. and Thrift, N. (eds). *Making sense of time; timing space and spacing time*. Vol. 1. London: Edward Arnold, p. 10.

Van Vuure, C. (2005). *Retracing the aurochs: History, morphology and ecology of an extinct wild ox*. Sofia, Moscow: Pensoft Publishers.

Vatcher, L. and Vatcher, F. (1973). Excavation of Three Postholes in Stonehenge Car Park. *Wiltshire Archaeological and Natural History Magazine*. Vol. 68: pp. 57–63.

Vera, F. (2000). Grazing ecology and forest history. Wallingford. Oxon. CABI Publishing. pp. 52–55.

Villa, P. and Lenoir, M. (2006). Hunting weapons of the Middle Stone Age and the Middle Palaeolithic: Spear points from Sibudu, Rose Cottage and Bouheben. *Southern African Humanities*. Vol. 18(1): pp. 89–122.

Villa, P. and Lenoir, M. (2009). Hunting and hunting weapons of the Lower and Middle Paleolithic of Europe. In: Richards, M. and Hublin, J. (eds). *The evolution of Hominid Diet*. Leipzig: Springer.

Viner, S., Evans, J., Albarella, U. and Parker Pearson, M. (2010). Cattle mobility in prehistoric Britain: Strontium isotope analysis of cattle teeth from Durrington Walls (Wiltshire, Britain). *Journal of Archaeological Science*. Vol. 37: pp. 2812–20.

Vines, G. and Page, S. (2012). *GIS-based spatial analysis for archaeological site prediction and evaluation*. Review Paper. Melbourne Geospatial Sciences. RMIT University Melbourne, Australia.

Waddington, C. (1999). *A landscape archaeological study of the Mesolithic-Neolithic in the Milfield Basin, Northumberland*. BAR British Series 291.

Waguespack, N., Surovell, T., Denoyer, A., Dallow, A., Savage, A., Hayneman, J. and Tapster, (2009). Making a point: Wood versus stone-tipped projectiles. *Antiquity*. Vol. 83: pp. 786–800.

Wainwright, G. (1960). Three microlith industries from south-west England and their affinities. *Proceedings of the Prehistoric Society*. Vol. 26: pp. 192–210.

Walker, M. (1999). Pollen analysis. In: Whittle, A., Pollard, J. and Grigson, C. (eds). *The harmony of symbols: The Windmill Hill causewayed enclosure*. Oxford: Oxford Books.

Waller, M. and Hamilton, S. (2000). Vegetation history of the English chalklands: A mid Holocene pollen sequence from the Caburn, East Sussex. *Journal of Quaternary Science*. Vol. 15: pp. 253–72.

Walters, M. (1992). *Principles of geoarchaeology. A North American perspective*. Arizona: University of Arizona Press.

Watson, A. (2001). Composing Avebury. *World Archaeology*. Vol. 33: pp. 296–314.

Webster, G., Hobley, B., Baker, A. and Pickering, J. (1964). Aerial reconnaissance over the Warwickshire Avon. *Archaeology Journal*. Vol. 121: pp. 1–32.

Wheatley, D. (1995). Cumulative viewshed analysis: A GIS based method for investigating intervisibility and its archaeological application. In: Lock, G. and Stancic, Z. (eds). *Archaeology and geographical information systems: A European perspective*. London: Taylor & Francis, pp. 171–85.

Wheatley, D. and Gillings, M. (2000). *Spatial technology and archaeology. The archaeological applications of GIS*. London: Taylor & Francis.

Whitehouse, N. (1998). *The evolution of the Holocene wetland landscape of the Humber: Head levels from a fossil insect perspective*. Unpublished PhD thesis, University of Sheffield.

Whitehouse, N. (2000). Forest fires and insects: Palaeoentomological research from a sub-fossil burnt forest. *Palaeogeography, Palaeoclimatology, Palaeoecology*. Vol. 164: pp. 231–46.

Whitehouse, N. and Smith, D. (2004). Islands' in Holocene forests: Implications for forest openness, landscape clearance and 'culture-steppe' species. *Environmental Archaeology*. Vol. 9: pp. 203–12.

Whittle, A. (1977). *The earlier Neolithic of southern Britain and its continental background*. Oxford: BAR.

Whittle, A., Brothwell, D., Cullen, R., Gardner, N. and Kerney, M. (1991). Wayland's Smithy, Oxfordshire: Excavations at the Neolithic Tomb in 1962–63 by R. J. C. Atkinson and S. Piggott. *Proceedings of the Prehistoric Society*. Vol. 57: pp. 61–101.

Whittle, A., Atkinson, R., Chambers, R. and Thomas, N. (1992). Excavations in the Neolithic and Bronze Age complex at Dorchester-on-Thames, Oxfordshire. 1947–52 and 1981. *Proceedings of the Prehistoric Society*. Vol. 58: Pp. 143–201.

Whittle, A., Pollard, J. and Grigson, C. (1999). *The harmony of symbols; the Windmill Hill causewayed enclosure.* Oxford: Oxbow Books.

Whittle, A., Healy, F. and Bayliss, A. (2011). *Gathering Time: Dating the Early Neolithic enclosures of Southern Britain and Ireland.* Oxford: Oxbow Books.

Wikars, L. and Schimmel, J. (2001). Immediate effects of fires severity on soil invertebrates in cut and uncut pine forests. *Forest Ecology and Management.* Vol. 141: pp. 189–200.

Wilkinson, N., Scaife, R. and Sidell, E. (2000). Environmental and sea level changes in London from 10500 BP to present: A case study from Silvertown. *Proceedings of the Geologists Association.* Vol. 111: pp. 41–54.

Wilson, R. (2003). Contributions In: Barclay, A., Lambrick, G., Moore, J. and Robinson, M. (eds). *Lines in the landscape.* Oxford: Oxford Archaeological Unit. Holywell Press.

Wood, J. (2007). A Mesolithic House, Saveock, Cornwall? Or from hooves to secret feather pits. In: Waddington, C. and Pedersen, K. (eds). *Mesolithic studies in the North Sea Basin and beyond.* Proceedings of a conference held at Newcastle in 2003. Oxford: Oxbow Books, pp. 96–100.

Wymer, J. and Bonsall, C. (1977) *Gazetteer of Mesolithic sites in England and Wales, with a gazetteer of Upper Palaeolithic sites in England and Wales.* Council for British Archaeology Research Report No. 20. Council for British Archaeology.

Young, R. (2017). *The secret life of cows.* London: Faber and Faber.

Zvelebil, M. (1994). Plant use in the Mesolithic and its role in the transition to farming. *Proceedings of the Prehistoric Society.* Vol. 60: pp. 35–74.

Zvelebil, M. and Rowley-Conwy, P. (1984). Transition to farming in northern Europe; a hunter-gatherer perspective. *Norwegian Archaeological Review.* Vol. 17(2): pp. 104–28.

Zvelebil, M., Green, S. and Macklin, M. (1992). Archaeological landscapes, lithic scatters and human behaviour. In: Rossignol, J. and Wandsnider, L. (eds). *Space, time and archaeological landscapes.* New York: Springer US, pp. 193–226.

Index

STUDIES IN THE BRITISH MESOLITHIC AND NEOLITHIC

Series Editors

David Jacques and Graeme Davis

Studies in the British Mesolithic and Neolithic presents the results of fieldwork and excavation as well as works of interpretation from all perspectives on the British Neolithic revolution. Archaeological methodology is augmented where appropriate with interdisciplinary techniques, reflecting contemporary practice in the discipline. Throughout the emphasis is on work which makes new contributions to the debate about the transition between hunter gatherer and farming cultures during this pivotal stage in British prehistory.

The series supports the archaeological community both in providing an appropriate forum for research reports as well as supporting interpretative work including cross-disciplinary research. It takes its inspiration from the work of the University of Buckingham's excavations at Blick Mead in the Stonehenge World Heritage Site.

Studies in the British Mesolithic and Neolithic is based at the Humanities Research Institute, University of Buckingham.

Volume 1 Blick Mead: Exploring the 'First Place' in the Stonehenge Landscape.
David Jacques, Tom Phillips and Tom Lyons
260 pages. 2018. 978-1-78707-096-7.

Volume 2 Stonehenge: A landscape through time.
David Jacques and Graeme Davis.
260 pages. 2018. 978-1-906165-85-7.

Volume 3 The Cursus Enigma: Prehistoric Cattle and Cursus Alignments.
David Saunders. 220 pages. 2020. 978-1-78997-745-5.

For the avoidance of doubt, input of the whole or any part of this publication into any generative artificial intelligence model, chatbot, program, software or technology is expressly prohibited without the prior written consent of the publisher.